T5-ARJ-761

Lecture Notes in Biomathematics

Managing Editor: S. Levin

64

Peripheral Auditory Mechanisms

Proceedings of a conference held at Boston University
Boston, MA, August 13–16, 1985

Edited by
J. B. Allen, J. L. Hall, A. Hubbard, S. T. Neely, and A. Tubis

Springer-Verlag
Berlin Heidelberg New York Tokyo

Editors

J.B. Allen
J.L. Hall
AT&T Bell Laboratories
Murray Hill, NJ 07974, USA

A.E. Hubbard
Department of Otolaryngology and Departments of Systems
Computer and Electrical Engineering, Boston University
110 Cummington Street, Boston, MA 02215, USA

S.T. Neely
Boys Town National Institute
555 North 30th Street, Omaha, Nebraska 68131, USA

A. Tubis
Department of Physics
Purdue University
West Lafayette, IN 47907, USA

Mathematics Subject Classification (1980): 92

ISBN 3-540-16095-7 Springer-Verlag Berlin Heidelberg New York Tokyo
ISBN 0-387-16095-7 Springer-Verlag New York Heidelberg Berlin Tokyo

Printing and binding: Beltz Offsetdruck, Hemsbach/Bergstr.
2146/3140-543210

PREFACE

How well can we model experimental observations of the peripheral auditory system? What theoretical predictions can we make that might be tested? It was with these questions in mind that we organized the 1985 Mechanics of Hearing Workshop, to bring together auditory researchers to compare models with experimental observations. The workshop forum was inspired by the very successful 1983 Mechanics of Hearing Workshop in Delft [1]. Boston University was chosen as the site of our meeting because of the Boston area's role as a center for hearing research in this country. We made a special effort at this meeting to attract students from around the world, because without students this field will not progress. Financial support for the workshop was provided in part by grant BNS-8412878 from the National Science Foundation.

Modeling is a traditional strategy in science and plays an important role in the scientific method. Models are the bridge between theory and experiment. They test the assumptions made in experimental designs. They are built on experimental results, and they may be used to test hypotheses and predict experimental results. The latter is the scientific method at its best.

Cochlear function is very complicated. For this reason, models play an important role. One goal of modeling is to gain understanding, but the necessary mathematical tools are often formidably complex. An example of this is found in cochlear macromechanics. The basic concept behind cochlear macromechanics is that the volume of fluid in each scala (vestibuli or tympani) must be constant in time. This is called Gauss' law, and it leads to some complicated mathematics. However, if we view the model in terms of its physical principles, then the mathematics can be viewed as secondary detail. From this global point of view, the important considerations are the model assumptions, their resulting predictions, and how they compare to the experimental observations. Such comparisons lie at the heart of this conference.

Why are we doing this research? This question brings forth many responses. Ours is not yet a science driven by a technology; however, such seeds are present because important human-serving applications lie along the research path. Consider a few questions which are related to hearing and speech technologies, but which will not be resolved without a good understanding of cochlear function:

(a) What is loudness, and what is its relation to deafness?

(b) By signal processing means, can we improve speech intelligibility for the hearing impaired?

(c) How is speech recognized by humans? How important is a cochlear "front end" processor to this end? Would a simple filter bank perform as well?

(d) What are masking and suppression? How important are they to speech and music perception?

Although hearing technologies presently do not drive our research, they could play a larger role in future funding. In the United States, it is estimated that the 10 to 20% of the population that have some type of hearing impairment would benefit from hearing support devices. Almost everyone could benefit from improvements in automatic speech recognition. Funding organizations are very receptive to broad-reaching goals, but they are not uniformly aware of them. It seems clear that a detailed cochlear model could contribute in a major way to the understanding and development of hearing and speech technologies.

The research reported on here will become historically significant as hearing and speech technologies mature. Plans are presently being made to have another Mechanics of Hearing Workshop in England in 1988. The main job ahead of us is to identify and model the principles of cochlear function. We should continue to work together toward our common goal of understanding of human oral-aural communication.

J. B. Allen
J. L. Hall
A. Hubbard
S. T. Neely
A. Tubis

1. *Mechanics of Hearing*, edited by E. de Boer and M. A. Viergever, Martinus Nijhoff Publishers, The Hague, and Delft University Press, Delft, 1983.

TABLE OF CONTENTS

OUTER AND MIDDLE EAR MECHANICS

COCHLEAR MACROMECHANICS

COCHLEAR MICROMECHANICS

ACTIVE FILTERING IN THE COCHLEA

NONLINEAR AND/OR ACTIVE PROCESSES

TRANSDUCTION IN THE COCHLEA

OUTER AND MIDDLE EAR MECHANICS

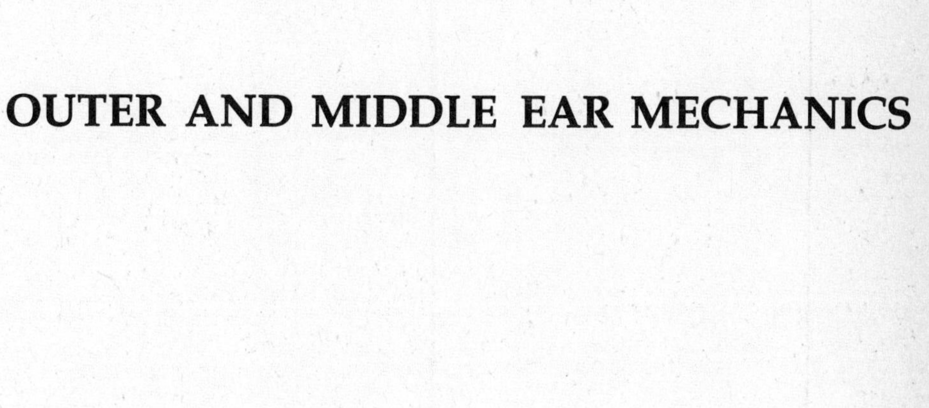

THE EFFECTIVENESS OF EXTERNAL AND MIDDLE EARS IN COUPLING ACOUSTIC POWER INTO THE COCHLEA

J.J.Rosowski, L.H.Carney*, T.J.Lynch III**, and W.T.Peake
Research Laboratory of Electronics, Massachusetts Institute of Technology, Cambridge, MA 02134 and Eaton-Peabody Laboratory, Massachusetts Eye and Ear Infirmary, Boston, MA 02114.

ABSTRACT

The aim of this paper is to evaluate measures of acoustic-power transfer for comparisons of the performance of the auditory peripheries of different species. To do so we will define three power transfer measures that can be computed from available data. The measures also separate the auditory periphery into functional subunits so as to enable estimations of the roles of different auditory specializations.

The three measures of power transfer are: the "Power Utilization Ratio at the TM" (PUR), the "Effective Area of the External Ear" (EA), and the "Middle Ear Efficiency" (MEE). The three power measures serve different purposes: PUR is an index of the impedance-matching performance of the external and middle ear; EA and MEE quantify power flow through the external and middle ears. The EA and MEE can be combined to obtain a single measure of the power into the cochlea that we call the "Net Effective Area" (NEA).

Our analysis suggests (1) the impedances of external and middle ears are poorly matched, (2) an appreciable fraction of the sound power which enters the middle ear is absorbed before it reaches the cochlea, (3) cochlear function at auditory threshold for pure tones can be roughly approximated by a power detector, and (4) the quantification of power transfer through the ears of different species is a useful comparative tool.

I. INTRODUCTION: Why Consider Power?

External and middle ears couple acoustic signals from the environment to the cochlea as schematized in Figure 1. A uniform acoustic plane wave, as defined by its free-field sound pressure, P_{ff}, and its direction, impinges on the peripheral parts of the system. The interaction of the impinging sound with the head, body, external ear and tympanic membrane, TM, generates a sound pressure, P_t, and volume velocity, U_t, at the TM and sound power is delivered to the middle ear. Some of this power is transmitted by the middle ear to the inner ear.

The performance of the ear as a coupler of acoustic power has been considered by many investigators (Siebert, 1970; Dallos, 1973; Zwislocki, 1975; Killion and Dallos, 1979; Shaw, 1979; Shaw and Stinson, 1983). Indeed, the common description of the middle ear as an "impedance-matching device" derives from the question: How can the middle ear extract the maximum power from a sound stimulus? (Wever and Lawrence, 1954 Chaps 5, 6 & 7). There is also evidence that the inner ear functions as a detector of acoustic power: the behavioral thresholds for tones correlate with the sound pressures needed to maintain a constant power input to the cochlea (Khanna and Tonndorf, 1969; Tonndorf and Khanna, 1976; Khanna and Sherrick, 1981).

---

* Present address: Department of Neurophysiology, University of Wisconsin, Madison, WI .
** Present address: MIT Lincoln Laboratory, 244 Wood St., Lexington, MA .

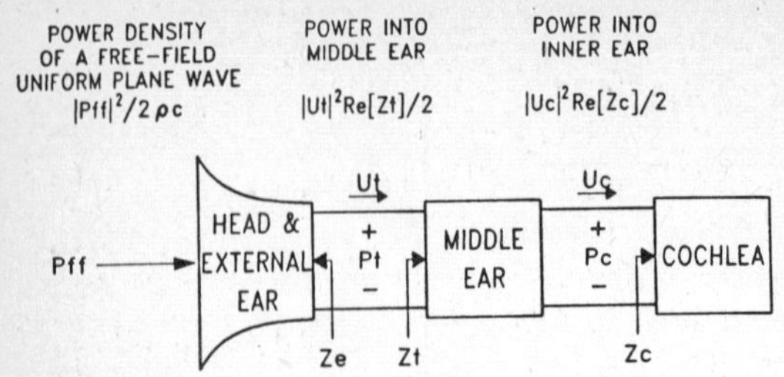

FIGURE 1. BLOCK DIAGRAM representing signals and power flow through the external and middle ear. The COCHLEA block is characterized by its acoustic input impedance, $Z_c = P_c/U_c$, where U_c is the volume velocity of the stapes footplate and P_c is the sound pressure in the perilymph at the input to the cochlea. The MIDDLE EAR is represented by a linear two-port network which when coupled to the cochlea has an acoustic input impedance, $Z_t = P_t/U_t$, where P_t is the sound pressure outside the TM, and U_t is the volume velocity of the TM. The HEAD AND EXTERNAL EAR component represents the transformation to P_t from a free-field, uniform plane-wave sound-stimulus, with sound pressure P_{ff} in a medium of characteristic impedance ρc, where ρ is the mass density of the medium and c is the propagation velocity of sound. Expressions for the average power (or average power/area) at the inputs of the three components are given at the top of the figure. Z_e is the impedance looking out from the tympanic membrane through the external ear. $Re[Z_t]$ = the real part of the complex impedance, Z_t.

Comparisons of power flow through different ears may be superior to comparisons of cochlear sound pressure or stapes velocity. The interpretation of measures of middle-ear motion or inner-ear pressure depends on knowledge of the input impedances of the middle and inner ear; these impedances are included in the quantification of power flow. Power related performance metrics can also be defined. For instance, it is possible to quantify (1) How well the external and middle ear match impedances? (2) How much of the power in a stimulus is extracted by the ear? (3) How much of the power which enters the middle ear reaches the cochlea?

II. POWER UTILIZATION RATIO AT THE TM: The Quality of Impedance Matching.

Siebert (1970) suggested that the quality of impedance matching in the ear can be conveniently determined by comparing the "output" impedance of the external ear with the input impedance of the middle ear. The middle-ear input impedance is Z_t, while the "output" impedance of the external ear, Z_e, is the impedance looking out through the external ear into free space (Figure 1). The Power Utilization Ratio, PUR, quantifies what fraction of the "available power" at the TM actually enters the middle ear, and is an index of the match between Z_e and Z_t. (The available power is the maximum power that the external ear could pass into a load).

$$\text{PUR} = \frac{\textit{POWER INTO MIDDLE EAR}}{\textit{POWER AVAILABLE AT TM}} = \frac{4\,Re[Z_t]\,Re[Z_e]}{|Z_t + Z_e|^2}.$$

When Z_t and Z_e are matched, i.e. when Z_t and Z_e are complex conjugates, PUR = 1.

In the cat, Z_t has been measured over a wide frequency range by Lynch (1981), and Carney has used Lynch's methods to measure Z_e (Figure 2). Both of these measurements are illustrated in Figure 3. Z_e is mass-like at frequencies below 2 kHz and primarily resistive at higher frequencies (Figure 3A), whereas Z_t is compliance-like at low frequencies and resistive at higher frequencies (Figure 3B). When the angle of Z_e or Z_t approach either $-90°$ or $+90°$, small errors in the angle cause the estimates of the real part of the impedances to become noisy, therefore theoretical estimates of the real parts are used in the low-frequency calculations.

The PUR of cat is highly frequency dependent (Figure 4A). At frequencies below 1 kHz PUR is much less than 1, i.e. there is a poor impedance match and the power that enters the middle ear is much less than the power available. At 3 kHz PUR is close to 1: Z_e and Z_t are matched (both are almost purely real and of equal magnitude). PUR is between 0.4 and 0.9 at frequencies between 4 and 10 kHz.

Using published measurements of Z_t and calculations of Z_e for appropriately sized exponential horns in baffles, (Malecki, 1969), we also estimate the PUR of the guinea pig and human ear (Figure 4B). The PURs of the three species have similar features. All are much less than 1 at frequencies below 1 kHz and all have maxima near 1 at higher frequencies. The multiple peaks of the human and guinea pig PURs result from resonances in the Z_es calculated for *lossless* exponential horns; such peaks are not seen in the cat Z_e measurement. Although we have displayed the computed PUR over as wide a range as was possible with the available data, it seems likely that appreciable errors occur at high frequencies (above 10 kHz) because neither our model nor the measurements take into account spatial variations in the sound field near the tympanic membrane (Khanna and Stinson, 1985).

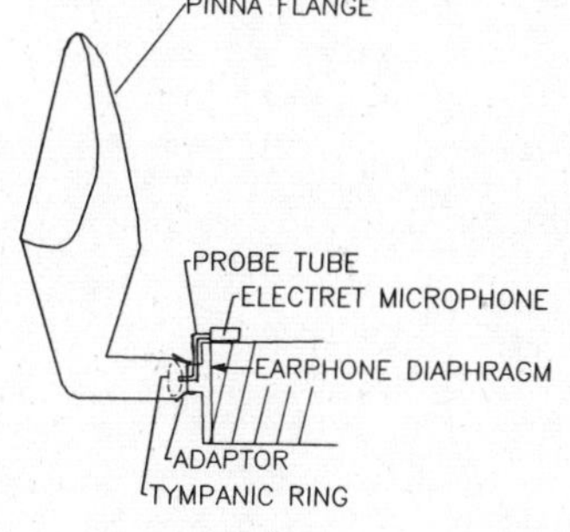

FIGURE 2. SCHEME FOR MEASURING Z_e. A calibrated earphone source was connected to the medial end of six excised external ears of cats, and a probe tube microphone was used to measure the pressure produced at the tympanic ring. The pressure measurements and the known source characteristics were used to calculate Z_e (Lynch, 1981).

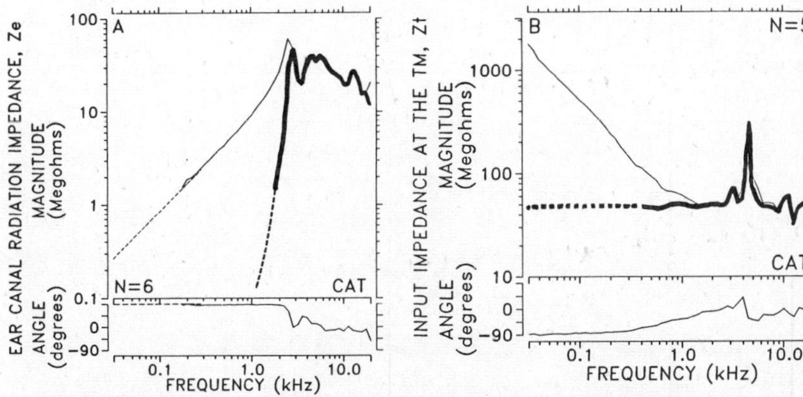

FIGURE 3. A) ACOUSTIC IMPEDANCE LOOKING OUT FROM THE TM OF CAT, Z_e, magnitude and angle (thin lines) and real part, $Re[Z_e]$, (thick lines). Data (solid lines) are means of six ears. The dashed lines are the low frequency Z_e and $Re[Z_e]$ calculated for an exponential horn (Malecki, 1969) of dimensions similar to those of the cat ear. B) MIDDLE EAR INPUT IMPEDANCE, Z_t, magnitude and angle and real part, $Re[Z_t]$. The thinner line represents the mean of 5 measurements from Lynch (1981). The thick line is the real part calculated from the data. The thick dashed line is based on an assumption that $Re[Z_t]$ is constant at low frequencies. The Z_t measurements were obtained in ears with the middle-ear cavity opened. In order to correct for this modification the mean measurements were multiplied by the ratio Z_t(intact cavities)/Z_t(opened cavities) obtained from a typical ear (Lynch, 1981). The large sharp peak in $|Z_t|$ near 4.5 kHz results from the middle-ear cavities. Impedance units are MKS.

The PUR calculation for cat, human and guinea pig all indicate that the ears of these animals approximately "match" impedances only in a narrow frequency range. The "match" is particularly poor at low frequencies: less than 1% of the available power enters the middle ear at frequencies less than 1 kHz.

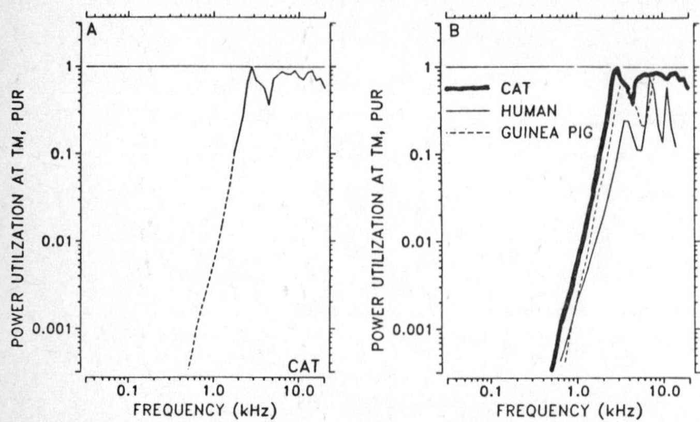

FIGURE 4. A) POWER UTILIZATION RATIO AT THE TM FOR CAT. Solid line shows the PUR calculated from the raw data of Figure 3. The dashed lines indicate the PUR calculated using the theoretical low-frequency real parts of Z_e and Z_t. B) THE PUR OF THE CAT, GUINEA PIG AND HUMAN EAR. The mean dimensions of four guinea pig external ears, and the human ear canal area function data of Johansen (1975) were used to calculate Z_e. The guinea pig Z_t was based on a typical ear of Mundie's (1963) (the Z_t angle data were modified so that no angle was less than $-90°$) and the human Z_t was a combination of the low and mid-frequency data of Rabinowitz (1981) (mean of 4 ears) and the high-frequency data of Hudde (1983) (mean of 6 ears).

III. EFFECTIVE AREA: Power Collection by the External Ear.

The efficacy of the "external-ear" (including the effects of the head, body and middle-ear load) as an acoustic power collector can be assessed by dividing the power input to the tympanic membrane by the power density (power/area) of the incident plane wave (Figure 1). This ratio has the dimensions of area, and is what we call the "Effective Area of the External Ear" (EA). EA can be computed from measurements of the pressure transformation of the external ear P_t/P_{ff} and the input impedance of the middle ear, Z_t, as follows

$$\mathrm{EA} = \frac{\textit{POWER INTO MIDDLE EAR}}{\textit{FREE-FIELD POWER DENSITY}} = \rho c \left| \frac{P_t}{P_{ff}} \right|^2 \frac{Re[Z_t]}{|Z_t|^2}.$$

(EA can be thought of as the area of a piston head whose specific impedance matches the characteristic impedance of air).

Measurements of P_t/P_{ff}, for three different plane wave directions (source positions), are pictured in Figure 5; Z_t and $Re[Z_t]$ are pictured in Figure 3B, and the results of the EA calculations are pictured in Figure 6.

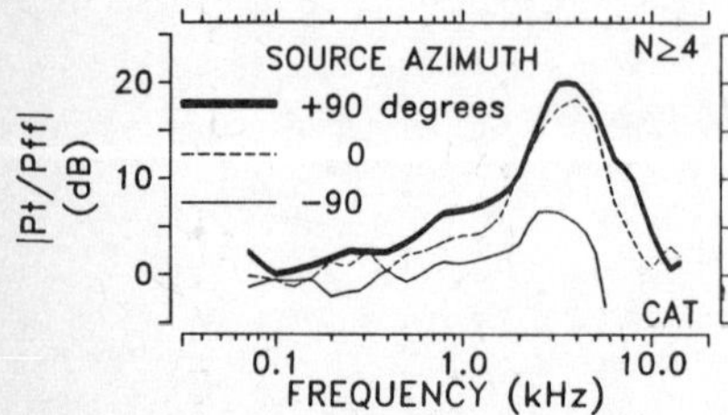

FIGURE 5. A) PRESSURE TRANSFORMATION OF THE EXTERNAL EAR OF CATS, P_t/P_{ff} for three horizontal azimuths, $+90°$, $0°$ and $-90°$ from the midline. Mean of 4 or 5 ears from Wiener, Pfeiffer and Backus (1965).

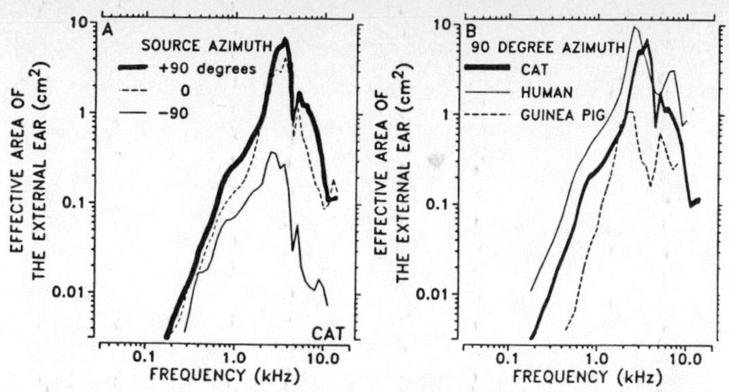

FIGURE 6. A) EFFECTIVE AREA OF THE CAT'S EXTERNAL EAR for three horizontal azimuths, +90°, 0° and −90° from the midline, calculated from the data of Figure 3B & 5. B) EA OF CAT, GUINEA PIG AND HUMAN EAR, at source azimuth of 90°. The guinea pig data were calculated from P_t/P_{ff} data of Sinyor and Laszlo (1973) (mean of 7 ears) and the Z_t data of Mundie (1963). The human data were the P_t/P_{ff} functions of Shaw (1974a) and the human Z_t data of Rabinowitz (1981) and Hudde (1983).

The EA magnitudes produced with a 90° azimuth are larger than those produced by 0° or −90° and are near the maximum EAs attainable at any azimuth (Figure 6) (We calculate slightly larger EAs when we use the P_t/P_{ff}s Wiener et al. measured at 45°.). This dependence of the Effective Area (EA) on stimulus azimuth reflects the directionality of the cat's head and pinna and is related to the animal's ability to localize sound. The EAs also vary with frequency, having a bandpass characteristic with a maximum value near 4 kHz. With the 90° stimulus, the EA maximum is almost 7cm^2. This is smaller than the 12 cm^2 area of the pinna flange (average of 6 ears) but larger than the 2.4 cm^2 cross-sectional area at the base of the flange. Between 1 and 10 kHz the dependence of EA on frequency is primarily determined by $|P_t/P_{ff}|$ except for the sharp minimum near 4.5 kHz, which is related to the effect of the middle-ear cavity resonance on $|Z_t|$. Below 0.5 kHz EA is controlled by Z_t and grows as the square of frequency (At low frequencies, $|P_t/P_{ff}| \approx 1$, Figure 5; $Re[Z_t]$ is approximated by a constant, dashed line of Figure 3B; $|Z_t| \approx 1/(\omega C_t)$, thin line of Figure 3B; thus EA $\approx \rho c\, \omega^2 C_t^2 R_t$.).

The EAs for cat, guinea pig and human ears are compared in Figure 6B. The maximum human EA ($\approx$10 cm^2) is larger than the cat maximum; the guinea pig maximum ($\approx$1 cm^2) is the smallest. This ranking of maxima is consistent with rankings of the dimensions of the external ears of the three species (Shaw, 1974b). At frequencies below 0.5 kHz, all three EAs are proportional to frequency squared and are controlled by the compliance at the tympanic membrane. The magnitude of the low-frequency EA is proportional to the square of the middle-ear compliance, and the ranking of the EA magnitudes below 0.5 kHz correlates with a ranking of the cat, guinea pig and human middle-ear compliance.

The interspecies comparison of EAs indicates that the size of the auricle and the compliance of the middle ear play roles in determining how much sound power enters the middle ear. The comparison is complicated by the azimuthal dependence of EA. We have attempted to overcome this complication by comparing EAs measured at a 90° azimuth. Slightly larger EAs could be calculated for man and cat using P_t/P_{ff} data at other azimuths. Shaw (1975) has calculated a "diffuse field sensitivity" measure of human external-ear function which is independent of azimuth and is closely related to our Power Utilization Ratio.

IV. MIDDLE-EAR EFFICIENCY: Power Flow Through the Middle Ear.

We define the efficiency of the middle ear (MEE) as the power into the inner ear divided by the power into the middle ear. This ratio can be calculated from measurements of the middle-ear transfer function, U_c/P_t (Figure 7A), the real part of the cochlear input impedance, $Re[Z_c]$ (Figure 7B), and Z_t (Figure 3B), as follows:

$$\text{MEE} = \frac{\textit{POWER INTO THE INNER EAR}}{\textit{POWER INTO THE MIDDLE EAR}} = \left|\frac{U_c}{P_t}\right|^2 \frac{|Z_t|^2 Re[Z_c]}{Re[Z_t]}.$$

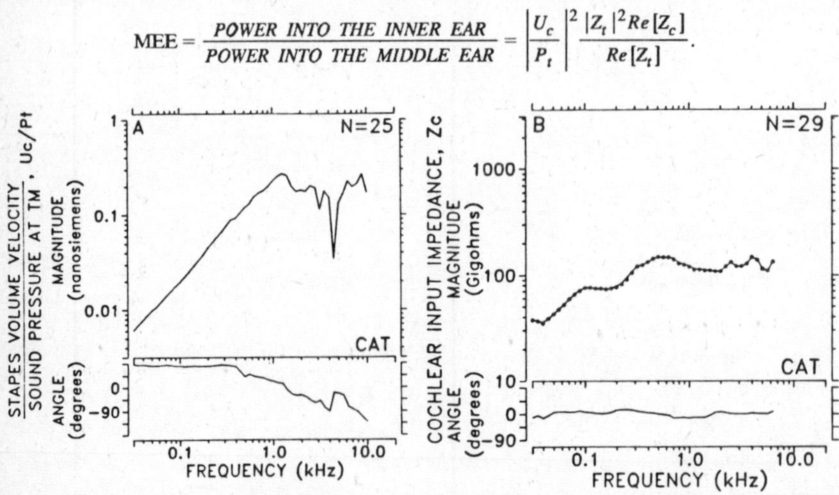

FIGURE 7. A) MIDDLE-EAR TRANSFER FUNCTION OF CAT, U_c/P_t, magnitude and angle from Guinan and Peake (1967). The mean from 25 ears measured with the middle-ear cavities opened and then corrected with the same middle-ear cavity data used to correct Z_t in Figure 3B. We have assumed an area of the stapes footplate of 1.25 mm^2. B) CAT COCHLEAR INPUT IMPEDANCE, Z_c, magnitude and angle and its real part, $Re[Z_c]$. The line is the mean Z_c from 29 ears (Lynch, Nedzelnitsky and Peake, 1982). The dots are the calculated real part. In the 0.03 - 8 kHz frequency range $|Z_c|$ and $Re[Z_c]$ are nearly identical.

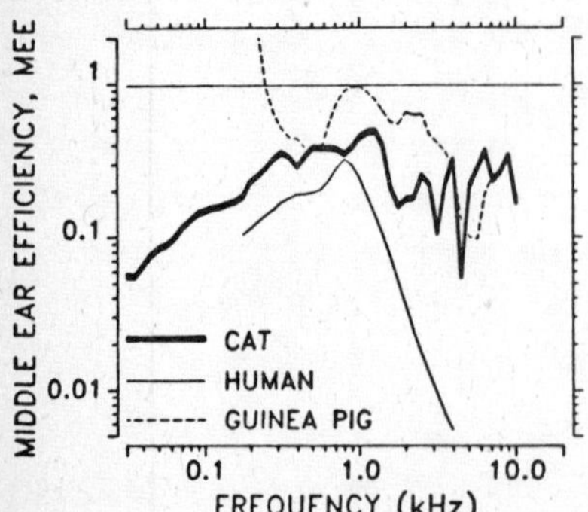

FIGURE 8. MIDDLE-EAR EFFICIENCY OF CAT, GUINEA PIG AND HUMAN. The MEE calculated for cat is based on the U_c/P_t and Z_c of Figure 7 and the Z_t of Figure 3B. The mean transmission data from 20 human temporal bones from Kringlebotn and Gunderson(1985) and the theoretical cochlear input impedance of Zwislocki(1975) have been combined with the human Z_t data (Hudde, 1983; Rabinowitz, 1981) to compute MEE for the human ear. The mean U_c/P_t from 9 guinea pig ears from Wilson and Johnstone(1975) (adjusted by the middle-ear cavity impedance of Zwislocki's model (1963) and assuming a 1 mm^2 footplate area) was used in conjunction with the mean Z_c estimate from 3 ears from Dancer and Franke (1980) and the Mundie Z_t data (1963) to calculate MEE for the guinea pig.

The results of MEE calculations show that even when MEE is at its largest, only half of the sound power which enters the cat middle ear reaches the inner ear (Figure 8). In cat, MEE maintains a magnitude near 0.4 between 0.3 and 2.0 kHz. At higher frequencies, MEE fluctuates rapidly with frequency. These fluctuations result from absorption of power by the middle-ear cavities: the minimum in MEE at 4.5 kHz is at precisely the same frequency as the middle-ear cavity induced dip in U_c/P_t and peak in Z_t. The low-frequency decrease in the cat MEE results from the decrease of the real part of Z_c and is clearly dependent on the *assumed* constancy of $Re[Z_t]$.

Data from the literature were used to calculate MEE in human and guinea pig (Figure 8). In human, the estimated MEE has its peak efficiency (≈0.3) at 0.9 kHz and falls off rapidly at higher frequencies. The estimated MEE of the guinea pig ear is near one at 1 kHz and falls at higher frequencies. At frequencies below 0.2 kHz the MEE calculated for guinea pig is greater than 1, and probably reflects errors in the data rather than active processes within the middle ear.

Small errors in any of the various impedances and transfer functions used in the calculations can greatly affect MEE. The middle-ear transfer function measurements used in the human MEE calculation were performed on cadaver ears and the relationship between these measurements and the transfer function of live ears is not clear. Nevertheless, the fact that most of the MEE estimates are substantially below 1 suggests that the middle ears of all three animals are far from perfectly efficient at most frequencies. The maximum middle-ear efficiencies of 50% estimated for the cat and 30% for the human middle-ear are similar to the 50% efficiency Shaw and Stinson (1983) calculated from a detailed model of the human middle ear.

V. NET EFFECTIVE AREA: Power Into the Cochlea.

The sound power that reaches the cochlea from the free-field is proportional to a "Net Effective Area" (NEA), where:

$$\mathrm{NEA} = \frac{\textit{POWER INTO THE INNER EAR}}{\textit{FREE}-\textit{FIELD POWER DENSITY}} = \mathrm{EA} \times \mathrm{MEE}.$$

The estimated NEAs of our three species are pictured in Figure 9. The NEAs of cat and human are smaller than their EAs because their estimated MEEs are less than 1, while the guinea pig NEA is similar to its EA, because its estimated MEE is close to 1. While the EA maxima can be ranked by size of the auricle, the ranking of NEA maxima is different. The guinea pig and cat maxima are more similar, and the human maximum is the smallest.

From NEA, the free-field sound pressure necessary to provide a constant sound power into the cochlea can be calculated. We have determined for each of the three species the iso-power contour (P_{ff} vs frequency) that most closely matches the behavioral thresholds. The sound power "threshold" of 6×10^{-19} watts into the cat cochlea (Figure 10A) is similar to the power threshold calculated by Khanna and Sherrick (1981). Also, the frequency dependence of the cat iso-power contour is grossly similar to the threshold curve. The differences in the low-frequency slopes of the two functions may reflect the absorption of low-frequency acoustic power by the helicotrema or other nonsensory cochlear structures. There are other differences between the curves, but it is difficult to determine their significance.

Iso-power contours for the guinea pig and human ear are also compared to the behavioral thresholds in Figure 10B & C. The magnitude of the constant power level which best fits the audibility data is similar for all three species.

Although inaccuracies in the data make interspecies comparisons questionable, we will briefly consider one issue suggested by the NEA calculations. Between 3 and 4 kHz, the effective area of the human external ear is 10 times larger than that of guinea pig (Figure 3B), and yet in this frequency range these animals have similar behavioral thresholds (≈ -8 dB SPL). If the power into the cochlea is indeed the relevant stimulus to the inner ear, the guinea pig could compensate for the smaller power flow into its ear canal by two methods: (1) its inner-ear could be a more sensitive power detector than the inner ear of man, or (2) its middle ear could be more efficient. The computations we have presented suggest the latter. If this conclusion were verified by more precise measurements, it would be an example of different species that have achieved similar auditory performances by "optimization" of different ear components.

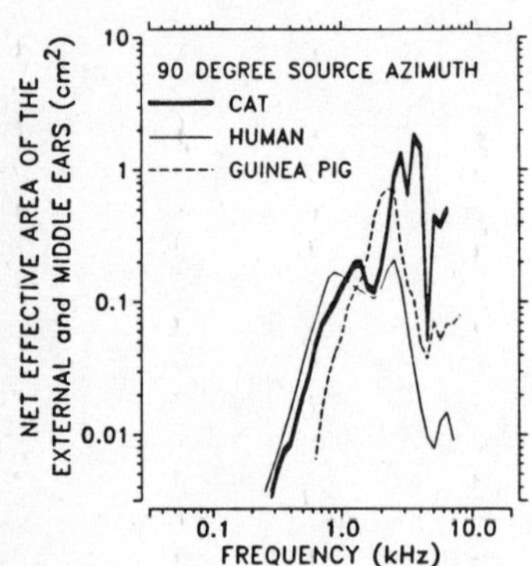

FIGURE 9. NET EFFECTIVE AREA OF THE EAR FOR CAT, GUINEA PIG AND HUMAN.

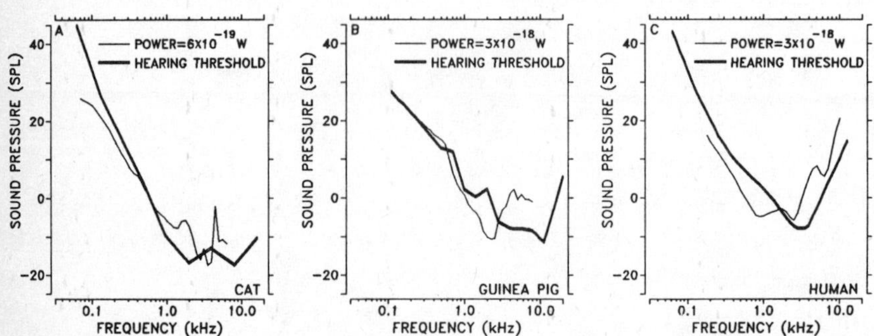

FIGURE 10. COMPARISON OF ISO-COCHLEAR POWER CONTOURS TO BEHAVIORAL THRESHOLDS. The power levels were chosen to minimize the mean difference between the 90° azimuth iso-power function and the threshold data. A) CAT. The auditory threshold curve (thick line) was determined by Miller et al. (1963). Also plotted is the free-field sound pressure necessary to deliver 6×10^{-19} watts of sound power to the cochlea. B) GUINEA PIG. The auditory threshhold curve from Prosen et al (1978) is plotted with the free-field sound pressure necessary to deliver 3×10^{-18} watts of power to the guinea pig cochlea. C) HUMAN. The auditory threshold curve from Sivian and White (1933) is plotted with the free-field sound pressure necessary to deliver 3×10^{-18} watts of power to the human cochlea.

VI. CONCLUSIONS

We need to mention two caveats related to the approach we have presented: (1) The model we have used may not apply to all species and environments. For instance, underwater sound reaches the inner ear of fishes through different pathways. To account for this difference our power transfer model would have to be generalized to allow for important alternate sound pathways to the inner ear. (2) In evaluating the ears' performance as a sound receiver one should undoubtedly take into account other key functions which are at odds with "perfect" sound reception, e.g. directionality. In general though, our model is relevant to the ears of most terrestrial vertebrates and the issue of directionality only has effects if we compare the performance of ears at non-optimal stimulus azimuths.

We suggest four general conclusions. (1) Before power coupling by the external and middle ears is understood, better measurements of middle-ear function are needed, particularly at low frequencies where estimates of acoustic power transfer are dependent on precise measurements of the real parts of the middle-ear and cochlear input impedances. (2) The middle ear neither extracts all the available power from the external ear nor delivers to the cochlea all the power it takes in. Therefore, describing the middle ear as an impedance matching device is not helpful. (3) As was suggested by Khanna and Tonndorf (1969) cochlear function for pure tones at auditory threshold can be roughly approximated by a power detector. (4) Comparisons of power-transfer through the ears of different species can lead to useful concepts. The idea that the effective area of the external ear is generally related to the size of the ear and the fact that ears tend to scale with body size (Khanna and Tonndorf, 1969) implies that the ears of small animals receive less acoustic power than those of large animals. If we assume that power handling by the rest of the ear is comparable, then small animals should have higher thresholds. Alternatively, smaller cochleas may require less power input or smaller middle ears may be more efficient. Thus, these measures of power coupling performance can provide a framework for functional comparisons across the structural variations of the vertebrate ear.

ACKNOWLEDGMENTS

We thank R.A.Eatock, D.M.Freeman, J.J.Guinan, T.F.Weiss for helpful comments, and D.B.Krakauer and the staff of the Eaton-Peabody Laboratory for their help in preparing this manuscript. This work was supported by NIH grants 5-P01-NS-13126 and 5-R01-NS-18682.

REFERENCES

Dallos, P. *The Auditory Periphery*. Academic Press, New York, 548 pg. , 1973.

Dancer, A. and Franke, R., "Intracochlear sound pressure measurements in guinea pigs." Hearing Res. *2*, pp. 191-206, 1980.

Guinan, J.J.,Jr and Peake, W.T., "Middle-ear characteristics of anesthetized cats." J. Acoust. Soc. Am. *41*, pp. 1237-1261, 1967.

Hudde, H., "Measurement of the eardrum impedance of human ears." J. Acoust. Soc. Am. *73*, pp. 242-247, 1983.

Johansen, P.A., "Measurement of the human ear canal." Acoustica *33*, pp. 349-351, 1975.

Khanna, S.M. and Sherrick, C. "The comparative sensitivity of selected receptor systems." In *The Vestibular System: Function and Morphology*, Ed. by T. Gualtierotti, Springer-Verlag, New York, pp. 337-348, 1981.

Khanna, S.M. and Stinson, M.R., "Specification of the acoustical input to the ear at high frequencies." J. Acoust. Soc. Am. *77*, pp. 577-589, 1985.

Khanna, S.M. and Tonndorf, J., "Middle ear power transfer." Arch. Klin. exp. Ohr.-, Nas. -u. Kehlk. Heilk. *193*, pp. 78-88, 1969.

Killion, M.C. and Dallos, P., "Impedance matching by the combined effects of the outer and middle ear." J. Acoust. Soc. Am. *66*, pp. 599-602, 1979.

Kringlebotn, M. and Gunderson, T., "Frequency characteristics of the middle ear." J. Acoust. Soc. Am. *77*, pp. 159-164, 1985.

Lynch, T.J.,III, *Signal processing by the cat middle ear: Admittance and transmission, measurements and models*. Ph.D. Thesis, Massachusetts Institute of Technology, Cambridge, MA., 256 pg., 1981.

Lynch, T.J.,III, Nedzelnitsky, V. and Peake, W.T., "Input impedance of the cochlea in cat." J. Acoust. Soc. Am. *72*, pp. 108-130, 1982.

Malecki, I., *Physical Foundations of Technical Acoustics*. Pergamon Press, Oxford. 743 pg., 1969.

Miller, J.D., Watson, C.S. and Covell, W.P., "Deafening effects of noise on the cat." Acta Oto-laryngol. Suppl. *176*, pp. 1-91, 1963.

Mundie, J.R., "The impedance of the ear--a variable quantity." U.S.Army Med. Res. Lab. Report No. 576. pp 63-85, 1963.

Prosen, C.A., Peterson, M.R., Moody,D.B. and Stebbins, W.C., "Auditory thresholds and kanamycin-induced hearing loss in the guinea pig assessed by a positive reinforcement procedure." J. Acoust Soc. Am. *63*, pp. 559-566, 1978.

Rabinowitz, W. M., "Measurement of the acoustic input immittance of the human ear." J. Acoust. Soc. Am. *70*, pp. 1025-1035, 1981.

Shaw, E.A.G., "Transformation of sound pressure level from the free field to the eardrum in the horizontal plane." J. Acoust. Soc. Am. *56*, pp. 1848-1861, 1974a.

Shaw, E.A.G., "The external ear." In *Handbook of Sensory Physiology, Vol V/1: Auditory System*. Ed. by W.D.Keidel and W.D.Neff, Springer-Verlag, New York, pp 455-490, 1974b.

Shaw, E.A.G., "Diffuse field sensitivity of external ear based on reciprocity principle." J. Acoust. Soc. Am. *60*, S102, 1975.

Shaw, E.A.G., "Performance of the external ear as a sound collector." J. Acoust. Soc. Am. *65*, S9, 1979.

Shaw, E.A.G. and Stinson, M.R., "The human external and middle ear: models and concepts." In *Mechanics of Hearing*, Ed. by E. de Boer and M. A. Viergever, Martinus Nijhoff Publishers, Delft University Press, Netherlands, pp. 3-10, 1983.

Siebert, W.M., "Simple model of the impedance matching properties of the external ear." Quarterly Progress Report No. 96, Research Laboratory of Electronics, Massachusetts Institute of Technology. pp. 236-242, 1970.

Sinyor, A. and Laszlo, C.A., "Acoustic behavior of the outer ear of the guinea pig and the influence of the middle ear." J. Acoust Soc. Am. *54*, pp. 916-921, 1973.

Sivian, L.J. and White, S.D., "On minimum audible fields." J. Acoust. Soc. Am. *4*, pp. 288-321, 1933.

Tonndorf, J. and Khanna, S.M., "Mechanics of the auditory system." In *Scientific Foundations of Otolaryngology*, Ed. by R. Hinchcliffe and D. Harrison, Yearbook Medical Publishers, Inc. Chicago. pp. 237-252, 1976.

Wiener , F.M., Pfeiffer, R.R. and Backus, A.S.N., "On the sound pressure transformation by the head and auditory meatus of the cat." Acta Otolaryngol. *61*, pp. 255-269, 1965.

Wever, E.G. and Lawrence, M. *Physiological Acoustics*. Princeton University Press, Princeton, NJ, 454 pg. , 1954.

Wilson, J.P. and Johnstone, J.R., "Basilar membrane and middle-ear vibration in guinea pig measured by capacitive probe." J. Acoust. Soc. Am. *57*, pp. 705-723, 1975.

Zwislocki, J.J., "Analysis of the middle ear function: Part II. Guinea-pig ear." J. Acoust. Soc. Am. *35*, pp. 1034-1040,1963.

Zwislocki, J.J., "The role of the external and middle ear in sound transmission." In *The Nervous System, Vol 3.: Human Communication and its Disorders*. Ed. by D.B. Tower, Raven Press, New York, pp 45-55, 1975.

SPATIAL DISTRIBUTION OF SOUND PRESSURE IN THE EAR CANAL

Michael R. Stinson
Division of Physics, National Research Council
Ottawa, Ontario, Canada K1A 0R6

ABSTRACT

For many experiments in physiological and psychological acoustics results are reported using, as a reference level, the sound pressure measured at the eardrum. However at higher frequencies there can be rather dramatic variations of sound pressure level within an ear canal and across the tympanic membrane. Different locations of a reference microphone can lead to quite different results, with the introduction of artifacts that relate only to peculiarities of the sound field. As a first step toward understanding this problem, measurements have been made of the spatial variation of sound pressure in scaled replicas of human ear canals and in the ear canals of live cats. The measured pressure distributions can be described reasonably well using a theoretical model that has been developed. This model is an extension of Webster's horn equation, taking into account the curvature and variable cross section of the ear canal and the absorption of acoustic energy at the eardrum. From both theory and experiment it is clear that variations in sound pressure level of over 20 dB can occur over the surface of the tympanic membrane.

1. INTRODUCTION

In the study of the hearing process a measured response is only meaningful when the applied acoustical stimulus is well-defined (Khanna and Stinson, 1985). This is the case in both physiological and psychological experiments, whether it be the measurement of basilar membrane tuning curves or subjective response to audiometric tones. A parameter often used to describe the applied stimulus is the sound pressure "at the eardrum". This reference signal is reasonably convenient to measure and is independent of the directional characteristics of the outer ear. However at higher frequencies large variations of sound pressure can occur over the eardrum surface (Stinson and Shaw, 1983), and a reference signal will depend critically on the location of the reference pressure probe.

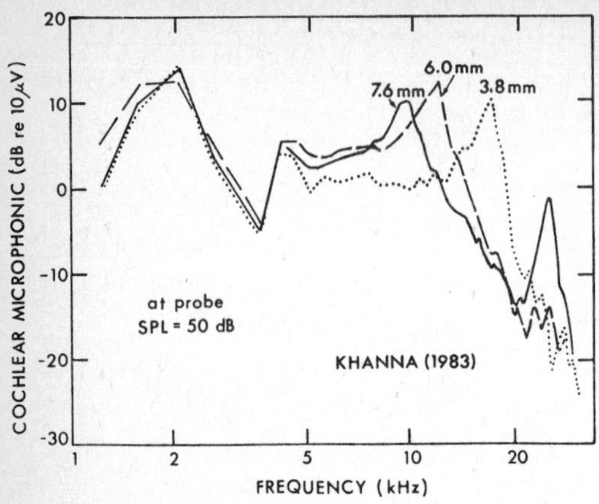

Figure 1. Measured CM response for an anesthetized cat, for three different positions of a reference pressure probe; an SPL of 50 dB was maintained at the probe for each curve. The labels give the distance between probe and innermost end of the ear canal.

This sensitivity to reference probe location is illustrated in Fig. 1. Measurements (Khanna, 1983) are shown of the round window cochlear microphonic, for a sound pressure level of 50 dB maintained at a probe microphone in the ear canal of an anesthetized cat. The three curves were produced under identical conditions except for the location of this probe. The differences can be due only to spatial variations of sound pressure in the ear canal. The sharp peak near 12 kHz is due, not to any property of the cochlea or middle ear, but the existence of a standing wave minimum in the ear canal. Figure 1 stands as a warning that "eardrum" pressure as a reference at high frequencies can be a misleading measure of the input stimulus. Many studies (particularly of cochlear dynamics) use other parameters (e.g. stapes velocity) for a reference. This is certainly preferable when a specific aspect of the hearing mechanism is being considered. Still, all such quantities must ultimately be related to the air-borne acoustical stimulus, and then an understanding of the spatial variation of canal sound pressure again is important.

In this paper a model is developed that is capable of predicting quantitatively sound pressure distributions in real ear canals, so that use of a reference eardrum pressure can be evaluated.

2. MODEL OF THE EAR CANAL

An acoustical description of the ear canal must account for several features indicated on Fig. 2(a). The cross-sectional area of the ear canal varies along its length especially at the inner end where the eardrum and adjacent canal wall form a wedge-shaped volume. The ear canal contains twists and turns along its length. While the canal walls

Figure 2. (a) Sketch of an ear canal, and (b) model used to calculate sound fields in the canal.

may be assumed to be acoustically hard significant absorption of acoustic energy at the tympanic membrane (TM) must be anticipated. The model that will be used is indicated in Fig. 2(b). A curved center axis is introduced in the ear canal and arc length s measured along this curve. At each s a cross section may be determined, normal to the axis, with area A(s). Provided that cross-sectional distances are small relative to the wavelength and to the local radius of axis curvature, the sound pressure p(s) is given by (Stinson and Shaw, 1983; Khanna and Stinson, 1985)

$$\frac{d}{ds}\left(A\frac{dp}{ds}\right) + k^2 Ap = 0, \tag{1}$$

where k is the wavenumber. This is the horn equation expressed in terms of the curvilinear coordinate s. The accuracy of this approximation is maximized by ensuring that the center axis passes through the <u>centroid</u> of each cross section.

To treat the absorption of acoustic energy at the eardrum the simplest approach is to assume that the effects of the eardrum may be represented by an acoustic impedance at a <u>single</u> location, perhaps somewhere near the physical center of the TM (Stinson, 1983). At the location of the impedance s_o there must be continuity of both pressure and volume velocity. This leads to the condition

$$\frac{dp}{ds}(s_o^+) = \frac{dp}{ds}(s_o^-) + \frac{i\rho\omega}{A(s_o)Z_{TM}}\, p(s_o) \tag{2}$$

where the supercripts (+) and (-) indicate outer and inner sides of s_o. The values for the complex eardrum impedance Z_{TM} can be taken from theoretical middle ear models (e.g., Shaw, 1977; Peake and Guinan, 1967) or, conversely, obtained by matching measured and predicted pressure distributions.

3. MEASUREMENTS

The formulation of the previous section has been tested in both replicas of human ear canals and the ear canals of live cats. In all cases the calculations first require a measurement of the area function A(s).

The human ear canal geometry was first investigated using a positive replica of a real canal, constructed using a polyester casting resin (Stinson and Shaw, 1983; Stinson, 1985). The basis for this procedure was an acrylic ear canal impression formed by casting in a human cadaver ear canal (cast supplied by P.A. Johansen, Denmark). The replica duplicates the shape of the original ear canal but is scaled up by a factor of 2.56 to increase the relative precision of measurements. Figure 3 shows a sketch of the replica and its location in the measurement system. Two channels allow a 1.2 mm probe to be inserted into the model canal, to survey the sound pressure either along the main body of the canal (x axis) or across the tympanic membrane (y axis).

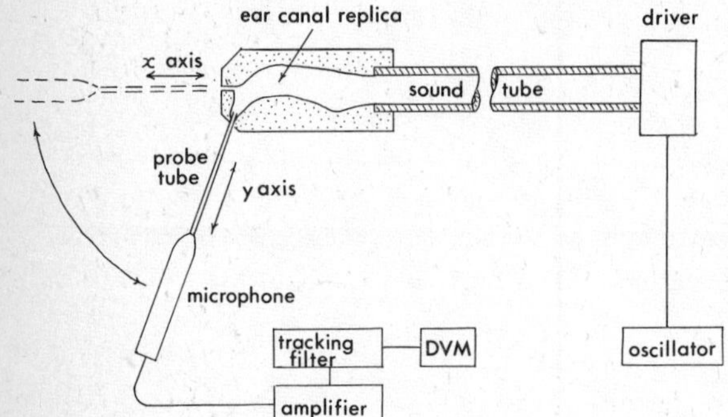

Figure 3. Measurement system used to survey the sound pressure distribution in replicas of human ear canals.

For a measurement frequency of 8 kHz (corresponding to about 20 kHz in a life-size human canal) the measured standing wave pattern shown in Fig. 4 is obtained; values along both x and y axes (circles and triangles, respectively) have been included. The smooth curve is the theoretical prediction based on the horn equation (1), using as input only the canal geometry. Agreement between theory and experiment is quite good; both show a regular series of maxima and minima of sound pressure, with the height of the maxima varying in conjunction with changes in cross-sectional area.

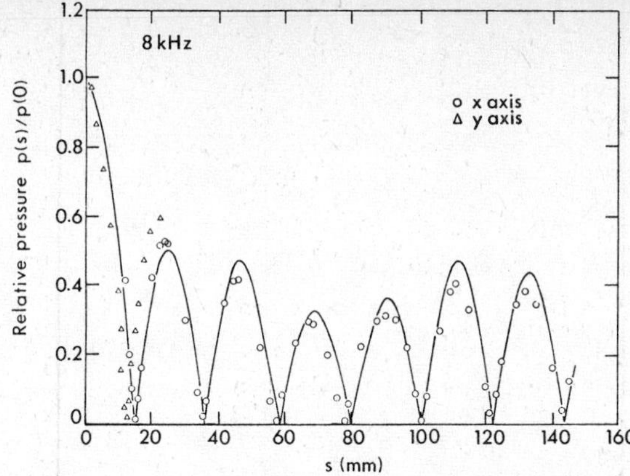

Figure 4. Measured pressure distribution in a scaled up replica model of a human ear canal, along the two axes defined in Fig. 3. The smooth curve is the theoretical prediction based only on the geometry (cross-sectional area along a curved center axis), with A(s) being measured independently. No absorption was allowed for.

To test the theory when there is absorption of energy, the replica canal was modified: at the eardrum location a hole was drilled through the casting material and a 7.6 m tube of 4.75 mm internal diameter attached. This provided a resistive load of 230 c.g.s. acoustic ohms at a position s_o = 10 mm. The measured pressure distribution at 6 kHz (Stinson, 1985) for this modified canal is shown in Fig. 5. The sound pressure is not zero at the minima, indicating absorption of energy. The ordinate shows sound pressure <u>level</u> to emphasize these minima. The solid curve again gives the theoretical prediction, this time incorporating the condition of Eq. (2) at the impedance location. The theory works quite well, predicting not only the location of extrema and height of maxima but the varying depths of the minima. In this example approximately 60% of the incident acoustic energy is being absorbed at the "eardrum" impedance.

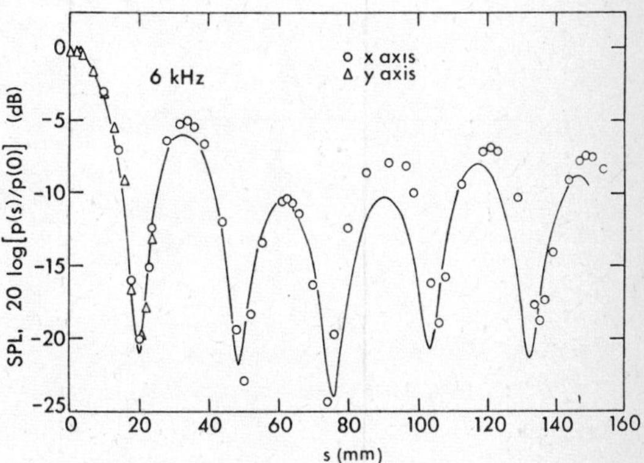

Figure 5. Measured and theoretical pressure distribution in the human replica canal, as in Fig. 4. A resistive load of 230 c.g.s. acoustic ohms has been placed 10 mm from the innermost end of the canal.

Recent experiments (Khanna and Stinson, 1985) provide pressure distributions in the ear canals of anesthetized cats. As in the replica experiments a movable probe surveys the sound field along a measurement axis. The measured standing wave patterns at three frequencies are shown for one animal in Fig. 6, by the open circles. At frequencies above about 10 kHz, relatively deep interference minima were evident in the patterns. This suggests that much of the incident energy is being reflected at these frequencies with little absorption at the eardrum. A calculation of the sound pressure using Eq. (1), with only the measured geometry [i.e. A(s)] as input (ignoring Z_{TM}), produced the solid curves shown in Fig. 6. The calculated pressure is normalized to unity at s = 0 and the experimental values have been adjusted vertically. The key features to compare are then the overall shape of the curves and location of extrema; on these counts the agreement is quite satisfactory. Between 5 and 10 kHz similar agreement was not obtained between theory and experiment indicating that there was significant absorption of sound energy at these frequencies.

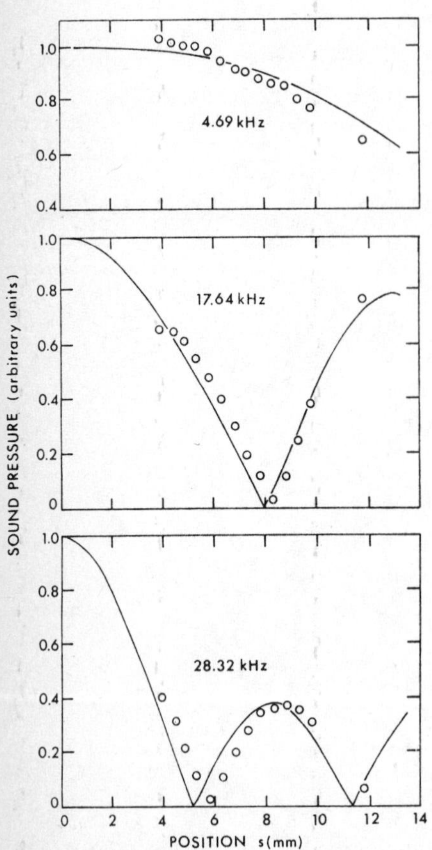

Figure 6. Measured standing wave patterns in the ear canal of a cat (open circles). The smooth curve shows a theoretical calculation that uses only the measured ear canal geometry and ignores absorption of energy at the eardrum. Experimental values have been scaled vertically.

The need to account for energy absorption was much more evident in measurements on some other animals. There seems to be quite a wide range in acoustic eardrum reflectivity between animals. Figure 7 shows a series of measurements (Khanna, 1984) for which experimental standing wave minima are relatively shallow at even quite high frequencies. It was necessary to include an eardrum impedance [through Eq. (2)] in the calculations to properly describe the measured pressure distributions. The best fit to the data were obtained using an impedance Z_{TM} (in c.g.s. acoustic ohms) of (429 + 233i) at 14.19 kHz and (1392 + 1006i) at 29.82 kHz; for both an eardrum location s_o = 2 mm was chosen arbitrarily. An attempt to apply the middle ear network model of Peake and Guinan (1967) did not work too well at these frequencies within the present formulation.

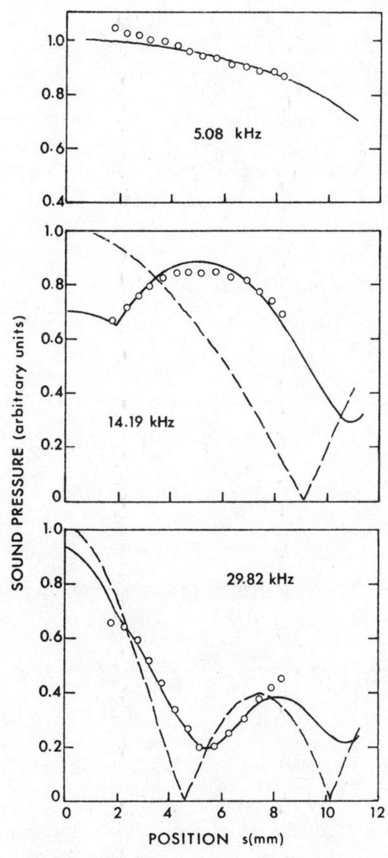

Figure 7. As in Fig. 6 but for a different animal. For frequencies of 14.19 kHz and 29.82 kHz the theoretical calculation ignoring absorption (dashed line) was not adequate and absorption via an eardrum impedance (solid line) had to be included.

4. CONCLUSION

To describe quantitatively the sound pressure distribution in real ear canals account must be made of both the ear canal geometry and the absorption of acoustic energy at the TM. The theory presented here models these factors by use of a horn equation formulation, defined in terms of a curved center axis, with energy dissipation provided by an effective eardrum impedance at a single position in the canal.

It is clear that significant variations of sound pressure level (over 20 dB at high frequencies) must be anticipated through the region adjacent to the eardrum. The use of a reference pressure probe can lead to anomalous results. It remains to be seen whether some more complicated characteristic of the sound field can be used to reliably characterize the acoustical input (e.g., SPL at innermost end of canal or at next maximum, or net acoustic power flowing into the ear canal), or if a different kind of quantity, such as malleus velocity, must be used.

Acknowledgement

The author would like to thank Shyam Khanna for the use of some of his measurements on cats.

REFERENCES

Khanna, S.M., Personal Communications, 1983 and 1984.

Khanna, S.M. and Stinson, M.R., "Specification of the Acoustical Input to the Ear at High Frequencies." J. Acoust. Soc. Am. 77, pp. 577-589, 1985.

Peake, W.T. and Guinan, J.J., Jr., "Circuit Model for the Cat's Middle Ear." Mass. Inst. Technol., Res. Lab. Electron. Quart. Progr. Rept. 84, pp. 320-326, 1967.

Shaw, E.A.G., "Eardrum Representation in Middle Ear Acoustical Networks." J. Acoust. Soc. Am. Suppl. 1 62, p. S12, 1977.

Stinson, M.R., "Implications of Ear Canal Geometry for Various Acoustical Measurements." J. Acoust. Soc. Am. Suppl. 1 74, p. S8, 1983.

Stinson, M.R., "The Spatial Distribution of Sound Pressure within the Human Ear Canal." Submitted to J. Acoust. Soc. Am., 1985.

Stinson, M.R. and Shaw, E.A.G., "Sound Pressure Distribution in the Human Ear Canal." J. Acoust. Soc. Am. Suppl. 1 73, pp. S59-S60, 1983.

THE IMPULSE RESPONSE VIBRATION OF THE HUMAN EAR DRUM

Viggo Svane-Knudsen[1] and Axel Michelsen[2]

1. ENT Department, Odense University Hospital, Odense, Denmark
2. Institute of Biology, Odense University, Odense, Denmark

ABSTRACT

Acoustical Dirac impulses (15 μsec duration, sound energy from 600 Hz to 25 kHz) are used under free field conditions for evoking impulse responses in the tympanic membranes of awake, unrestrained human volunteers. The impulse responses are analysed in the time- and frequency domains. The impulse responses within different frequency bands add to a total impulse response of complex shape (Fig. 3). The frequency spectrum of the vibration velocity of the mallear handle (umbo) shows several maxima in amplitude and corresponding deflections of phase, but the maxima are mainly caused by changes in the spectrum of the sound travelling to the ear drum. In contrast, the transfer function of the umbo (sound pressure to vibration) is smooth with only little variation in amplitude and a gradual development of phase. It is argued that the common notion of the middle ear as a low-pass filter is misleading.

1. INTRODUCTION

Most studies of the vibrations of the human tympanic membrane have been restricted to frequencies below a few kHz, and only little is known about the behaviour of the ear drum in the time domain (review: Funnell and Laszlo, 1982). In theory, the impulse response of the ear drum may be calculated from the amplitude- and phase spectrum, and *vice versa* but for such a calculation accurate data are required, and these are not easily obtained without interacting with the properties of the ear. Measurements of the absolute phase of sound are difficult at high frequencies. Furthermore, the acoustical properties of the ear canal are complex, and the driving force on the membrane may not be represented by the sound pressure at any single location in the canal (Khanna and Stinson, 1985). Calculations of the behaviour of tympanic membranes in the time domain may therefore not be realistic at present.

An experimental determination of the impulse response requires that the tympanum is activated by an ideal "acoustical Dirac impulse" and that the vibrational response is measured without any interference of the apparatus with the acoustics or mechanics of the ear. This paper is a report of our efforts in achieving this goal in awake and unrestrained human subjects.

2. MATERIALS AND METHODS

We produce very short sound impulses of moderate intensity and measure the vibrational response of the ear drum to these sounds. The vibration is measured by directing a laser beam into the ear canal and analysing the frequency shifts in the reflected light.

Unipolar acoustic impulses of about 15 sec duration are produced by letting a light membrane perform a quick jump towards the ear. This is achieved by rapidly short-circuiting the dc-voltage of a home-built electrostatic loudspeaker. Because of the limited size of the membrane, the amplitude spectrum falls off towards low frequencies, but for the impulses used here it is almost flat above 600-800 Hz, and the phase spectrum is fairly simple (Fig. 1). The loudspeaker is placed at

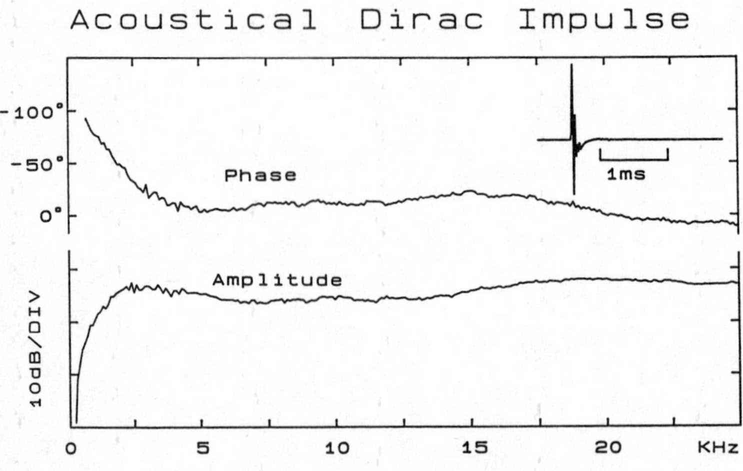

Figure 1. The sound impulse (low-pass filtered: upper right) and its amplitude and phase spectra. The phase is calculated relative to the initial rise of the sound impulse. The 90° deviation at low frequency is caused by the dipole nature of the loudspeaker.

a distance of about 30 cm from the ear and approximately in the direction of the ear canal. The surroundings of the head are arranged so as to approach the conditions of a free acoustic field for about 5 msec after each impulse (because of the limited duration of the tympanal responses, reflected sounds arriving later can be ignored). The peak sound pressure of the impulses measured at 30 cm distance is about 103 dB (re 20 μPa). Because of the short duration of the impulses, the sound level is well below the threshold for evoking a reflex contraction of the middle ear muscles.

The principle of laser vibrometry has been discussed elsewhere (Michelsen and Larsen, 1978). In short, a laser beam is divided in a measuring beam and a reference beam by a beam splitter. One of the beams is frequency shifted 40 MHz in a Bragg cell. The measuring beam is focused at the vibrating surface, and reflected light is picked up by the optical system and mixed with the reference beam. The resulting beat frequency is 40 MHz plus the Doppler frequency caused by the velocity of movement of the vibrating surface. The frequency modulation is transformed into an analog signal, which is further processed by signal averaging and FFT-frequency analysis (HP 3582A, controlled by a HP85 computer) as well as band-pass and band-reject frequency filtering of the time signals using digital finite-impulse-response (FIR) filters with "ringing" of short duration (compared with the vibrational impulse responses).

The use of short sound impulses for activating the ear drum make phase measurements relatively simple. The phase angles indicated in this paper are defined relative to the time when the ear drum starts to move when acted upon by an impulse sound. We are not certain how this phase angle relates to those measured by previous investigators (by using probe microphones in the ear canal for measuring the timing of the driving force acting on the ear drum), but our method can be used also at high frequencies where probe measurements are doubtful.

The measuring beam is focused on the ear drum by means of optics with 50 cm focal length, which allows the optical system to be placed far away from the ear as not to disturb the sound field. Only a small part of the ear drum reflects light in the direction of the ear canal. The reflection elsewhere on the ear drum is improved by placing 5 to 6 highly reflecting spheres at the points selected for measurements. The spheres (obtained from 3M reflective tape) each have a weight of about 0.5 μg. No gluing is necessary. Measurements with and without one of these spheres in the area with natural light reflection demonstrate that a sphere does not affect the vibration of the ear drum, and that a sphere vibrates with the same amplitude and phase as the ear drum.

The volunteer subjects have normal pure-tone audiograms and normal tympanograms. The ear drums are inspected with otomicroscopy and with Siegle´s pneumatic speculum. The subjects are laying on a heavy couch, and the head is supported by a sand bag. The tragus is pulled aside by a thin ear-mould-like cast in the cavum concha and the most lateral parts of the ear canal. A horn-shaped probe microphone (B&K 4170) is used for monitoring the sound entering the ear canal, either before or after the laser measurements.

3. RESULTS

The impulse responses obtained vary in duration, shape and frequency spectrum (amplitude and phase), both between subjects and between different areas of the tympanum. Anterior and posterior to the mallear handle we observe vigorous and complex vibrations which have been described by Tonndorf and Khanna (1972). The vibration of the umbo is more simple, both with respect to amplitude and phase. In this short paper we will concentrate on the vibration of the umbo.

The erratic movements of the ear drum cause a substantial noise level in recordings from living subjects, even when using reflective spheres and after averaging 256 times. The signal-to-noise ratio of the spectra calculated from the measured impulse responses may be improved by limiting the calculation to the part of the time window occupied by the main part of the impulse response. The amplitude maxima thus determined can then be used for designing band-pass FIR-filters suitable for improving the signal-to- noise ratio in the time domain. These techniques were used in producing the results of Figs. 2-3.

The frequency spectrum calculated from the impulse response of the umbo is generally not smooth (although more smooth than those obtained elsewhere on the ear drum). The amplitude spectrum has maxima and minima, which are reflected by the phase spectrum (Fig. 2). Small changes of the position of the head relative to the loudspeaker may cause the maxima and minima to move to other frequencies. Our data support the hypothesis that the transfer function (sound pressure at the ear drum to vibration velocity of the umbo) is smooth with only little variation in amplitude and a gradual development of phase, and that most of the maxima and minima observed represent the spectrum of the sound pulses at the ear drum. The umbo appears to move as a piston in the entire frequency range of the ear, and higher modes of vibration seem to occur in other areas of the ear drum.

The impulse response of the umbo appears to vary much from

subject to subject. One may easily be misled by differences in the shape of the impulse response, which only reflect minor differences in the relative phase of the frequency components. However, genuine differences are seen between subjects, not only in the raw (averaged) impulse responses, but also in the filtered impulse responses in different frequency bands (Fig. 3). At present, we are not sure how much of the inter-subject variation in the amplitude at high frequencies may be ascribed to differences in the outer ear (affecting the spectrum of the driving force on the ear drum). The differences in the effective damping (rate of decay of the impulse response), however, appear to reflect real differences between the middle ear mechanics in different subjects.

The raw impulse responses may look as composed by a "head" of high-frequency vibration followed by a "tail" of lower frequency. However, the band-pass analysis demonstrates that this impression is wrong. The "tail" is part of an impulse response with maximum energy around 3 kHz, and this impulse response starts at the same time as the "head", which mainly contains energy from higher frequencies.

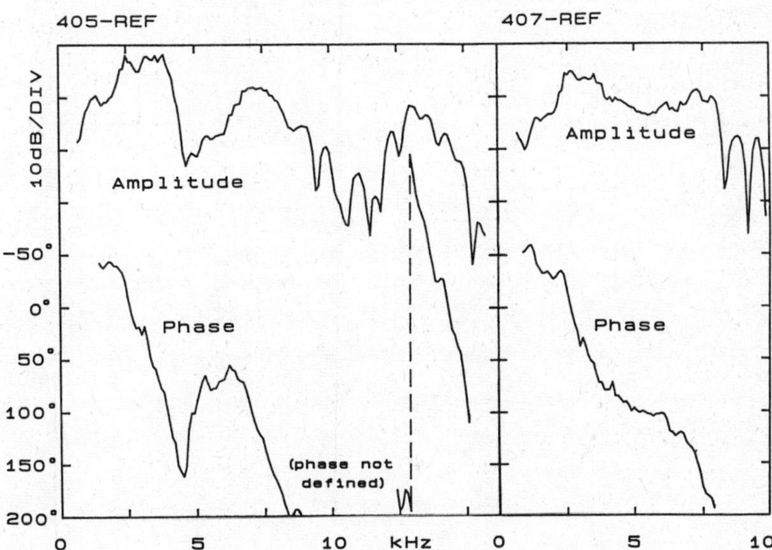

Figure 2. Two measurements of the vibration velocity of the umbo in the same person. A slight difference in the position of the loudspeaker relative to the ear canal during the two measurements is reflected in both amplitude and phase spectra. The spectra have been corrected for the spectra of the sound impulse (Fig. 1).

4. DISCUSSION

The present study includes a frequency range (6-20 kHz), which has seldom been covered by previous investigators. The behaviour of the ear drum is measured directly in the time domain, and the methods used allow for an analysis of the shape of the impulse response within different frequency bands. The sound impulses used here are close to ideal Dirac impulses (their duration is short relative to the evoked impulse responses; the amplitude- and phase spectra are known and are sufficiently smooth that deviations from a flat spectrum can be corrected for when computing the spectra of the ear vibrations; the sound level is moderate). The main shortcoming of the sound impulses is that the lowest frequencies are not represented.

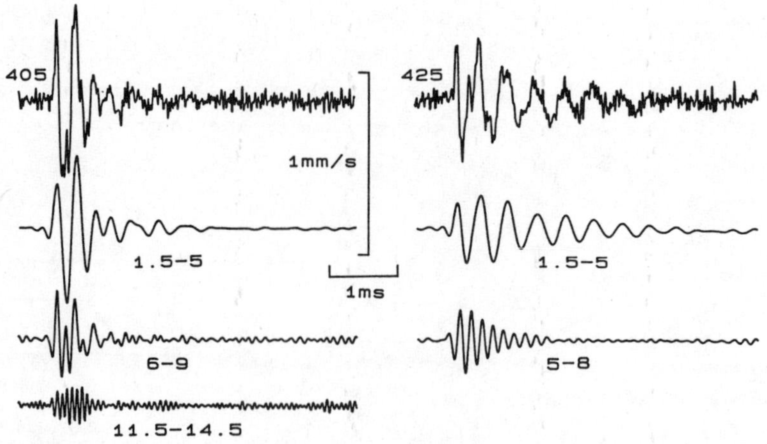

Figure 3. The impulse response of the umbo in two different subjects (left and right columns). Upper line: the original data (0.6-25 kHz). Lower lines: band-pass filtered versions (-6 dB cut-off at the frequencies indicated). Note the different amounts of damping in the two individuals. The upper-left impulse response (405) was used for computing the "405-ref" spectra in Fig. 2.

The middle ear is often referred to as a low-pass filter. This is partly a misunderstanding caused by the habit of some investigators to plot amplitude curves as displacement curves. When indicated as velocities (the correct measure when the transfer of sound energy is concerned), we observe a maximum of vibration amplitude around 3 kHz, but apart from this the amplitudes do not fall off very much with frequency. In an experiment on a guinea pig with most of the ear canal removed, we observed an amplitude spectrum which was flat within 10 dB up to 22 kHz (confirming the trend observed by Wilson and Johnstone, 1975). A part of the fall-off we observed in human ear drums may be due to our failure to select the best direction of sound to the ear.

Although more data from the ear drum and from the stapes is needed in order to establish the actual transfer function, our preliminary data suggest that the neural threshold curve with its fall-off towards high frequencies may perhaps not represent the transfer of energy to the inner ear. This assumption has been made by the manufacturers of measuring equipment for industrial noise when selecting the time constants of the instruments for impulsive sounds (Brüel, 1977; Brüel and Baden-Kristensen, 1985). We should like to stress, however, that more solid evidence is needed before the concepts behind the present procedures for noise measurements can be properly evaluated.

ACKNOWLEDGEMENTS

This work is supported by the Danish Medical and Natural Science Research Councils and by the Oticon, Novo, and Carlsberg Foundations. We are most grateful to a large number of colleagues for help on many aspects of the experimental technique.

REFERENCES

Brüel, P.V., "Do we measure damaging noise correctly?" Noise Control Engineering March/April 1977.

Brüel, P.V., and Baden-Kristensen, K., "Time constants of various parts of the human auditory system and some of their consequences". In: Time Resolution in Auditory Systems. Ed. by A. Michelsen, Springer Verlag, pp. 205-214, 1985.

Funnell, W.R.J., and Laszlo, C.A., "A critical review of experimental observations on ear-drum structure and function." ORL 44, pp. 181-205, 1982.

Khanna, S.M., and Stinson, M.R., "Specification of the acoustical input to the ear at high frequencies." J.Acoust.Soc.Am. 77, pp. 577-589, 1985.

Michelsen, A., and Larsen, O.N., "Biophysics of the Ensiferan ear. I. Tympanal vibrations in bushcrickets (Tettigoniidae) studied with laser vibrometry." J.comp.Physiol. 123, pp. 193-203, 1978.

Tonndorf, J., and Khanna, S.M., "Tympanic-membrane vibrations in human cadaver ears studied by time-averaged holography". J.Acoust.Soc.-Am. 52, pp. 1221-1233, 1972.

Wilson, J.P., and Johnstone, J.R. "Basilar membrane and middle-ear vibration in guinea pig measured by capacitive probe." J.Acoust.-Soc.Am. 57, pp. 705-723, 1975.

FORMULATION AND ANALYSIS OF A DYNAMIC FIBER COMPOSITE CONTINUUM MODEL OF THE TYMPANIC MEMBRANE

Richard D. Rabbitt and Mark H. Holmes
Rensselaer Polytechnic Institute
Troy, New York 12180

ABSTRACT

A dynamic continuum model of the tympanic membrane is formulated by accounting for its fibrous structure and including membrane type restoring mechanisms, internal structural damping, curvatures of the drum, and spatially varying properties. Accepted experimental observations are combined with the ultrastructure to argue that bending, torsional, and shear restoring forces are secondary at moderate to high sound pressure levels. The resulting model is sufficiently simple that closed form asymptotic solutions can be found which contain adequate physical content to address questions of the vibrational shape, the transient response, impulse failure, tympanoplasty effects, Eustachian tube coupling and similar related problems.

INTRODUCTION

The relatively complex geometry and fiber composite construction of the tympanic membrane (TM) have presented researchers with a formidable modeling task. Models appearing in the literature over the last century can be grouped into five basic catagories depending on the type of physical mechanisms they contain. The groups are: lumped parameter models [14,15], stiff plate models [1], curved lever models [3,7,8,9], membrane models [3,4,5,] and curvilinear shell models [4,5,6]. Even the simplest lumped parameter models are able to describe some of the behavior of the TM, however, in order to describe actual stresses, detailed vibrational shapes, and the effects of structural changes it is necessary to formulate a distributed parameter model. This requirement eliminates the first two groups. The remaining continuum type models can be discussed within the general theory of nonlinear composite shells. These models are distinguished by their ability to identify and describe the physical mechanisms associated with inertia (transverse, rotary, etc.), restoring forces (bending, membrane, shear, etc.), and damping (transverse, bending, shear, etc.).

The primary inertia mechanism can be found by inspecting the vibratory behavior of the TM. From time averaged holographic inferferometry it appears that the primary accelerations are transverse, which indicates that transverse inertia is the dominant inertia term [2,9]. Similarly, evidence of internal structural damping is easily seen by studying the manubrium amplitude over a range of forcing frequencies. Bending of the TM causes the mucous and epidermal layers to change shape at a rate equal to the rate of change in local curvature. Since the mucous and epidermal layers are water intensive cellular structures, the force required to change their shape will increase with the speed of deformation. This dissipative force is a damping mechanism and can be written as a series in the rate of change of two principal curvatures. The first term in the series is a linear bending type damping term.

Existence of a dominant restoring force is not as direct as the inertia and damping mechanisms. The three primary restoring forces that exist in a shell are bending, membrane, and shear forces. Most models appearing in the literature consider bending or membrane forces for an isotropic drum [4,5,6,7]. An exception is Funnell and Laszlo [5,6] who briefly mention a finite element single phase orthotropic shell model. Helmholtz's original curved lever work, along with Khanna and Tonndorf's [9] variation thereof, are among the few to directly discuss the structure and its relationship to deformation. To address this problem, a thin composite shell, consisting of a set of locally orthogonal fibers imbedded in a base material, is used to model the TM ultrastructure (Fig. 1). By applying the Kirchoff hypothesis and integrating the stress over the thickness, it is possible to derive formuli giving membrane bending, and shear stiffnesses in terms of the constituent components [13]. One finds that for a tightly packed fiber spacing the bending stiffness in a fiber direction is approximately 1.37 dyn-cm while the membrane stiffness is roughly 6.9×10^{6} dyn-cm. This value for the bending stiffness is essentially identical to the measured values [1,5,10]. An interesting feature of this result is that the derived stiffnesses are incompatible with isotropic materials. In fact, if an isotropic material is selected to have a bending stiffness equal to the estimated eardrum value then the membrane stiffness will be an order of magnitude less than the fiber composite eardrum material. Hence, an isotropic, or even a single phase orthotropic model may underpredict the membrane stresses by an order of magnitude if the bending stiffness is matched exactly.

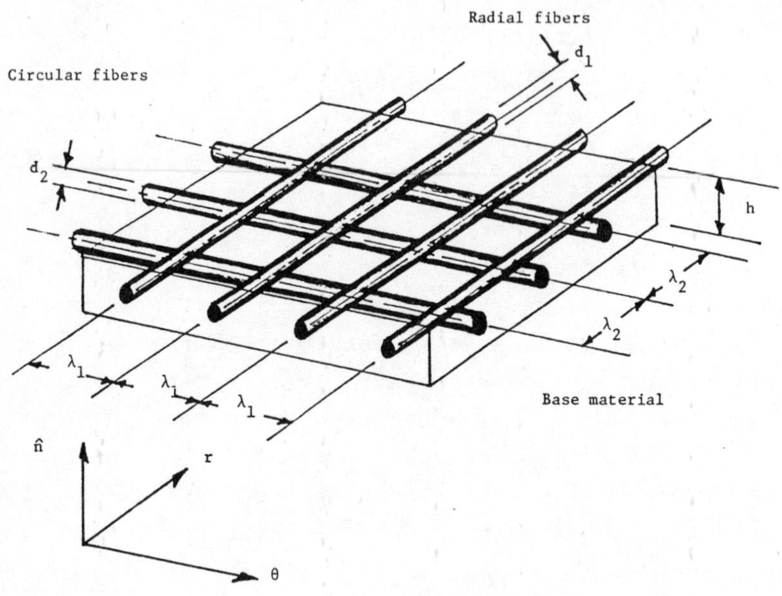

Fig. 1. Orthogonal fiber composite model of the TM (fiber spacing and size not to scale).

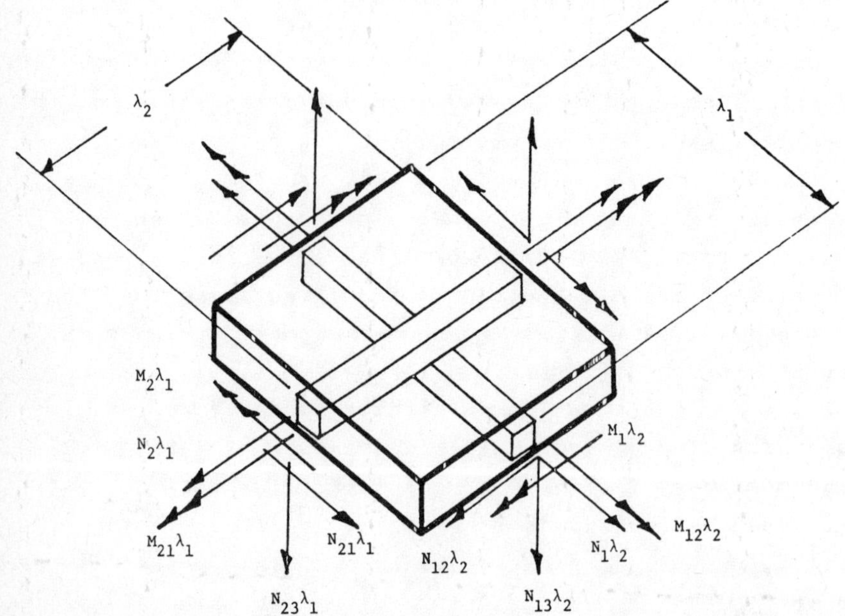

Fig 2. Equilibrium of an element of composite material.

FORMULATION OF MODEL

With this introduction and motivation, the task is to formulate a model of the eardrum including transverse inertia, bending dissipation, and membrane stresses. These features will be described in the context of a curvilinear shell theory such that the actual drum geometry is included. A model of this type is consistent with curved lever concepts as well as composite thin shell models.

By applying conservation of momentum to an element of drum material (Fig. 2), the following curvilinear membrane force equilibrium equations are obtained [12]

$$\frac{1}{H_1H_2}\{\frac{\partial}{\partial q_1}(H_2N_1) + \frac{\partial H1}{\partial q_2}N_{12} + \frac{\partial}{\partial q_2}(H_1N_{21}) - \frac{\partial H2}{\partial q_1}N_2\} + P_1 = 0 \quad (3)$$

$$\frac{1}{H_1H_2}\{\frac{\partial}{\partial q_2}(H_1N_2) + \frac{\partial H2}{\partial q_1}N_{21} + \frac{\partial}{\partial q_1}(H_2N_{12}) - \frac{\partial H1}{\partial q_2}N_1\} + P_2 = 0 \quad (4)$$

$$\frac{1}{H_1H_2}\{\frac{\partial}{\partial q_1}(H_2(\theta_1N_1 + \theta_2N_{12})) + \frac{\partial}{\partial q_2}(H_1(\theta_2N_2 + \theta_1N_{21}))\}$$

$$- \frac{N1}{R_1} - \frac{N2}{R_2} + P_3 = 0 \ , \quad (5)$$

where N_{ij} are the products of membrane stresses with local thickness, R_i are radi of curvature, q_i are the curvilinear coordinates, P_i are generalized D'Alembert pressures, H_i are coordinate normalizers, and θ_i are transverse deformation gradients. As written, damping and inertia terms are contained in P_i.

We describe the eardrum shape as a "perturbed cone" where departure from the perfect shape is given by $\varepsilon g(r,\theta)$, and the resting position is defined by $f(r,\theta) = \alpha r+\varepsilon g(r,\theta)$(see Fig. 3). Taking $q_1 = r$ and $q_2 = \theta$ then

$$H_2 = r + O(\varepsilon^2)$$

$$H_1 = (1+\alpha^2)^{1/2} + \frac{\varepsilon\alpha g_r}{2(1+\alpha^2)^{1/2}} + O(\varepsilon^2) \quad (6)$$

$$H_{12} = \varepsilon\alpha g_\theta + O(\varepsilon^2) \ .$$

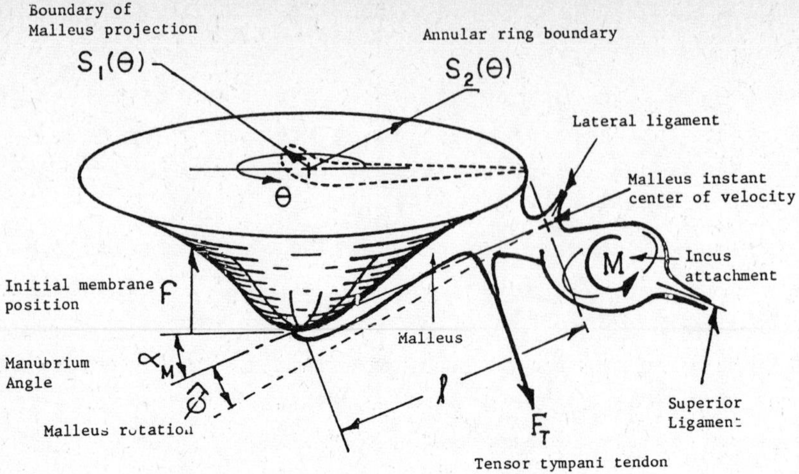

Fig. 3. Geometrical sketch of the TM.

The radi of curvature are

$$\varepsilon R_1 \simeq \frac{-1}{n_o g_{rr}} + \varepsilon \frac{r(1+\alpha^2)}{n_o \alpha} + O(\varepsilon^2) ,$$

$$R_2 \simeq \frac{r(1+\alpha^2)}{n_o \alpha} + O(\varepsilon) , \text{ and} \tag{7}$$

$$n_o = (1+\alpha^2)^{-1/2} - \varepsilon \frac{\alpha g_r}{2(1+\alpha^2)^{3/2}} + O(\varepsilon^2) .$$

Letting $w(r,\theta,t)$ denote the transverse deformation, then

$$\theta_1 = \frac{\partial w}{\partial r} \{(1+\alpha^2)^{-1/2} - \varepsilon g_r \frac{\alpha}{2(1+\alpha^2)^{3/2}} + O(\varepsilon^2)\} , \tag{8}$$

$$\theta_2 = \frac{\partial w}{\partial \theta} \{\frac{1}{r} + O(\varepsilon^2)\} .$$

ANALYSIS

The fiber composite structure of the drum is such that in-plane shear (N_{12}, N_{21}) is much smaller than the tensile forces (N_1, N_2). This feature can be seen by testing materials consisting of stiff orthogonal fibers imbedded in a flexible base matrix, or by using analytical methods. In addition to small in-plane shear, calculations based on holographic measurements show that stress in the radial direction greatly exceeds stress in the circumferential direction ($N_1 \gg N_2$, maintaining $N_2 \gg N_{12}$, N_{21}). This can be shown by measuring strain in the fibers and applying a constitutive law to calculate stress [13].

There are three small parameters in the resulting problem for the displacement $w(r,\theta,t)$, the ratio of the average circumferential stress to the average radial stress, the perturbation (ε) from the perfect cone shape, and the ratio (γ) of the amplitude of the TM to the amplitude of the malleus. One finds, after expanding in these parameters, that a first order approximation of the transverse displacement has the form

$$w(r,\theta,t) \sim \frac{[r-S_2(\theta)]\gamma\phi}{S_1(\theta)-S_2(\theta)} + v_o(r,\theta,t) , \qquad (9)$$

where the boundary $r = S_1(\theta)$ is the malleus projection, $r = S_2(\theta)$ is the annular ring, and $\phi[v_o]$ is the operator representing the coupling of the TM with the malleus. In the case of when the TM is radially homogenous and for harmonic forcing with frequency ω and amplitude $q(r,\theta)$ the function v_o in (9) is

$$v_o = \sum_{n=1}^{\infty} \Bigg\{ \frac{c^2 \, q_n \sin(\omega t+\psi_n)}{\omega_n^2\sqrt{\left[1+\left(\frac{\omega}{\omega_n}\right)^2\right]^2 + (\zeta\omega)^2}}$$

$$+ c_n e^{-\zeta\omega_n^2 t/2} \sin\left[\sqrt{1+\tfrac{1}{4}(\zeta\omega_n)^2}\, \omega_n t + \xi_n\right]\Bigg\} \sin\left[n\pi\left(\frac{r-S_1}{S_2-S_1}\right)\right] ,$$

where ζ is the damping coefficient, c is the sound speed, c_n, ξ_n are constants determined from the inital conditions,

$$\psi_n(\theta) = \mathrm{Arctan}\left(\frac{\zeta\omega_n^2\omega}{\omega_n^2-\omega^2}\right) , \qquad \omega_n(\theta) = \frac{n\pi c}{S_2-S_1}$$

and

$$q_n(\theta) = \frac{-2}{S_1-S_2} \int_{S_2}^{S_1} q(r,\theta) \sin\left[n\pi\left(\frac{r-S_1}{S_1-S_2}\right)\right]dr \ .$$

We now have a closed form asymptotic description of the motion of the TM coupled to a general middle ear and cochlear displacement operator. In order to obtain numerical results from the asymptotic solution the detailed geometry, damping, material, middle ear and cochlear constraints must be specified. To test the model, without optimizing parameters, some calculations were done for the geometry of a cat. Using a simple angular spring as a constraint on the malleus to define ϕ and very small structural damping the following experimental phenomena were reproduced [13]: low frequency vibration shapes and amplitudes as reported by Khanna and Tonndorf [9]; transition to complex vibration shapes at frequencies above 3 KHZ [9]; peaks on the "TM/malleus amplitude vs. frequency" curve including location and spacing [9,10]; and decrease in malleus amplitude with frequency having a shape as reported by Manley & Johnstone [11]. These features were obtained directly from the asymptotic solution. It should be noticed that coupling with the middle ear air chambers, Eustachian tube, and external ear has not been addressed directly. All three of these items appear as a perturbation on the excitation pressure so it is not difficult to include these features in our continuum model [13].

To summarize, a curvilinear membrane model of the general TM including structural damping has been presented. Three small parameters appearing in the formulation have been used to obtain an approximate closed form solution describing vibratory response of the eardrum. The work that remains is to optimize and physically select correct parameters, and couple the model with adjoining continuum models. The nature of the model allows direct address of questions concerning tympanoplasty, impulse failure, transient response, impedance analysis, Eustachain tube function, and related continua problems.

ACKNOWLEDGEMENT

This work was supported, in part, by the U.S. Army Research Office Grant DAAG29-83-K-0092.

REFERENCES

1. Bekesy, G. von, "The structure of the middle ear and the hearing of one's own voice by bone condition," J. Acoust. Soc. 21, 217-232, 1949.

2. Dancer, A.L., Franke, R.B., Smigielski, P., Albe, F., and Fagot, H., "Holographic interferometry applied to the investigation of tympanic membrane displacements in guinea pig ears subjected to acoustic impulses," J. Acoust. Soc. 58, 223-228, 1975.

3. Esser, M.H.M., "The mechanism of the middle ear: part II. The drum," Bulletin Math. Biophysics 9, 75-91, 1947.

4. Frank, O., "Sound conduction in the ear," Sitzungsbeer Math.-Physikal. Klass, Bayerischen Akad. Wiss. Munchen, 1923.

5. Funnell, W.R.J, "A theoretical study of eardrum vibrations using the finite-element method," Ph.D thesis, McGill University, Montreal, 1975.

6. Funnell, W.R.J. and Laszlo, C.A., "Modeling of the cat eardrum as a thin shell using the finite-element method," J. Acoust. Soc. 63, 1461-1467, 1978.

7. Gran, S., "The analytical basis of middle-ear mechanics. A contribution to the application of the acoustical impedance of the ear," Dissertation, University of Oslo, 1968.

8. Helmholtz, H.L.F., "The mechanism of the middle-ear ossicles and of the eardrum," Hinton J. Publ., New Sydenham Soc., #62, 97-155, 1874.

9. Khanna, S.M., and Tonndorf, J., "Tympanic membrane vibrations in cats studied by time-averaged holography," J. Acoust. Soc. 51, 1904-1920, 1972.

10. Kirikae, I., The structure and function of the middle ear, University of Tokyo Press, Tokyo, 1960.

11. Manley, G.A. and Johnstone, B.M.,"Middle-ear function in the guinea pig," J. Acoust. Soc. 56, 571-576, 1974.

12. Novozhilov, V.V., The theory of thin shells, P. Noordhoff Ltd., Groningen, Netherlands, 1959.

13. Rabbit, R.D., "A dynamic fiber composite continuum model of the tympanic membrane. Part I: Model formulation," Rensselaer Polytechnic Institute, Mathematics Report No. 151, Troy, N.Y. 1985.

14. Zwislocki, J., "Analysis of the middle ear function, Part I: Input impedance," J. Acoust. Soc. 34, 1514-1523, 1962.

15. Zwislocki, J., "Analysis of the middle ear function, Part II: Guinea pig ear," J. Acoust. Soc. 35, 1034-1040, 1963.

How Do Contractions of the Stapedius Muscle Alter the Acoustic Properties of the Ear?

X.D. Pang and W.T. Peake

Res. Lab. of Electronics and Dept. of Elec. Eng. & Comp. Sci., MIT, Cambridge, MA 02139
and Eaton-Peabody Lab. of Aud. Physiol., Mass. Eye & Ear Infirmary, Boston, MA 02114

Abstract

We describe our investigations of the *mechanisms* through which contractions of the stapedius muscle in the cat cause alterations in acoustic transmission through the middle ear. We have observed that stapedius contractions displace the stapes head along the direction of the stapedius tendon, which is perpendicular to the direction of stapes motion in response to sound. This stapes-head displacement (SHD) occurs *without* detectable displacement of the incus or malleus. This result suggests that the changes in transmission are solely caused by changes in the stapes impedance due to the SHD. Measurements of SHD were made together with the associated transmission changes. For SHDs up to 40 μm, the transmission was reduced up to 10 dB in the frequency range below 1.5 kHz with little change for higher frequencies. For SHDs larger than 40 μm, reductions in transmission up to 30 dB were observed in the low frequency range and up to 15 dB for high frequencies. We explore the possibility that changes in the configuration of the annular ligament at the stapes footplate are the source of the acoustic changes by comparing our results with other measurements of changes of stapes impedance. We conclude that this hypothesis is tenable.

I. Introduction

The effects of the stapedius-muscle contractions on the acoustic transmission through the middle ear have been studied in human, cat and other species (e.g., see Møller, 1984). However, the mechanisms by which the stapedius contractions alter properties of specific middle-ear structures are not clear. From the anatomy of the middle ear (Fig. 1) one can suggest several possibilities for changes in the configuration of the ossicular chain that might result from contraction of the stapedius muscle. The simplest possibility is that the stapes alone is displaced without moving the incus and malleus. On the other hand, all three ossicles and the tympanic membrane could be displaced. The primary goal of our study has been to understand the mechanism through which the stapedius muscle modulates the acoustic properties of the middle-ear. Our approach was first to observe and quantify the displacements of the ossicular chain in response to stapedius contractions. Our second step was to relate the ossicular displacements to changes in acoustic transmission of the middle ear. The third step was to analyse the mechanical consequences of the ossicular displacements so as to identify the middle-ear structures that alter the transmission.

Because of differences among published measurements of stapedius-muscle effects (e.g., see reviews by Rabinowitz, 1977, 1981), we also wished to determine how large an effect the stapedius-muscle contraction can have on the middle-ear transmission (in cat). In previous studies in the cat, the muscle contractions elicited by an intense sound (acoustic reflex) had rather small effects (e.g., Møller, 1965), whereas electric stimulation near the muscle motoneurons in the brainstem produced larger effects (Teig, 1973). In neither case is it clear whether all muscle fibers contracted. Even larger effects have been demonstrated when stapedius contractions were simulated by an external force applied on the stapedius tendon (Wever & Bray, 1942), but it is not clear whether the muscle by itself is able to generate those forces. We have used electric stimulation of the stapedius muscle (as did Nedzelnitsky, 1979) with the goal of producing maximum possible contraction of the muscle.

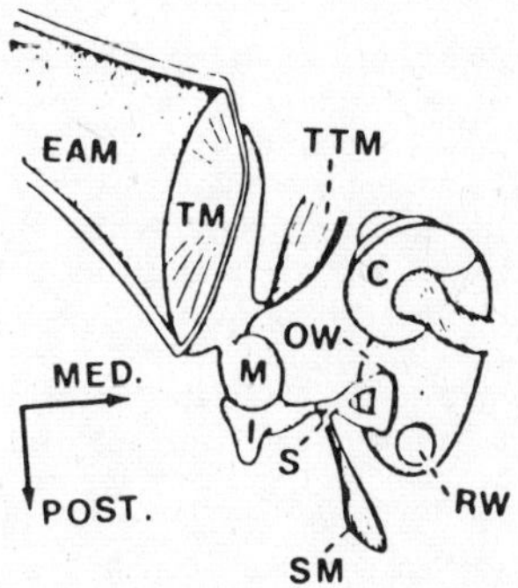

Fig. 1. A somewhat schematic view of the cat middle ear. In response to sound the stapes vibrates in the medial-lateral direction. The stapedius tendon pulls in a posterior direction. The symbols are: C: Cochlea; EAM: External Auditory Meatus; I: Incus; M: Malleus; OW: Oval Window; RW: Round Window; S: Stapes; SM: Stapedius Muscle; TM: Tympanic Membrane; TTM: Tensor Tympani Muscle. MED. and POST. stand for the medial and posterior directions, respectively.

II. Methods

Experiments were conducted on seventeen adult cats whose tensor tympani tendons were cut and whose acoustic reflex was blocked by anesthesia. The stapedius muscle was stimulated with a sinusoidal electric current (67 Hz) through a pair of electrodes placed either on or near the muscle. When the electric stimulation was turned on it produced a steady contraction of the muscle that was maintained for 30 seconds or longer. The resultant ossicular chain displacement was observed and measured during this interval through a dissecting microscope with an eyepiece micrometer. With tonal stimuli ranging from 0.1 to 10 kHz, the input and output of the middle ear were monitored through measurements of sound pressure at the tympanic membrane and cochlear potential (CP) from the round-window (through a narrow-band filter). Moderate sound pressure levels were used (e.g., 70 dB SPL) so as to avoid the noise floor and saturation of CP. Control experiments showed that the transmission-change measurements were not particularly sensitive to the sound levels for the range 65-85 dB SPL. Transmission changes during electric stimulation were eliminated after the stapedius tendon was cut. This demonstrates that the observed transmission changes were caused by the stapedius-muscle contractions.

III. Results

A. Ossicular displacement

When the stapedius muscle contracted, the stapes head was pulled posteriorly by the stapedius tendon in a direction perpendicular to that of stapes' motion in response to sound (see Fig. 1). No displacement of the incus or malleus (detection threshold 1 μm) was observed for stapes-head displacements up to 65 μm. Thus the incudo-stapedial joint appeared to be perfectly flexible in allowing the posterior sliding motion of the stapes head with respect to the incus. This is a sharp contrast with the rigidity of this joint in coupling the incus and stapes motion (in the medial-lateral direction) in response to sound (Guinan & Peake, 1967). The restriction of the ossicular displacement to the stapes allows us to characterize the ossicular displacement in response to stapedius contractions with one variable, the stapes-head displacement (SHD) in the direction of the stapedius tendon (Fig. 2).

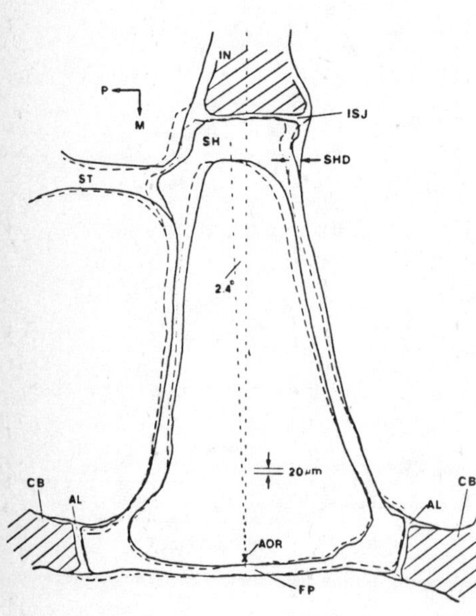

Fig. 2. Representation of stapes displacement in response to contraction of the stapedius muscle. The undisplaced configuration (solid lines) was obtained by tracing a histological section of the cat middle ear. In the experiments the region of the incudo-stapedial joint and the tendon was viewed from a direction perpendicular to this plane. We have assumed that the stapes rotated as a rigid body about an axis at the footplate (AOR) that is perpendicular to this plane. This notion is consistent with our observation that there was no stapes-head displacement in the medial-lateral direction in response to stapedius contractions. The "displaced" tracing (dashed contour) was designed to mimic a stapes-head displacement (SHD) of 60 μm, with a 2.4 degree angle of rotation. No displacement of the lenticular process of the incus is shown, which is consistent with our observations. For comparison of displacement amplitudes, the peak-to-peak displacement of the stapes for a sound input of 140 dB SPL at the tympanic membrane (20 μm) is shown. Symbols: AL: Annular Ligament; AOR: Axis Of Rotation; CB: Cochlear Bone; FP: Foot-Plate; IN: INcus (lenticular process); ISJ: Incudo-Stapedial Joint; M: Medial; P: Posterior; SH: Stapes Head; SHD: Stapes-Head Displacement; and ST: Stapedius Tendon.

Measures of SHD (the displacement from the normal position of the stapes head) were repeatable with a given electric stimulation level of the muscle. The SHD increased monotonically with increasing muscle stimulation level until the stimulation level was so high that jaw muscles contracted. In this range the SHD increased up to about 60 μm, which is about one-tenth of the diameter of the incudo-stapedial joint (Fig. 2).

B. Dependence of transmission changes on the stapes-head displacement

We have used the fundamental component of the cochlear potential as a measure of the output of the middle ear. Thus the transmission through the middle ear is proportional to the ratio of the cochlear potential to the sound pressure at the tympanic membrane; the change in transmission (ΔT) was defined as the ratio of the transmission with SHD to the transmission without SHD and we express it in decibels.

Figure 3 shows the magnitude of the change in transmission as a function of SHD and acoustic frequency for a typical ear. Transmission was generally attenuated by the stapedius contractions. For small SHDs (i.e., < 40 μm), the transmission change occurred primarily in the low frequency range (< 1.5 kHz), and the amount of attenuation was 7 dB or less. At larger SHDs, the low frequency transmission decreased by as much as 25 dB (in this case), and the high frequency transmission also decreased by as much as 14 dB at 7 kHz. In general, the attenuation increased monotonically with SHD. For frequencies above 1 kHz the attenuation was not monotonic with frequency. These general features of the data in Fig. 3 were seen in the 13 preparations in which transmission changes were measured. In several ears the attenuation at low frequencies was as much as 30 dB.

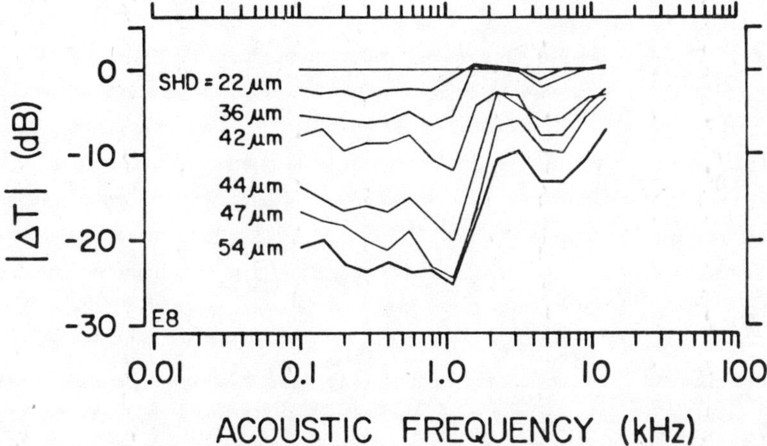

Fig. 3. Change in transmission in one middle ear as a function of acoustic frequency for six different stapes-head displacements (SHDs). Measurements at ten other SHDs are not shown to avoid congestion of the figure. Each curve is an average of six measurements. Each measurement was made during an interval (≈ 30 seconds) after the onset of the electric stimulation of the muscle. During this interval SHD, cochlear potential, and sound pressure at the tympanic membrane were constant. Measurements were made at equal intervals of 1/7 decade in log frequency. The horizontal line marks 0 dB. The sound pressure level used for the measurements was about 70 dB SPL.

C. Mechanical consequences of the stapes-head displacement

Because the stapedius contractions apparently did not displace the incus or malleus, the change in transmission appears to be caused by a change in the stapes impedance, that is, the impedance of the stapes and its attachments. The circuit model in Fig. 4 illustrates this idea, which has been previously developed by Rabinowitz (1977, 1981). Thus, the hypothesis is that the stapes impedance, Z_S, is a variable that modulates the acoustic transmission from the tympanic membrane to the cochlea.

This hypothesis is attractive because it is generally consistent with a previous demonstration that the acoustic impedance of the stapes is substantially altered by changes in the static pressure-difference across the stapes footplate (Lynch, Nedzelnitsky & Peake, 1982, Fig. 10). We will compare these published measurements of changes in Z_S with our measurements of transmission change as a function of SHD (Fig. 3) to see whether they can be jointly consistent with the hypothesis.

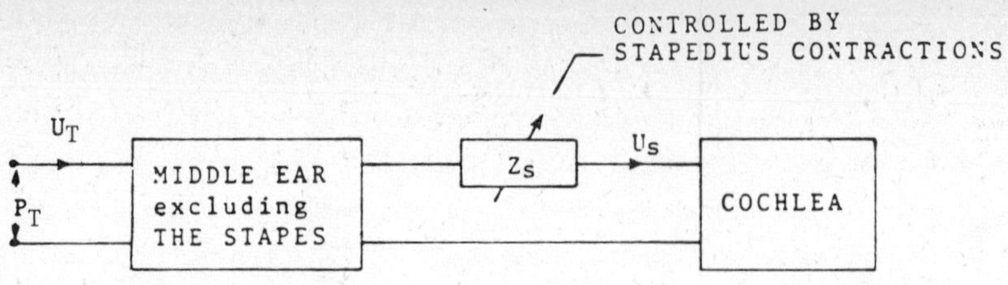

Fig. 4. Circuit model illustrating the stapedius modulation of the middle-ear transmission. The impedance of the stapes, Z_S, is shown in series with the cochlear input impedance, Z_C. Middle-ear transmission = U_S/P_T, where P_T = sound pressure at the tympanic membrane, and U_S = volume velocity of the stapes.

Lynch, et al. (1982, Fig. 10) measured the acoustic impedance of the stapes at a low acoustic frequency as a function of a static pressure on the stapes. The impedance was measured by delivering sound directly to the stapes with the other parts of the middle ear removed. Lynch, et al. had concluded that the stapes impedance was primarily determined by the annular ligament and they hypothesized that the static pressure caused a static displacement of the stapes footplate and therefore a deformation of the annular ligament. The low frequency acoustic admittance can be considered as proportional to the slope (derivative) of the displacement-pressure curve and therefore integration of the admittance measurement with a proper offset provides a relation between the pressure and the displacement (Lynch, 1978). From this displacement-pressure relation we have constructed Fig. 5A, in which the change in stapes admittance (ΔY_S) is plotted as a function of the static displacement of the annular ligament (The annular ligament was presumably displaced equally at all locations around the footplate.). The curve shows that the admittance is reduced to 1/10 of its maximum magnitude by a static displacement of 9.5 μm.

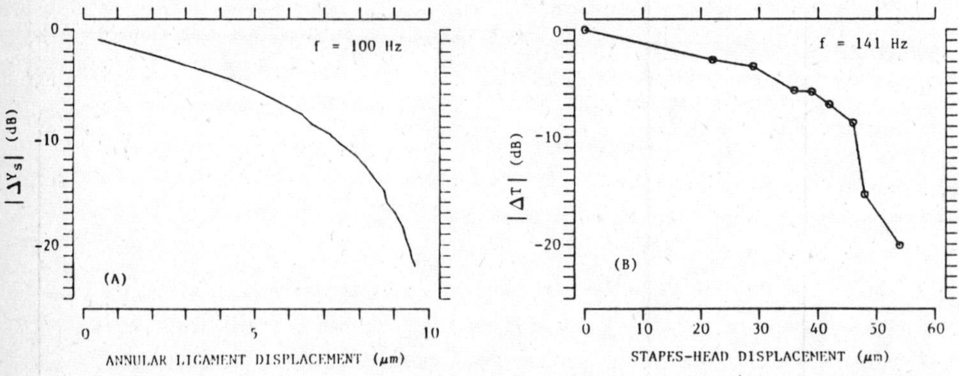

Fig. 5. (A) Magnitude of the change in acoustic admittance of the stapes (Y_S = $1/Z_S$) vs. calculated static annular-ligament displacement. The measurements of $|Y_S|$ vs. pressure were approximately symmetrical about the maximum value of $|Y_S|$, thus this curve applies to displacements in either direction. The change in admittance is defined as the ratio of the admittance (Y_S) with static displacement to the admittance without displacement. (B) Magnitude of change in middle-ear transmission vs. stapes-head displacement (SHD) from data similar to those shown in Fig. 3. These data are typical of those in the low frequency region.

For comparison with Fig. 5A we have juxtaposed (in Fig. 5B) some of our results of the effects of the stapes-head displacement caused by stapedius contractions on middle-ear transmission (at a low frequency). In this case we see that a transmission reduction of 20 dB occurred with an SHD of 54 μm. Our working hypothesis is that both of these results (Fig. 5A and 5B) are caused by the change in the configuration (mechanical strain) of the annular ligament.

Are the available data consistent with the hypothesis? In exploring this question we first deal with a qualitative issue which may appear to invalidate the hypothesis. Lynch, et al. (1982, p.115) reported that, although static pressure variations produced large alterations in admittance magnitude at 0.1 kHz, they produced small alterations (3 dB) at 1.0 kHz. A possible explanation for this is that at 1 kHz the effects of the displacements produced by their static pressure differences were acoustically "equivalent" to those of the smaller SHDs that have effects only at low frequencies (Fig. 3). If we adopt this possibility as a part of our working hypothesis, we then expect that larger displacement than those of Fig. 5A would produce changes in Z_S at high frequencies.

We now take a closer look at the low frequency measurements. The change in middle-ear transmission can be related to the change in stapes admittance from the model:

$$\frac{T'}{T} = \frac{1 + B}{1 + \dfrac{B}{Y_S'/Y_S}} \tag{1}$$

where T = middle-ear transmission, Y_S = stapes admittance, the primed variables apply to the stapedius contracted situation, $B = Z_S/(Z_O + Z_C)$, Z_O is the output impedance of the middle ear block, and Z_C is the input impedance of the cochlea. At low frequencies both Y_S' and Y_S are compliance-like (Lynch, et al., 1982) and their ratio is real. From a low-frequency model by Lynch (1981), we conclude that B is also real and is approximately equal to one for low frequencies. Thus, Eq.(1) allows one to determine T'/T as a function of Y_S'/Y_S and each $[T'/T, Y_S'/Y_S]$ pair can be transformed into a [stapes-head displacement (SHD), annular-ligament displacement] pair through Fig. 5A and 5B. The result of this transformation is shown in Fig. 6. In so far as the resulting curve is a straight line through the origin, each SHD is "equivalent" to a uniform annular ligament displacement of SHD/5.4 at low acoustic frequencies.

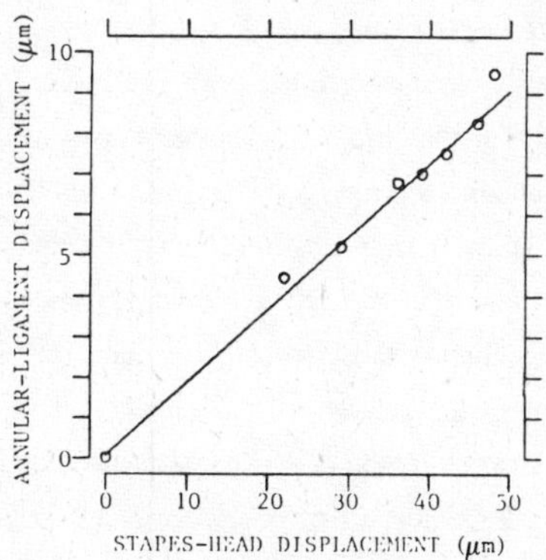

Fig. 6. Relationship between "equivalent" stapes-head displacements and the uniform annular-ligament displacements for low acoustic frequencies. Equation (1) with B = 1 was used along with Fig. 5A and 5B to determine "equivalent" displacements. The points were determined at the stapes-head displacements for which ΔT measurements were made (Fig. 5B). The straight line is a regression line with slope = 0.18; the correlation coefficient of the data = 0.99.

Let us now consider how the above "equivalence" can come about and whether the factor of 5.4 is reasonable. From Fig. 2 we see that a stapes-head displacement causes an annular-ligament displacement. Since the long-axis of the footplate (parallel to the stapedius tendon) is approximately equal to the height of the stapes from the footplate to the tendon, the *maximum* displacement of the annular ligament is about half of the SHD. With the assumed rotation of the stapes that is shown, the resultant displacement of the annular ligament is spatially *nonuniform* with its maximum at the poles of the long-axis of the footplate and minimum (zero) along the axis of rotation. Thus, only a fraction of the ligament is strained in the SHD case as compared to the uniform displacement of the ligament in the case of static pressure. It is conceivable then that this difference in spatial uniformity can contribute the other factor of 5.4/2 = 2.7.

From the above comparisons, we conclude that the available data are consistent with the hypothesis that the annular ligament controls the stapes impedance through some process in which its acoustic (or incremental) "stiffness" is altered by the static strain of the ligament.

IV. Discussion

On the basis of the observed motion of the stapes head caused by stapedius-muscle contraction and the *absence* of simultaneous motion of other middle-ear ossicles, we have pursued the hypothesis that the stapedius muscle produces its effects through an increase of the stapes impedance caused by mechanical strain of the annular ligament. Our test of this hypothesis could be improved (especially for high frequencies), but with a few reasonable assumptions we concluded that the hypothesis is tenable. One way to provide a good quantitative test would be to measure the impedance of the stapes, Z_S, e.g. in the manner of Lynch, et al. (1982), as a function of SHD. In addition the value of B (or Z_O) at high frequencies needs to be determined. With these quantities it would be possible to test the hypothesis by predicting the T'(SHD) from Eq.(1). In the meantime we can speculate about some larger implications of the observations.

How large can the effects of physiological contractions of the middle-ear muscles be? In human and animals, stapedius contractions evoked through the acoustic reflex produce transmission attenuations of about 10 dB for acoustic frequencies below 1 kHz and little effect for higher frequencies (Mϕller, 1984; Rabinowitz, 1977). However, our data and those obtained by Teig (1973) suggest that at least in the cat much larger effects can occur. It may be that the measurements of acoustic reflex have not evoked the maximum attenuation, because experimenters have not used stimuli that evoke maximum activity of the motoneurons involved. Also, recent experiments (McCue and Guinan, 1983) suggest that only a fraction of the stapedius motoneurons is involved in the acoustic reflex. Thus, it is conceivable that the brain can cause much stronger stapedius muscle contractions than those evoked in the reflex measurements. It is then conceivable that the largest effects that we have observed (including those for high frequencies) with electric stimuli (Fig. 3) can also occur with physiological activation of the muscle.

The mechanical behavior of the incudo-stapedial joint and the hypothesized consequences suggest some general speculations. The role of this joint in localizing the mechanical effect of stapedius contractions to *only* the stapes suggests a general function for ossicular joints and thereby suggests an advantage for middle ears with three ossicles as found in mammals, compared to the one-ossicle model found in birds, reptiles, and most amphibians. If, in general, a joint allows one ossicle *only* to move when a muscle contracts, this provides the ear with a system in which a muscle can have a "localized" effect through ligaments that are attached to only one ossicle. The advantage of this arrangement might be that the other muscle (the tensor tympani) can exert its force on ligaments that are unstrained by the stapedius. Thus, two muscles can have relatively independent effects on the middle ear by having them attached to separate ossicles that are effectively "uncoupled" for the motions introduced by the muscles. This is consistent with our (not extensive) observation that electric stimulation of the tensor tympani in cat causes displacement of the malleus but not the incus or stapes at the incudo-stapedial joint. This suggests the further speculation that it is possible for the stapedius and the tensor tympani *together* to produce significantly larger effects on the middle-ear transmission than those that have been observed.

Acknowledgments

We thank M.F. Bourgeois for surgical preparations, J.W. Larrabee for figure preparation, and J.J. Guinan, Jr. and J.J. Rosowski for much assistance. R.A. Eatock, D.M. Freeman, J.J. Guinan, Jr., W.M. Rabinowitz, J.J. Rosowski and T.F. Weiss provided helpful comments on the manuscript. This research was supported by NIH grant NS-13126.

References

Guinan, J.J., Jr. & Peake, W.T. (1967) Middle-ear characteristics of anesthetized cats. J. Acoust. Soc. Am., 41(5): 1237-1261.

Lynch, T.J., III (1978) Personal communication.

Lynch, T.J., III (1981) Signal processing by the cat middle ear: admittance and transmission, measurements and models. Ph.D. thesis, Dept. of Elec. Eng. & Comp. Sci., MIT, Cambridge, MA.

Lynch, T.J., III, Nedzelnitsky, V. & Peake, W.T. (1982) Input impedance of the cochlea in cat. J. Acoust. Soc. Am., 72(1): 108-130.

McCue, M.P. & Guinan, J.J., Jr. (1983) Functional segregation within the stapedius motoneuron pool. Soc. Neurosci. Abstr., 19: 1085.

MΦller, A.R. (1965) An experimental study of the acoustic impedance of the middle ear and its transmission properties. Acta Oto-laryng., 59: 1-19.

MΦller, A.R. (1984) Neurophysiological Basis of the Acoustic Middle-ear Reflex. In *The Acoustic Reflex* ed. by S. Silman, Academic Press.

Nedzelnitsky, V. (1979) Effects of middle-ear muscle contraction on transmission of sound to the inner ear (cat): some direct measurements. J. Acoust. Soc. Am., 65 Suppl(1): S10.

Rabinowitz, W.M. (1977) Acoustic-reflex effects on the input admittance and transfer characteristics of the human middle-ear. Ph.D. thesis, Dept. of Elec. Eng. & Comp. Sci., MIT, Cambridge, MA.

Rabinowitz, W.M. (1981) Acoustic-reflex effects on middle-ear performance. J. Acoust. Soc. Am., 69 Suppl(1): S44.

Teig, E. (1973) Differential effect of graded contraction of middle ear muscles on the sound transmission of the ear. Acta Physiol. Scand., 88: 382-391.

Wever, E.G. & Bray, C.W. (1942) The stapedius muscle in relation to sound conduction. J. Exp. Psychol., 31: 35-43.

MEASUREMENT OF EARDRUM ACOUSTIC IMPEDANCE

J. B. Allen
AT&T Bell Laboratories
Murray Hill, NJ 07974

ABSTRACT

In this paper we describe a system which we have developed to measure cat ear canal specific acoustic impedance Z_{sp}, magnitude and phase, as a function of frequency, for frequencies between 200 Hz and 33 kHz, and impedance magnitudes between 4.0 to 4.0×10^5 rayles (MKS). The object to be measured is placed at the end of a 3.5 mm diameter sound delivery tube. After a simple calibration procedure, which determines the Thévenin parameters for the acoustic source transducer, the impedance may be calculated from the pressure measured at the orifice of the delivery tube with the unknown load in place. This procedure allows for a fast but accurate measure of a specific acoustic impedance. The system has been tested by measuring the impedance of a long cavity and comparing this response to the exact solution of the linearized Navier Stokes equations (acoustic equations including viscosity and thermal conduction). We have used this system to measure the impedance of the normal cat tympanic membrane in more than 30 cats. Healthy animals were found to have a real input impedance of ρc between 0.3 to 20.0 kHz. When the scala vestibuli was drained, the real part of the impedance dropped to less than $\rho c/10$ for frequencies less than 3.0 kHz. Above 3 kHz, the impedance for the drained cochlea is best described by an open circuited transmission line.

1. INTRODUCTION

As in the case of electrical networks, impedance is an important characterization of an acoustical network. However, unlike the electrical case, no convenient commercial system is available for acoustical impedance measurement. Reasons for this lack include the unavailability of low distortion (0.005% distortion) acoustic sources, the unavailability of precisely calibrated acoustic impedances, and the complications introduced by the wave-like nature of sound (due to the relatively slow sound speed). Therefore the need exists for a fast, precise, automatic (computer) method of linear system identification. In this paper we shall describe a method of acoustical measurement which address all of the above problems. The impedance measurement technique used here is based on accurately estimating the Thévenin equivalent parameters for a sound source, namely the open circuit pressure and the source impedance, as functions of frequency. If one knows the Thévenin equivalent source parameters, then the impedance of any acoustic load may be calculated given the pressure at its input. This technique has been previously applied [Beranek (1949); Lynch (1974); Mawardi (1949); Tonndorf and Khanna (1967)]; however, the implementation discussed here differs in several ways. First, we use an electret push-pull (Hunt, 1954) low distortion sound source that has a uniform frequency response up to 30 kHz. Second, this sound source is connected to a uniform diameter (3.5 mm) tube through a matching acoustic resistor which is used to reduce reflections at the source end of the sound delivery tube, thereby minimizing standing waves in the delivery tube [Sokolich, G. W. (1977)]. As a result of the matching resistor, the Thévenin source impedance at the sound delivery end is close to the characteristic impedance of the acoustic transmission line.

The transducer's open circuit pressure and source impedance are computed from four different pressure responses, (measured via a single microphone at the system's orifice) which result from its being terminated by four different acoustic loads. The precise procedure for doing this is believed to be both novel and considerably more accurate than previous published methods. Based on the assumptions made in the calibration procedure, the calibration is insensitive to temperature changes and positioning of the probe microphone. Also, the method does not require the use of a calibrated probe microphone, since the measurement method (as we shall show) is independent of the probe transfer function.

We have developed this system to measure the specific acoustic impedance looking into the cat ear canal. Our results differ from those of others in significant ways. First, we find that the eardrum (TM) is matched to the impedance of air over the frequency range from 300 Hz to 20 kHz. However, significant animal variability was observed. For those animals for which the TM was clear (transparent), which we took as a measure of a healthy middle ear, the measured impedance was uniform over frequency and closest to ρc. The impedance was usually measured within 5 mm of the TM, with the bulla and septum widely opened. A closed bulla and septum strongly modified the measured impedance at certain frequencies, such as at the bulla resonance frequency of 3 kHz.

2. THEORETICAL METHOD

In Fig. 1 we show an equivalent circuit for our sound delivery system, loaded by an unknown load impedance $Z_x(\omega)$. The output response pressure $P_x(\omega)$ across the load $Z_x(\omega)$ is measured through a probe tube having transfer function $H_p(\omega)$. From Fig. 1

$$R_x(\omega) = H_p(\omega)P_x(\omega) \,, \tag{1}$$

where $R_x(\omega)$ represents the response measured through the probe tube at radian frequency $\omega = 2\pi f$. Note that Z_x, R_x, H_p, and P_x are all complex functions of frequency representing the Fourier transforms of time functions $z(t)$, $r(t)$, $h(t)$, and $p(t)$.

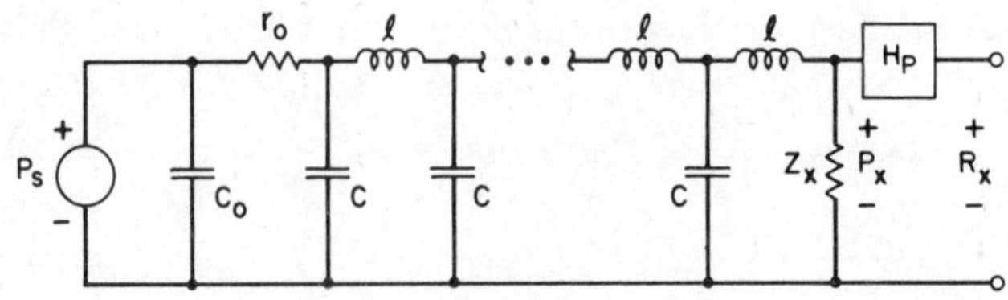

Fig. 1

From Fig. 1, the input sound source P_s is connected to the sound delivery tube, represented here as a mass-compliance transmission line, through a porous screen which acts as an acoustic resistor r_0. The screen is created by cascading several screens together which terminate the transmission line at its driven end. The proper screen resistor is determined by setting Z_x to an acoustic open circuit, (zero volume velocity via a rigid wall condition at the probe microphone) and then measuring the standing wave ratio (SWR) with the probe as a function of frequency. The screen is chosen to minimize the resulting SWR.

We define $Z_0(\omega)$ to represent the Thévenin complex source impedance and $R_0(\omega)$ to represent the Thévenin open circuit source pressure. For reasons which will become clear, we shall define all impedances in terms of the *specific acoustic impedance* [Beranek (1954)], which is defined as the ratio of the pressure to the particle velocity, and which is measured in MKS rayles, or equivalently Pa sec/m, where one Pa is the pressure in Pascals (1 nt/m^2).

The relation between $R_x(\omega)$ and $Z_x(\omega)$ in terms of $R_0(\omega)$, and $Z_0(\omega)$ is,

$$R_x = \frac{Z_x R_0}{(Z_0 + Z_x)} . \tag{2}$$

Given four known load impedances $Z_1(\omega)$, $Z_2(\omega)$, $Z_3(\omega)$, and $Z_4(\omega)$, which correspond to measured pressure responses R_1, R_2, R_3, and R_4, one may solve for $Z_0(\omega)$ and $R_0(\omega)$ in terms of the knowns R_1, R_2, R_3, R_4, Z_1, Z_2, Z_3 and Z_4 by solving the over specified system of equations:

$$\begin{bmatrix} Z_1 & -R_1 \\ Z_2 & -R_2 \\ Z_3 & -R_3 \\ Z_4 & -R_4 \end{bmatrix} \begin{bmatrix} R_0 \\ \\ \\ Z_0 \end{bmatrix} = \begin{bmatrix} R_1 Z_1 \\ R_2 Z_2 \\ R_3 Z_3 \\ R_4 Z_4 \end{bmatrix} \tag{3}$$

by least squares methods, resulting in the solution:

$$\begin{bmatrix} R_0 \\ \\ Z_0 \end{bmatrix} = \frac{1}{\Delta} \begin{bmatrix} \Sigma|R_i|^2 & -\Sigma Z_i^* R_i \\ \\ \Sigma R_i^* Z_i & -\Sigma|Z_i|^2 \end{bmatrix} \begin{bmatrix} \Sigma|Z_i|^2 R_i \\ \\ \Sigma|R_i|^2 Z_i \end{bmatrix} \tag{4a}$$

where

$$\Delta = \left(\Sigma|Z_i|^2\right) \left(\Sigma|R_i|^2\right) - \left(\Sigma R_i^* Z_i\right) \left(\Sigma Z_i^* R_i\right) \tag{4b}$$

Once R_0 and Z_0 have been determined, Z_x may be found from R_x using the relation

$$Y_x = Y_0 \left(\frac{R_0}{R_x} - 1 \right) , \tag{5}$$

where Y_x and Y_0 are $1/Z_x$ and $1/Z_0$ respectively.

The above analysis assumes that we have four standard impedances Z_1, Z_2, Z_3, and Z_4 which are known. We next discuss the choice of these standard impedances.

3. CALIBRATION IMPEDANCES

For an acoustic transmission line closed at the far end, the specific input impedance is [Beranek, (1954)]

$$Z(\omega) = -i\,\rho c\,cot(kL)\,. \tag{6}$$

In air, $\rho c = 412.5$ rayles, $i = \sqrt{-1}$, $k = \omega/c$ and L is the length of the tube. Since any transmission line model assumes uniform (plane-wave) flow, the specific impedance is independent of the cross-sectional area of the tube. Equation (6) is therefore a one parameter model of the impedance, where the length L may be approximately determined from the first antiresonance of the impedance (that frequency f_0 where Z first becomes zero). The length L is related to f_0 by the relation $L = c/4f_0$. Thus the specific acoustic impedance Eq. (6) is completely determined if we know the length, or equivalently the frequency f_0 of the first impedance zero. This frequency is easily estimated from the pressure response since the pressure has its zeroes at the impedance zeroes.

These lengths may be very precisely determined by minimizing the norm over frequency of the residual error of the over determined equations (Eq. (3)), with respect to the unknown cavity lengths. The resulting estimated lengths then provide the best overall fit to the equations. This procedure is used to improve the accuracy of the estimate of L_1, L_2, L_3, and L_4. It was found to be necessary to include damping in the model for Z_1, Z_2, Z_3, and Z_4. This was done by using the 'exact' solution to the transmission line equations with viscous and thermal conduction losses included rather than Eq. (6) [White et al., (1982); Zuercher et al., (1977)].

In Fig. 2 we show the open circuit pressure and source impedance magnitude as estimated by the above defined procedure for our transducer system. The units for pressure are in A/D volts, with

Fig. 2

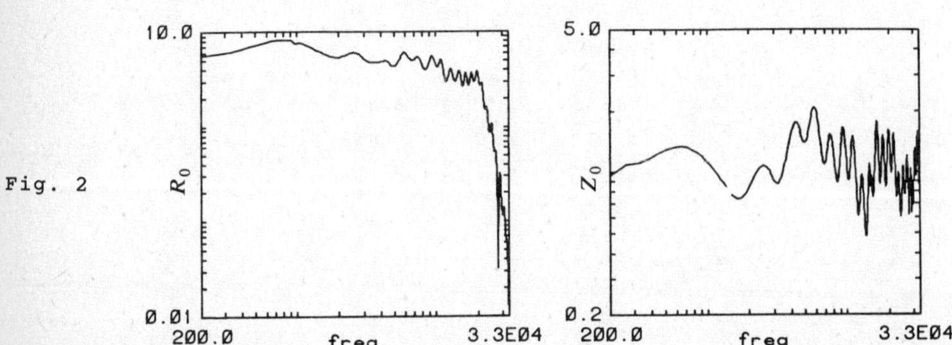

10 volts applied to the transducer, while the units for the source impedance are normalized by ρc, the impedance of air.

4. CAT IMPEDANCE MEASUREMENTS

Next we present our experimental results. Periodically, prior to use, the system was recalibrated to reduce the effects of temperature and system variations. The calibration procedure only took a few minutes and was not inconvenient. In order to estimate the accuracy of the calibration of the system, we initially measured the impedance of a 2.35 cm uniform cavity. Since the exact solution, including losses, is known for this case, it is possible to compare the measured impedance to the exact results. We show this comparison in Fig. 3. In this figure, the exact numerical solution is shown as a dashed line, and the experimental result is shown as the solid line. In the left panel, the magnitudes of the two impedances are shown, while in the right panel, the real part of the impedances are compared. The experimental and theoretical results are in very good agreement, with the two curves almost totally overlapping over three orders of magnitude.

Fig. 3

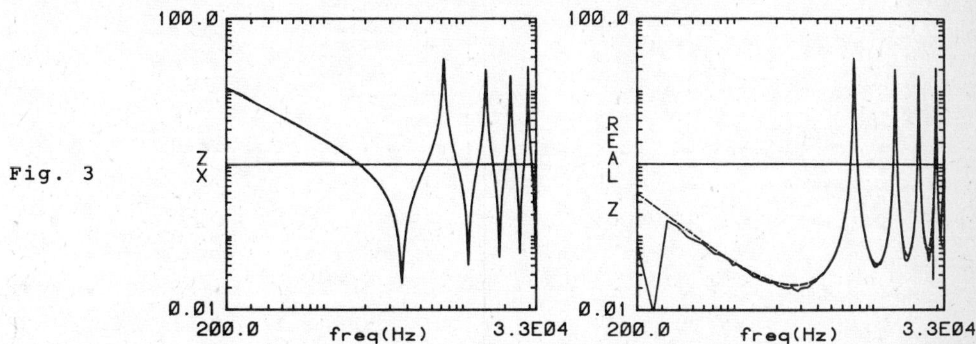

We measured the impedance looking into a cat ear with the probe tube tip between 2.5 to 5.0 mm from the ear drum. In Fig. 4, curve 1, we see the measured impedance of a healthy cat ear over the frequency range from 200 Hz to 33 kHz. Over this frequency range, the input impedance is nearly real and is equal to ρc. The majority of healthy cat ears showed similar results, with damaged ears showing impedance frequency dependent variations of between 6 to 10 dB. While it is esthetically pleasing to see such a uniform impedance match, it is also surprising.

The remaining curves of Fig. 4 show the results of a systematic damage experiment, where the cochlea was progressively removed from

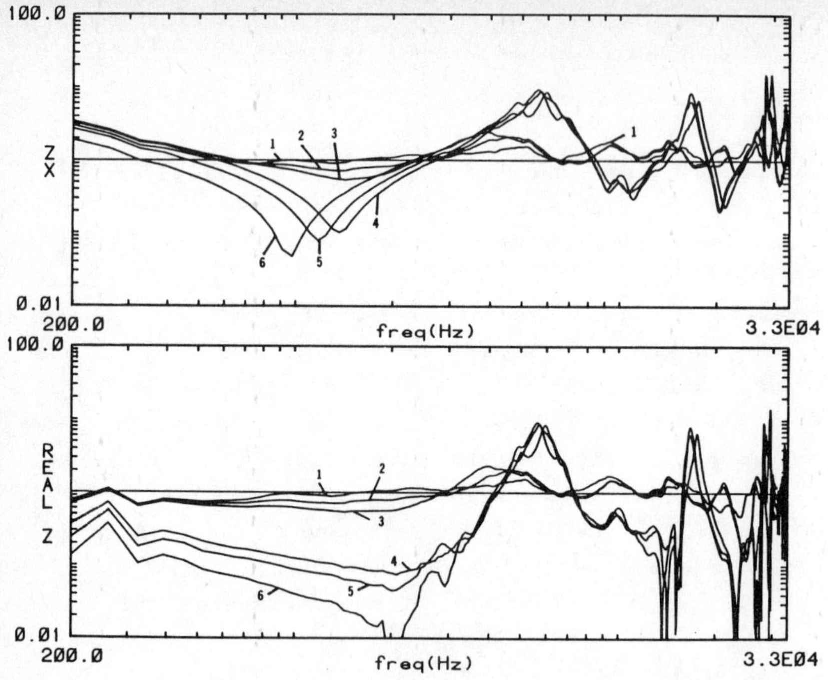

Fig. 4

the system while the eardrum impedance was monitored. Curve 1 shows the normal eardrum impedance prior to damage. In curve 2, we show the measured impedance after touching the basilar membrane with a glass probe. In curve 3, the basilar membrane has been punctured with the glass probe, producing a small hole. Note that after touching the BM and after creating the hole, the eardrum had a small, but measurable change in the input impedance. Relative to the ear canal pressure, this difference was only 2 and 4 dB at 1.8 kHz, while the impedance change was 3 and 6 dB respectively. After removal of the round window, no observable impedance change was observed. Next, the fluid was drained from the scala tympani. Again no observal change was seen. However, when the BM was removed, and the scala vestibuli was drained, a dramatic change in impedance was observed, as may be seen in curve 4. Below 3.5 kHz the resistance decreased by 20 dB. Above that frequency a standing wave pattern is observed in the magnitude response, which is similar to the cavity response seen in figure three, in that the resistance increases at the pole frequencies and decreases at the zero frequencies. From curve 4, it is clear that the largest component of the resistance is due to the cochlea.

In curve 5 we see the effect of cutting the Tensor Tympani. Its removal reduced the eardrum stiffness and slightly reduced the resistance. Finally cutting the stapes free (curve 6) from the incus further reduced the stiffness and resistance. From these results it appears that the tensor tympani and annular ligament have only a small resistive component, and that the major resistive component in the eardrum impedance is due to the cochlea. In summary, it seems likely that for the normal ear, the cochlear loss is largely due to the basilar membrane resistance (since the fluid resistance is believed to be small, based on theoretical estimates).

From these experimental results, it appears feasible that basilar membrane viability might be estimated from the ear canal by measuring the real component of the cochlear input impedance, in human hearing impair subjects. Such a correlation would have important clinical diagnostic applications.

REFERENCES

Beranek, L. L. (1949). *Acoustic Measurements,* (Wiley, New York).

Hunt, F. V. (1954). *Electroacoustics, The Analysis of Transduction, and its Historical Background,* Harvard Univ. Press, John Wiley and Sons, Inc., New York, pp. 187-212.

Kinsler, L. E., Frey, A. B., Coppens, A. B., and Sanders, J. V. *Fundamentals of Acoustics,* J. Wiley, 1982, third edition p. 206, Section 9.5.

Lynch, T. J., III (1974). ``Measurements of Acoustic Input Impedance of the Cochlea in Cats,'' S. M. Thesis, Massachusetts Institute of Technology, Cambridge, Ma., pp. 1-180.

Mawardi, O. K. (1949). ``Measurement of acoustic impedance,'' J. Acoust. Soc. Am. 21, 84-91.

Møller, A. R. (1972). The middle ear. In J.V. Tobias (Ed.), *Foundations of modern auditory theory.* New York: Academic Press, 1972.

Sokolich, G. W. (1977), ``Improved acoustic system for auditory research,'' J. Acoust. Soc. Am. Suppl. 1 *61,* S12.

Tonndorf, J., and Khanna, S. M. (1967). ``Some properties of sound transmission in the middle and outer ears of cats,'' J. Acoust. Soc.

White, R. A., Studebaker, G. A., Levitt, H., and Mook, D. (1980). "The Application of Modeling Techniques to the study of Hearing and Acoustic Systems." In *Acoustical Factors Affecting Hearing Aid Performance,* G. A. Studebaker and Hockberg, Eds., Univ. Park Press, Baltimore.

Zuercher, J. C., Carlson, E. V., and Killion, M. C. (1977). "The calculation of isothermal and viscous effects in acoustical tubes," presented at the 94th Meeting of the Acoustical Soc. Am.

MIDDLE EAR RESEARCH USING A SQUID MAGNETOMETER

I. MICRO- AND MACROMECHANICAL SELECTION OF POLYMER MATERIALS FOR ARTIFICIAL TYMPANIC MEMBRANES

W.L.C. Rutten+, D. Bakker, J.H. Kuit*, M. Maes** and J.J. Grote
ENT department, University Hospital, Leiden, The Netherlands

+present address: dept. of Electrical Engineering and * dept. of Chemical Technology, Twente University of Technology, Enschede, The Netherlands.
**dept. of Exp. Physics, University of Antwerp, Belgium.

1. INTRODUCTION

In developing an artificial membrane in a total alloplastic middle ear (TAM) prosthesis (Grote 1984) one of the main research goals is mechanical compatibility (besides biological compatibility). First, micromechanically ($\approx 10^{-9}$-10^{-6} m), vibration amplitude spectra in response to sound must match that of the natural membrane in a sufficiently wide frequency range (between 200 and 10000 Hz). Second, macromechanically ($\approx 10^{-3}$ m), the elasticity modulus must be in the natural range for purposes of epithelial overgrowth and ingrowth. As ingrowth requires a porous material structure, porosity is a relevant structural variable, to be monitored by use of scanning electron micrographs. Finally, the material must sustain sterilization, at a temperature of about 120 ^{o}C.

In this paper we present vibration measurements of six polymers and also of two natural membranes in temporal bones in which the middle ear ossicles were removed. Also elasticity data are presented. Materials which did not sustain sterilization are omitted.

2. METHODS AND MATERIALS

To measure microscopic vibrations contact-free and in three dimensions we developed a method, based on the use of a SHE-330 r.f. SQUID magnetometer and a second order gradiometer, which measures flux variations of a vibrating tiny magnetic dipole ($SmCo_5$, typical mass 1 mg) glued to the structure of interest. Principles and details are given elsewhere (Rutten et al. 1982, 1985). The set-up (fig. 1) measures amplitude and phase spectra down to 10 Ångström between 200 and 10000 Hz at levels from 60 to 90 dB SPL. Sensitivity is by far the largest along the gradiometer axis and amounts to 2.12×10^{-8} Tesla/Volt, or 1.1 μV/Ångström.

Polymers are stretched slightly in a perspex/aluminium holder (fig. 2). Six polymers were tested, three has a porous structure with a porous-to-dense volume ratio of 0.6, a macroporediameter of 10-100 μm and micropores of 3 μm. Three polymers were totally dense. The three porous polymers were polyure-

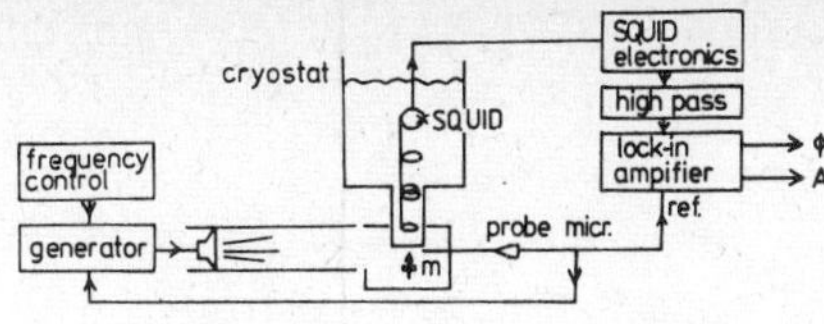

Figure 1. Sketch of the set-up. m is the vibrating magnet, ϕ and A are phase and amplitude output (rms value). Frequency is swept logarithmically from 200 Hz to 14 kHz in five minutes. Integration time of the lock-in amplifier is 300 msec. Sound level is held automatically constant, range 60-90 dB SPL re 20 μPa.

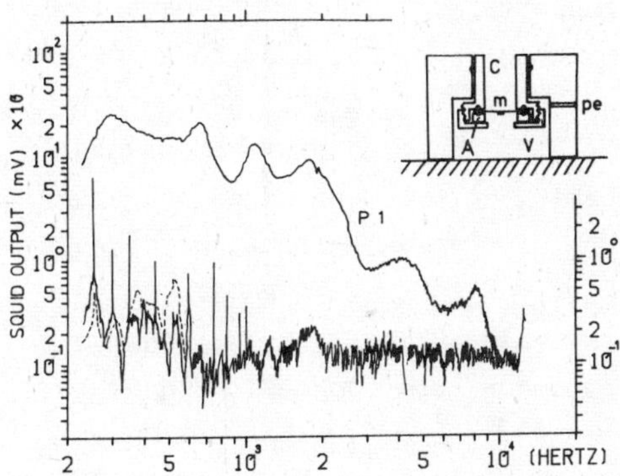

Figure 2. Amplitude spectrum for polyurethane 1 (p1) at 80 dB SPL, with the magnet in center of the membrane. Lower curve: baseline in absence of a magnet, representing mainly environmental magnetic noise. Spikes at 50 Hz or 100 Hz intervals between 250 and 1000 Hz are harmonics of the mains' magnetic field. Dashed baseline curve: as p1 but with the magnet glued to the top of the perspex holder. Inset upper right: cross-section of the artificial membrane holder. The membrane is stretched between O-ring and aluminium ring A. Dimensions: V is 2 cm^3, diameter C is 0.6 cm. Membrane surface is 0.86 cm^2.

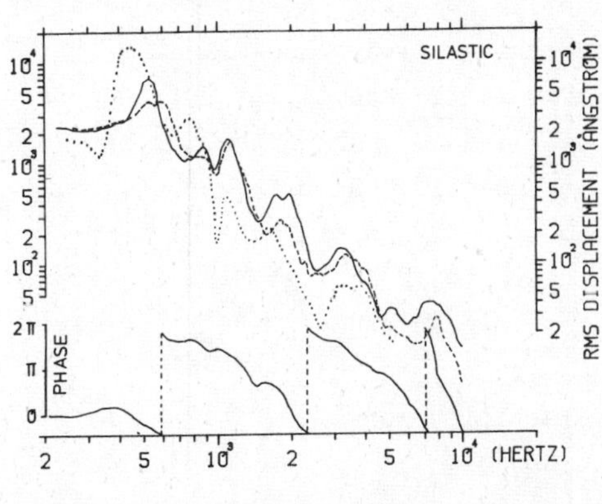

Figure 3. Effect of varying the magnet mass. Amplitude and phase spectra of silastic for 80 dB SPL. Dashed curve: magnet mass is 0.45 mg. Drawn curve: magnet mass is 0.9 mg. Dotted curve: magnet mass is 1.95 mg. Above 0.9 mg mass-loading can be observed to alter the system. Note the downard shift of "resonance frequencies" with increasing mass. Membrane mass is 15 mg.

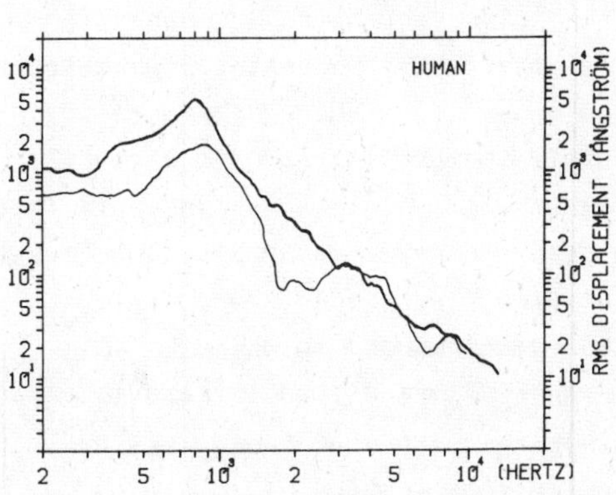

Figure 4. Amplitude spectra of two freely vibrating natural tympanic membranes in temporal bones in which the ossicles have been removed and the middle ear cavity closed (about 22 hours after death). The magnet is glued on the umbo. Low-pass cut-off frequency is about 900 Hz, slope is about 12 dB/octave.

thanes (Goodrich Esthane 5714 F1), to be called polyuretheane 1 (p1, micro/ macroporous surface, macroporous body), p2 (same as p1, but a dense body) and p3 (dense/microporous surface, open body). The three dense materials were a polyester-polyether copolymer (pe), silastic^{R} (sil) and Dow Corning MPX-4-4210 elastomore (ela). All membranes had a thickness of about 100 μm.

3. RESULTS

Figure 2 shows an example of an amplitudespectrum, i.e. displacement-rms-amplitude (volts at output A in figure 1) versus frequency for the p1 material. The lower curve is the baseline, measured without a magnet on the membrane, representing background magnetic activity and electrical instrumental noise. The dashed baseline is for the situation in wich the magnet has been glued on top of the holder, it indicates that mechanical vibrations of holder and/or set-up are negligible.

Mass-loading effects of the finite magnet-mass upon the vibrating membrane are investigated in figure 3. Effects occur upon increasing the mass beyond 0.9 mg.

Figure 4 shows human tympanic membrane data for two freely vibrating membranes. Note the approximate low-pass behaviour with cut-off frequencies of about 900 Hz and slopes of about 12 dB/octave. Figure 5 summarizes the 80 dB SPL amplitude spectra of the six polymers (after sterilization, the effect of which was showed elsewhere, Rutten et al. 1985), some of them being shifted horizontally. Average behaviour is low-pass, cut-off frequency is between about 800 Hz (ela and sil) and 2000 Hz (p2 and pe). The high frequency slope gradually slow down from 30 dB/octave to 12 dB/octave at 10 kHz.

Figure 6 sketches the macroscopic stress-strain behaviour in a stress-length diagram for the polymers as well as for the natural membrane. Slopes of the polymer curves, i.e. the Young's elasticity moduli, of p2 and pe resemble most closely that of the natural membrane.

From figures 5 and 6 it is obvious that with respect to low- and high-frequency vibratory properties as well as elasticity the best porous material is polyurethane 2 (p2) and the best dense material is the copolymere (pe).

It is of interest to model the artificial membrane system as a circular plate or, alternatively, as an electrical LCR series circuit, since all parameters are known. Conventional circular plate theory (see Kinsler and Frey, Fundamentals of Acoustics) yields for the first resonance frequency
$f_1=0.47\ t\ a^{-2}\ Y^{1/2}\ \rho^{-1/2}\ (1-\sigma^2)^{-1/2}$ in which t=membrane thickness, Y=Young's modulus, ρ=specific mass and σ=Poisson's ratio (assumed=1/2). For silastic one calculates f_1=100 Hz. This value is also found in the LCR model, using $f=(2\pi)^{-1}(LC)^{-1/2}$ in which LC is mass/tension [kg/ m Pa]. A tension-value of 2N/m has been used.

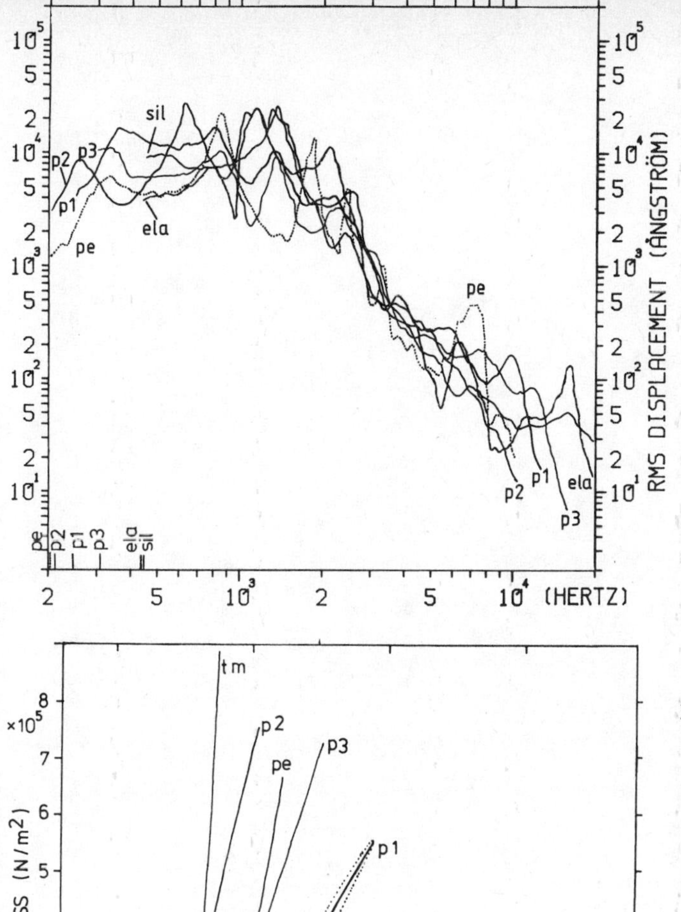

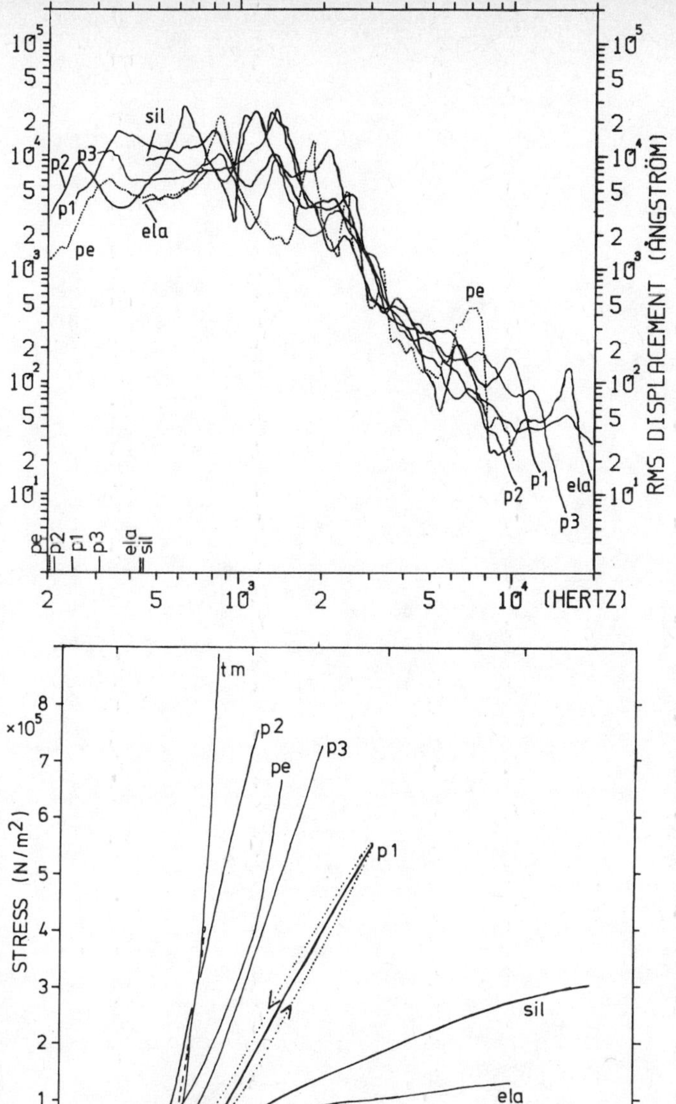

Figure 5. Amplitude spectra for the six polymer membranes p1,p2,p3, pe, sil and ela. Curves for p1,p3, sil and ela have been shifted to the right in order to let all curves coincide in the 2-5 kHz interval. The starting point of each curve is 200 Hz, the endpoint is at 10 kHz. The applied shift can also be seen in the left/under corner of the figure. All membranes show a low-pass character with a cut-off frequency of about 800 Hz (ela and sil) to about 2000 Hz for the stiffest materials (p2 and pe). High frequency roll-off changes from 30 dB/octave gradually to 12 dB/octave.

Figure 6. Uniaxial tensile stress versus length of the six polymers (see fig. 5) and of a natural tympanic membrane in vitro. Dashed tm curve is taken from Decraemer et al. 1980. For clarity the actual "hysteresis loops" have been given only for p1, see the dotted curve with the arrows indicating the loop direction.

REFERENCES

Decraemer, W.F., Maes, M.A. and Vanhuyse, V.J., "An elastic stress-strain relation for soft biological tissues." J. Biomech. 13, pp. 463-468, 1980.

Rutten, W.L.C., Peters, M.J., Brenkman, C.J., Mol, H., Grote, J.J. and v.d. Marel, L.C., "The use of a SQUID magnetometer for middle ear research." Cryogenics 22, 457-461, 1982.

Grote, J.J., Ed. Biomaterials in otology, Martinus Nijhoff, The Hague, 1984.

Rutten, W.L.C., v. Blitterswijk, C.A., Brenkman, C.J. and Grote, J.J., "Vibrations of natural and artificial middle ear membranes," in Biomagnetism, Ed. Weinberg, H., Pergamon press, New York, pp. 461-465, 1985.

MIDDLE EAR RESEARCH USING A SQUID MAGNETOMETER
II. TRANSFER CHARACTERISTICS OF HUMAN MIDDLE EARS

C.J. Brenkman, W.L.C. Rutten[xxx] and J.J. Grote, ENT-dept.,
University Hospital Leiden, 2333 AA Leiden, The Netherlands

I. INTRODUCTION

Using a tiny magnet and a commercial SHE r.f. SQUID magnetometer set-up, supplemented by an adjustable second order gradiometer, amplitude and phase spectra of vibrations of human middle ears were determined without disturbing the anatomy of middle ear and intact cochlea. The magnet had a mass of 1.5 mg and was positioned at the tip of the malleus (umbo), halfway the malleus near the processus brevis and on the anterior crus of the stapes. Post mortem changes of vibratory umbo displacements were measured in human temporal bones and guinea pigs. In a group of twelve ears the mean and standard deviation were determined at the three positions as mentioned above at 80 dB SPL stimulation level. These amplitude and phase spectra yielded information about the transfer characteristics of the middle ear and about the projection (into the tympanic plane) of a rotation axis.
In the past, one axis was assumed (v. Bekesy, Tonndorf and Khanna). Later, Gundersen observed an axis the position of which changed with frequency. His measuring method however, disturbed the middle ear anatomy and gave insight only in a small anatomical area.

II. MATERIALS AND METHODS

The experiments were carried out on 12 fresh temporal bones, the age of death varying between 6 and 80 years. The external ear was removed, leaving about 5 mm intact, not disturbing the fibrous annulus.
The middle ear was reached by widening the Eustachian tube, thus making an entrance of about 4 mm diameter.In this way the condition of the middle ear could be checked (in case of severe middle ear pathology the temporal bones were excluded). When the magnet was positioned the temporal bone was closed and encapsulated in modelling clay, the ear drum lying in the upper plane. The measurements took place immediately afterwards and no special preserving precautions were taken (constant room temperature about 20°C, normal air humidity). The guinea pigs in this study were healthy adult animals sedated by an i.p. injection of 0.5 ml/100 g body weight of a 20% aqueous solution of urethane. The death of the animal was caused by an overdosis urethane in combination with

Norcuron^R as a muscle relaxant.

Measurements were performed in a sound proof room which was not magnetically shielded. Stimulation was carried out in a semi closed sound system (2-way) at a 80 dB SPL stimulation level in the frequency range of 200 up till 10.000 Hz. Sound pressure levels were measured at the site of the annulus by a carefully calibrated probe-tube microphone. The SQUID output was high pass filtered (48 dB/oct.) above 200 Hz and led to a Brookdeal 9505 two-phase lock in amplifier (integration time 300 ms). The probe microphone signal served as reference input. Amplitude and phase spectra between 200 Hz and 10 kHz were written on a X-Y writer and off-line processed by a DEC PDP-11 computer system with a Houston HIPAD digitizer and a Houston HIPLOT plotter. The latter system corrected for the not perfectly flat amplitude response of the probe microphone as well as for phase shifts in this microphone and in the high pass filter.

III. RESULTS

Mass loading effects of the magnet

Experiments were performed to investigate which mass of the magnet was still acceptable, giving enough flux variation at 80 dB SPL not causing serious mass loading effects. Figs. 1 and 2 show the results for three different magnet masses on the umbo and the stapes. The figures indicate that up to 2.26 mg (2.16 for the stapes) no serious mass loading effects occur.

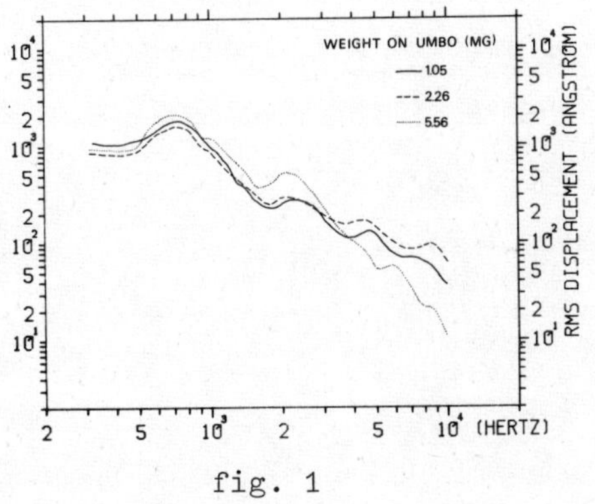

fig. 1

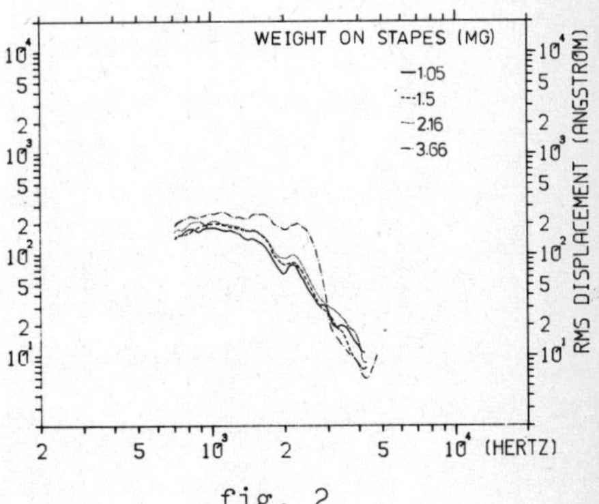

fig. 2

Post mortem changes

The umbo amplitude of two temporal bones was measured directly after preparation (12-24 hours post mortem) and at regular intervals up till 60 hours post mortem. Fig. 3 shows only few changes at three frequencies (5000 Hz, 2000 Hz and 600 Hz, see fig. 4).

To gain more insight in possible immediate post mortem changes an animal model was chosen. The umbo amplitudes of three alive guinea pigs were recorded before, during and after sacrifycing up till 48 hours post mortem. Fig. 4 shows a few measurements when the animal is alive and then (marked with an arrow) the first measurement after death. When changes occur, they do this predominantly in the lower frequencies. Nevertheless, changes occur gradually and especially also after death.

fig. 3

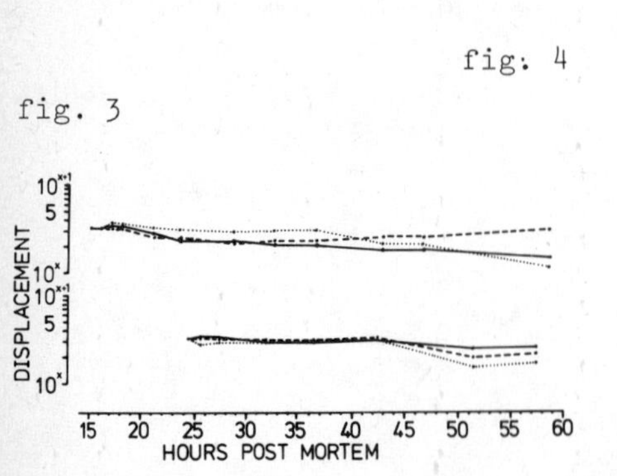

fig: 4

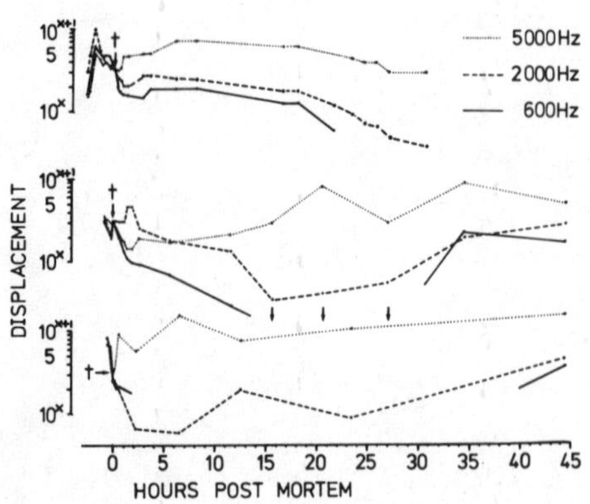

Mean displacement and standard deviation

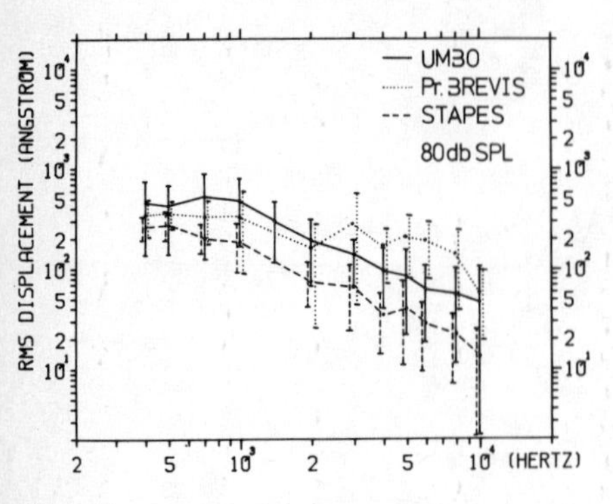

fig. 5

For 12 human temporal bones displacement was measured at two positions on the malleus (umbo and near processus brevis) and on the anterior crus of the stapes. The vertical lines represent the standard deviation (fig. 5).

The combination of amplitude and phase spectra measured at three positions of the middle ear chain

Combining the results of amplitude and phase spectra of the three positions mentioned above it is possible to determine three main groups:

Group I In the low frequencies all the three positions are in-phase (fig. 6).

Group II In the low frequencies only the stapes differs 180° from the two other positions (fig. 7).

Group III In the low frequencies only the pr. brevis position differs 180° from the two other positions (fig. 8).

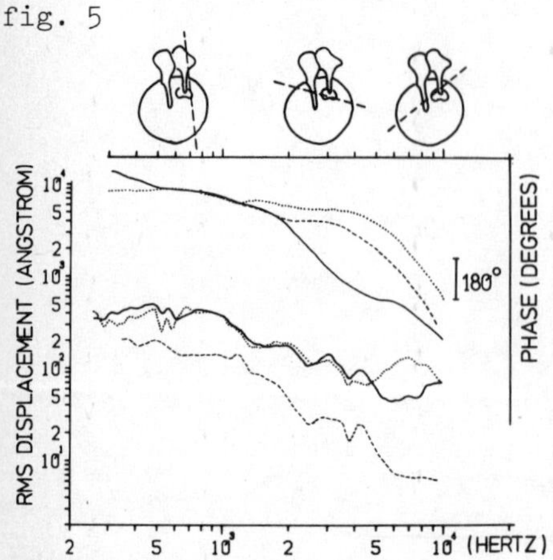

fig. 6

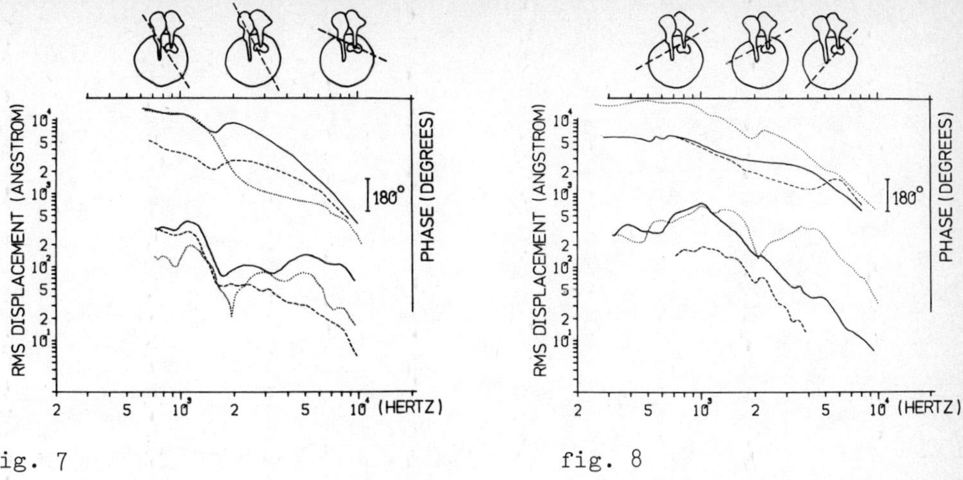

fig. 7

fig. 8

Using these measurements it is possible to draw the projection of the axis in the plane of the tympanic membrane, as indicated on top of each figure.

ACKNOWLEDGEMENTS

We thank the Heinsius Houbolt Fund for financial support.

REFERENCES

Bekesy, G. v. "Uber die Messung der Schwingungsamplitude der Gehörknöchelchen mittels einer kapazitiven Sonde," Akust. Z. 6, 1 (1941).

Gundersen, T. and Høgmoen, K. "Holographic Vibration Analysis of the Ossicular Chain," Acta Otolaryngol. 82: 16-24, 1976.

Rutten, W.L.C., Peters, M.J., Brenkman, C.J., Mol, H., Grote, J.J. and van der Marel, L.C. "The Use of a SQUID-Magnetometer for Middle Ear Research," Cryogenics, Sept. 1982; 457-460.

Tonndorf, J. and Khanna, S.M. "Tympanic-Membrane Vibrations in Human Cadaver Ears Studied by Time-Averaged Holography," J. Acoust. Soc. Amer. 52, 1221-1233, 1972.

COCHLEAR MACROMECHANICS

COCHLEAR MACROMECHANICS - A REVIEW

Max A. Viergever
Delft University of Technology - The Netherlands

ABSTRACT

This paper reviews macromechanical models of the cochlea. The emphasis is on two questions: (i) which geometrical and mechanical features should be included, and (ii) which experimental results can be matched.

1. MACROMECHANICAL MODELS

Since it has become clear that the behaviour of the cochlea is for a large part determined by mechanical processes at hair cell level, cochlear mechanics has known a division into two classes: cochlear *micromechanics*, which refers to the mechanics of the organ of Corti and the tectorial membrane, and cochlear *macromechanics* which is most concisely defined as the complement of micromechanics.

One characteristic feature of macromechanical models is that the various parts of the organ of Corti/tectorial membrane complex are not considered separately, but move in unison with the basilar membrane (BM). The cochlear partition is thus a system having one degree of freedom.

The second characteristic feature derives from the nowadays commonly accepted view that cochlear nonlinearity and activity (i.e., the creation of mechanical energy) originate in hair cell processes. Nonlinear and active behaviour does manifest itself at the level of BM vibration, but this is attributed to a - hitherto unknown - feedback mechanism. Consequently, macromechanical models are linear and passive.

2. ONE-DIMENSIONAL MODELS

The first macromechanical models of the cochlea leading to realistic results are the so-called one-dimensional (1D) models [60, 35, 11, 16]. The simplest geometry for such models is shown in Fig. 1. The characteristic assumption of 1D models is that the fluid pressure is uniform over a channel cross-section and thus depends only on the coordinate x.

The equation describing the harmonic behaviour of the model is [55]

$$\frac{d^2P}{dx^2} - \frac{16j\omega\rho\beta}{\pi^2 AZ} P = 0, \qquad (1)$$

with boundary conditions

$$\frac{dP}{dx} = - j\omega\rho\frac{A_{st}}{A} \quad \text{at } x = 0, \qquad (2)$$

$$\frac{dP}{dx} = 0 \quad \text{or } P = 0 \qquad \text{at } x = L. \tag{3}$$

Here, $P(x;j\omega)$ is the pressure in the upper channel, defined in the frequency domain; ω is the radian frequency of the input signal. Furthermore, ρ is the fluid density, A_{st} is the area of the stapes footplate, and L is the length of the cochlea. The BM impedance Z has the form

$$Z(x;j\omega) = j\omega\, M(x) + R(x) + \frac{S(x)}{j\omega}, \tag{4}$$

with M, R and S mass, resistance and stiffness of the BM per unit area. The relation between the pressure P and the velocity V of the BM centreline is

$$2P = ZV. \tag{5}$$

The boundary value problem (1)-(3) admits of an analytic solution only for specific choices of the parameters. For example, when M/β and R/β are constant and S/β depends exponentially on x, the solution can be written in terms of hypergeometric functions [25].

The solution of the 1D model without parameter constraints may be obtained either directly by numerical means or, in an approximate manner, by applying an asymptotic technique. For the latter purpose, the Liouville-Green (LG) method [59, 53] is the most suitable.

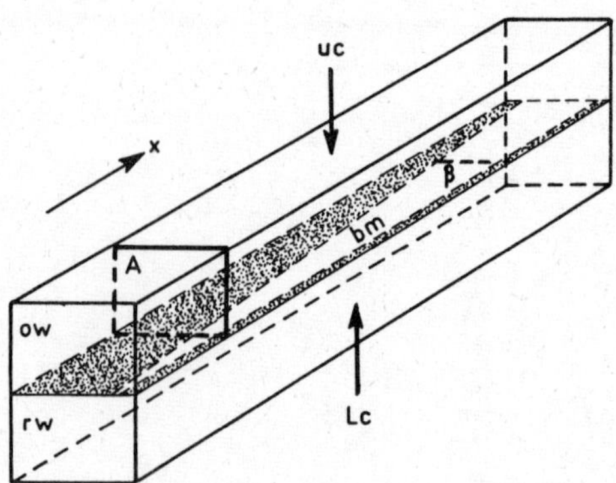

Fig. 1. Geometry of the one-dimensional cochlear model. The cross-sections of the channels have been drawn as rectangular, the membrane as linearly tapered. These structures may have any shape in the model, however; only the parameters A and $\beta(x)$, denoting the (constant) cross-sectional area of the channels and the width of the membrane, respectively, appear in the equations.

3. EXAMINATION OF THE SIMPLIFICATIONS

The 1D model of the previous section is based on geometrical and mechanical simplifications which will now be examined briefly. The discussion of the assumptions a-l is not restricted to a 1D model but valid as well for 2D and 3D formulations.

a. The cochlea is an isolated cavity, communicating with its surroundings only through the windows.

At its basal end, the actual cochlea has several mechanical leaks, by which I understand the contact of the cochlear fluids with other spaces. The presence of the perilymphatic and endolymphatic ducts presumably has little significance for the mathematical formulation. The connection of scala vestibuli with the vestibule, on the other hand, influences the BM response near the basal end at high input frequencies. It seems useful to study the extent to which this occurs in more detail than done so far [12].

b. The cochlear fluid is incompressible.

Several authors [27, 53] have shown that fluid compressibility has no effects on BM motion, possibly except at the highest frequencies in the range of hearing.

Compressibility is important with regard to the fluid pressure, however [35, 28]. The pressure can be divided into two components, one related to fluid compressibility which is uniform over each cross-section of the cochlea (and hence is irrelevant to BM motion), the other related to the flexibility of the cochlear partition. It is the latter component which is usually termed pressure. This is admissable only in models which are symmetrical with respect to the partition.

c. The fluid behaves linearly.

The Navier-Stokes equation describing an incompressible, Newtonian fluid contains a number of linear terms, including one featuring linear convection, and one quadratic convection term. An estimate of the ratio of the convection terms [53] shows that the nonlinearity is negligible even at the highest intensity levels.

d. The fluid is inviscid.

The influence of viscosity can be studied using a boundary layer approach. The thickness of the boundary layer is about 100 μm for very low frequencies (10 Hz), and about 10 μm for middle frequencies (1 kHz). From these values it can already be concluded that viscosity effects are negligible in cochlear macromechanics, a characteristic dimension of the scalae being 1 mm. Boundary layer calculations [53, 33] support this conclusion.

Viscosity does play a significant, or even dominant, role in micro fluid mechanics of the cochlea, however, where characteristic length dimensions are of the order of 10 μm.

e. The geometry of the cochlea does not change by stimuli.

In response to a stimulus, flexible structures in the cochlea move and thereby change the geometry. Maximum BM displacements at very high intensity levels (120 dB SPL) are of the order of 1 µm, which is still negligibly small with respect to scala dimensions. Consequently, the geometry of the model may be taken as time-invariant.

f. The spiral coiling is ignored.

The role of the spiral coiling in cochlear macromechanics can be studied by formulating the equations for fluid motion and BM motion in a spiral-shaped model. An analysis of the difference in formulation to that of a straight cochlear model shows that BM motion is only slightly affected by the spiral [52]. This conclusion has recently been confirmed by calculations [29, 48].

g. The presence of Reissner's membrane is neglected.

Reissner's membrane is substantially more flexible than the BM, except in the vicinity of the helicotrema [3]. Since moreover perilymph and endolymph are mechanically alike [37], Reissner's membrane is unlikely to serve mechanical purposes. Its main function probably is that of an ionic barrier between the perilymphatic and endolymphatic fluids.

h. The structures inside scala media are not taken into account.

This is the basic feature of macromechanics, see section 1. The complex consisting of organ of Corti and tectorial membrane is 'lumped' onto the BM. In other words, the cochlear partition has only one degree of freedom.

i. The scala walls are rigid.

The mechanical effects of periosteum, ligamentum spirale and stria vascularis are ignored. This assumption remains undiscussed; suffice it to say that it is the least disputed of the simplifications made.

j. The helicotrema is disregarded.

The helicotrema is not modelled as an opening in the cochlear partition. Instead, the BM runs the length of the cochlea. This hardly influences the BM response, even at the lowest audiofrequencies [12].

k. The oval and round windows have equal cross-sections and move as pistons.

The motion of the stapes is approximately piston-like [17, 39]. The round window must follow this behaviour owing to fluid incompressibility. The presence of the vestibule partly destroys the uniformity of the input over the channel cross-section, but this has only a very limited effect on BM motion [53].

l. The cross-sectional areas of the two channels are equal to another and constant along the length of the cochlea.

Outside the region of the vestibule, the cross-sectional area of the upper channel (scala vestibuli plus scala media) is approximately equal to that of the lower channel (scala tympani). Furthermore, the areas show only moderate variations along the length of the cochlea as compared with the mechanical properties of the

BM [56].

The BM width is not assumed to be constant, although this quantity also varies much less than the BM impedance. The reason is that the impedance components depend greatly on the BM width; for instance, BM stiffness is proportional to β^{-5}. Consequently, keeping Z x-dependent while making β constant would be rather artificial.

The model now is symmetrical with respect to the cochlear partition. The description may, therefore, be confined to one channel.

m. The cochlear partition is represented by a linear, passive point-impedance function.

Linearity and the restriction to passive mechanical properties are characteristic of macromechanical models, as has been argued in section 1. The point-impedance is the analogue in 1D and 2D models of the fairly general characterization of the BM by an orthotropic visco-elastic plate of variable width and thickness [51].

n. The fluid flow is one-dimensional.

It has been shown repeatedly that the 1D model provides for a qualitatively good BM response [35, 16, 61]. The picture is that of a travelling wave, starting from the basal end. As it progresses, the speed of propagation gradually diminishes and the vibration amplitude increases. At a certain point the maximum is attained and the wave appears as rapidly attenuated beyond this point. The position of the maximum is frequency-dependent: the higher the stimulus frequency, the more basal lies the maximum. Such a response is completely in agreement with all experimental data of BM vibration.

Quantitatively, however, the 1D model shows quite large discrepancies both with 2D and 3D models [46, 53, 4] and with measurement results [59]. A closer inspection of the differences with multi-dimensional models reveals that their nature is not purely quantitative. For instance, it can be shown that the restriction to a uniform pressure distribution is equivalent to considering only long waves on the BM: the 1D approach is valid so long as $\lambda >> 2\pi h$, where λ is the wavelength of the velocity wave propagated along the BM and h is the channel height. From measurements it can be deduced that λ is approximately equal to h in the region of maximum BM amplitude [38, 57]. Consequently, a multi-dimensional treatment of cochlear macromechanics is mandatory.

4. TWO-DIMENSIONAL MODELS

Figure 2 shows the description of the standard 2D model in concise form; the pressure P now depends on both x and y. The formulation is due to Lesser and Berkley [26], who were the first to give a full 2D treatment of cochlear mechanics. Earlier attempts were limited to short waves ($\lambda << 2\pi h$) [36] or long waves ($\lambda >> 2h$), the latter being a refinement of the approach leading to 1D models [34]. Other formulations than that of Fig. 2 have also been used, viz. in the spatial frequency domain [42] and in the time domain [2].

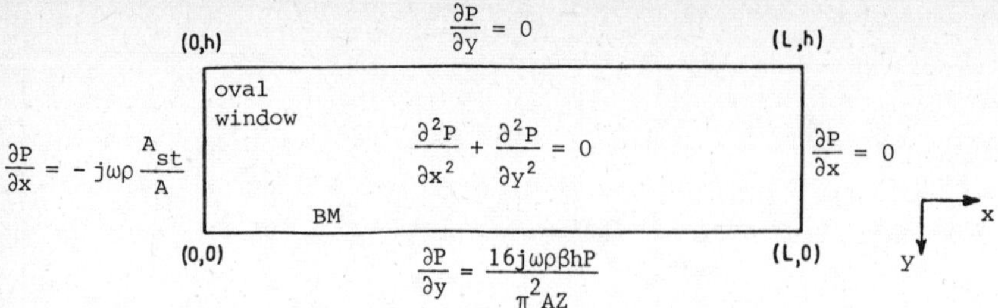

Fig. 2. The boundary value problem describing the two-dimensional model.

Efficient techniques to solve the 2D model have either been based on dimension reduction, notably to a 1D integral equation [1, 43, 2], or on a straightforward numerical approach [31, 53]. Recently, the two ideas have been fruitfully combined [30]. This last algorithm is so fast that it allows for extensive numerical experimentation with the parameters. Consequently, one no longer has to rely on asymptotic approximations in order to fit measurement data.

Asymptotics will remain important, however, in clarifying the physical mechanisms that govern the response. The LG method, or rather a 2D generalisation of it, is pre-eminently suited for this purpose [45, 53, 9, 10].

5. THREE-DIMENSIONAL MODELS

In 3D models of the cochlea, fluid flow variations in the direction lateral to the BM are also taken into account. Steele and Taber [46] solve a 3D cochlear model using the LG approximation and show that the fluid flow is fully three-dimensional. Nonetheless, the BM response differs only marginally from its 2D counterpart for the parameter sets they considered.

De Boer formulates a 3D model equation in the spatial frequency domain and solves it using *ad hoc* approximate methods [4, 8]. The formulation also allows the derivation of a 'correspondence principle' indicating for which choices of the parameters the 3D model response is satisfactorily approximated by a 2D or even by a 1D response [5].

Apart from assessing the influence of the vestibule (see section 3), the only unsolved problem in cochlear macromechanics is obtaining a numerically exact solution of a realistic 3D model. The storage capacity and the speed of present-day computers have brought realization of this well into reach.

6. COMPARISON WITH EXPERIMENTS

Both 2D and 3D model calculations have been compared with experimental data of BM motion. Most of the early in vivo data [20, 21, 58] can be matched quite well using either a 2D or a 3D model [47, 54], see for example Fig. 3.

On the other hand, the responses recorded by Rhode [38, 39] and the recent data

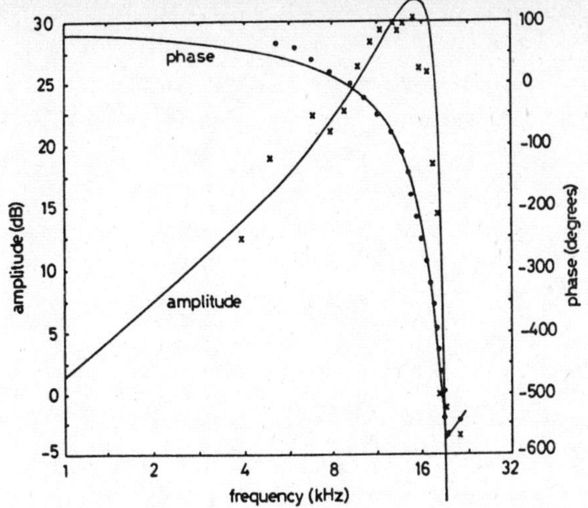

Fig. 3. Comparison of 2D model response with BM/stapes transfer ratio (crosses: amplitude, dots: phase) observed in the guinea pig by Johnstone and Yates [21]. For details of calculations and parameter values, see [54].

of BM vibration exhibiting an even sharper tuning [22, 41, 40] cannot satisfactorily be matched by macromechanical model results [55]. This is not really surprising since in the measurements concerned strong nonlinear effects have been observed. Moreover, it can be shown that the sharp tuning of the responses is most likely due to active mechanical processes [6, 13].

7. FROM MACROMECHANICS TO MICROMECHANICS

The comparison with the experimental data of BM vibration shows that macromechanical models of the cochlea fully meet the purpose for which they were originally developed: to describe the behaviour of the linear and passive cochlea.

It is known now that a cochlea exhibiting such properties is pathological. Intact cochleae have much sharper tuning characteristics and behave in a nonlinear fashion. Macromechanical models are not able to match responses of intact cochleae, let alone describe the underlying physics.

The obvious next step towards a more complete understanding of cochlear functioning is to include active and nonlinear features in the models, although the necessity for this has not rigorously been proved. It is, for instance, possible to match the recent data of BM motion with a passive and linear cochlear model, provided this model is allowed to have many degrees of freedom [13]. The physics of the problem, however, do not seem to suggest this type of model. The highest order realistic cochlear models that have been proposed have four degrees of freedom [44, 49, 50], but their responses differ only marginally from the ones discussed in this paper.

Active and nonlinear features may be included either by considering processes at hair cell level, or by representing these processes phenomenologically in an otherwise macromechanical model. The latter approach can give an insight into the *effects* on BM vibration of mechanical activity [24, 6, 7], nonlinearity [19, 23, 18], or both [32, 14, 15]. Insight into the *origin* of active and nonlinear behaviour can only be gained from truly micromechanical models. These are still at a quite primitive stage, though, mainly for lack of experimental findings to verify or disprove model hypotheses.

REFERENCES

1. Allen, J.B. (1977). Two-dimensional cochlear fluid model: New results, J. Acoust. Soc. Am. 61, 110-119.
2. Allen, J.B. and Sondhi, M.M. (1979). Cochlear macromechanics: Time domain solutions, J. Acoust. Soc. Am. 66, 123-132.
3. Békésy, G. von (1960). *Experiments in Hearing*, McGraw-Hill, New York.
4. Boer, E. de (1981). Short waves in three-dimensional cochlea models: solution for a 'block' model, Hearing Res. 4, 53-77.
5. Boer, E. de (1982). Correspondence principle in cochlear mechanics, J. Acoust. Soc. Am. 71, 1496-1501.
6. Boer, E. de (1983). No sharpening? A challenge for cochlear mechanics. J. Acoust. Soc. Am. 73, 567-573.
7. Boer, E. de (1983). Wave reflection in passive and active cochlea models. In: *Mechanics of Hearing*, E. de Boer and M.A. Viergever (eds), Martinus Nijhoff Publishers/Delft Univ. Press, 135-142.
8. Boer, E. de and Bienema, E. van (1982). Solving cochlear mechanics problems with higher order differential equations, J. Acoust. Soc. Am. 72, 1427-1434.
9. Boer, E. de and Viergever, M.A. (1982). Validity of the Liouville-Green (or WKB) method for cochlear mechanics, Hearing Res. 8, 131-155.
10. Boer, E. de and Viergever, M.A. (1984). Wave propagation and dispersion in the cochlea. Hearing Res. 13, 101-112.
11. Bogert, B.P. (1951). Determination of the effects of dissipation in the cochlear partition by means of a network representing the basilar membrane, J. Acoust. Soc. Am. 23, 151-154.
12. Borsboom, M.J.A. (1979). Linear and non-linear one-dimensional cochlear models, Eng. D. thesis, Dept. of Mathematics, Delft Univ. of Technology.
13. Diependaal, R.J., Boer, E. de and Viergever, M.A. (1985). Determination of the cochlear power flux from basilar membrane vibration data. These proceedings.
14. Diependaal, R.J. and Viergever, M.A. (1983). Nonlinear and active modelling of cochlear mechanics: A precarious affair. In: *Mechanics of Hearing*, E. de Boer and M.A. Viergever (eds), Martinus Nijhoff Publishers/Delft Univ. Press, 153-160.
15. Duifhuis, H., Hoogstraten, H.W., Netten, S.M. van, Diependaal, R.J. and Bialek, W. (1985). Modelling the cochlear partition with coupled Van der Pol oscillators. These proceedings.
16. Fletcher, H. (1951). On the dynamics of the cochlea, J. Acoust. Soc. Am. 23, 637-645.
17. Guinan, J.J., Jr. and Peake, W.T. (1967). Middle-ear characteristics of anesthetized cats, J. Acoust. Soc. Am. 41, 1237-1261.
18. Hall, J.L. (1974). Two-tone distortion products in a nonlinear model of the basilar membrane, J. Acoust. Soc. Am. 56, 1818-1828.
19. Hubbard, A.E. and Geisler, C.D. (1972). A hybrid computer model of the cochlear partition, J. Acoust. Soc. Am. 51, 1895-1903.
20. Johnstone, B.M., Taylor, K.J., and Boyle, A.J. (1970). Mechanics of the guinea pig cochlea, J. Acoust. Soc. Am. 47, 504-509.
21. Johnstone, B.M. and Yates, G.K. (1974). Basilar membrane tuning curves in the guinea pig, J. Acoust. Soc. Am. 55, 584-587.
22. Khanna, S.M. and Leonard, D.G.B. (1982). Basilar membrane tuning in the cat cochlea, Science 215, 305-306.
23. Kim, D.O., Molnar, C.E., and Pfeiffer, R.R. (1973). A system of nonlinear

differential equations modeling basilar-membrane motion, J. Acoust. Soc. Am. 54, 1517-1529.
24. Kim, D.O., Neely, S.T., Molnar, C.E., and Matthews, J.W. (1980). An active cochlear model with negative damping in the partition: Comparison with Rhode's ante- and post-mortem observations. In: *Psychophysical, physiological and behavioural studies in hearing*, G. van den Brink and F.A. Bilsen (eds), Delft Univ. Press, 7-15.
25. Kok, L.P. and Haeringen, H. van (1979). Reflections on 'Reflections on reflections', Internal report, Inst. of Theoretical Physics, Univ. of Groningen.
26. Lesser, M.B. and Berkley, D.A. (1972). Fluid mechanics of the cochlea. Part 1, J. Fluid Mech. 51, 497-512.
27. Lien, M.D. (1973). A mathematical model of the mechanics of the cochlea, Ph.D. thesis, Washington Univ. School of Medicine, St. Louis.
28. Lighthill, J. (1981). Energy flow in the cochlea, J. Fluid. Mech. 106, 149-213.
29. Loh, C.H. (1983). Multiple scale analysis of the spirally coiled cochlea, J. Acoust. Soc. Am. 74, 95-103.
30. Miller, C.E. (1985). VLFEM analysis of a two-dimensional cochlear model, J. Appl. Mech., in press.
31. Neely, S.T. (1981). Finite difference solution of a two-dimensional mathematical model of the cochlea, J. Acoust. Soc. Am. 69, 1386-1393.
32. Netten, S.M. van and Duifhuis, H. (1983). Modelling an active, nonlinear cochlea. In: *Mechanics of Hearing*, E. de Boer and M.A. Viergever (eds), Martinus Nijhoff Publishers/Delft Univ. Press, 143-151.
33. Neu, J.C. and Keller, J.B. (1985). Asymptotic analysis of a viscous cochlear model, J. Acoust. Soc. Am., in press.
34. Oetinger, R. und Hauser, H. (1961). Ein elektrischer Kettenleiter zur Untersuchung der mechanischen Schwingungsvorgänge im Innenohr, Acustica 11, 161-177.
35. Peterson, L.C. and Bogert, B.P. (1950). A dynamical theory of the cochlea, J. Acoust. Soc. Am. 22, 369-381.
36. Ranke, O.F. (1950). Hydrodynamik der Schneckenflüssigkeit, Z. Biol. 103, 409-434.
37. Rauch, S. (1964). *Biochemie des Hörorgans*, Thieme, Stuttgart.
38. Rhode, W.S. (1971). Observations of the vibration of the basilar membrane in squirrel monkeys using the Mössbauer technique, J. Acoust. Soc. Am. 49, 1218-1231.
39. Rhode, W.S. (1978). Some observations on cochlear mechanics, J. Acoust. Soc. Am. 64, 158-176.
40. Robles, L., Ruggero, M., and Rich, N.C. (1984). Mössbauer measurements of basilar membrane tuning curves in the chinchilla, J. Acoust. Soc. Am. 76, S35.
41. Sellick, P.M., Patuzzi, R., and Johnstone, B.M. (1982). Measurement of basilar membrane motion in the guinea pig using the Mössbauer technique, J. Acoust. Soc. Am. 72, 131-141.
42. Siebert, W.M. (1974). Ranke revisited - a simple short-wave cochlear model, J. Acoust. Soc. Am. 56, 594-600.
43. Sondhi, M.M. (1978). Method for computing motion in a two-dimensional cochlear model, J. Acoust. Soc. Am. 63, 1468-1477.
44. Steele, C.R. (1974). Behavior of the basilar membrane with pure-tone excitation, J. Acoust. Soc. Am. 55, 148-162.
45. Steele, C.R. and Taber, L.A. (1979). Comparison of WKB and finite difference calculations for a two-dimensional cochlear model, J. Acoust. Soc. Am. 65, 1001-1006.
46. Steele, C.R. and Taber, L.A. (1979). Comparison of WKB calculations and experimental results for three-dimensional cochlear models, J. Acoust. Soc. Am. 65, 1007-1018.
47. Steele, C.R. and Taber, L.A. (1981). Three-dimensional model calculations for guinea pig cochlea, J. Acoust. Soc. Am. 69, 1107-1111.
48. Steele, C.R. and Zais, J.G. (1985). Effect of coiling in a cochlear model, J. Acoust. Soc. Am. 77, 1849-1852.
49. Taber, L.A. (1979). An analytic study of realistic cochlear models including three-dimensional fluid motion. Ph.D. thesis, Stanford Univ.
50. Taber, L.A. and Steele, C.R. (1981). Cochlear model including three-dimensional fluid and four modes of partition flexibility, J. Acoust. Soc. Am. 70, 426-436.

51. Viergever, M.A. (1978). On the physical background of the point-impedance characterization of the basilar membrane in cochlear mechanics, Acustica 39, 292-297.
52. Viergever, M.A. (1978). Basilar membrane motion in a spiral-shaped cochlea, J. Acoust. Soc. Am. 64, 1048-1053.
53. Viergever, M.A. (1980). *Mechanics of the inner ear - a mathematical approach.* Delft Univ. Press.
54. Viergever, M.A. and Diependaal, R.J. (1983). Simultaneous amplitude and phase match of cochlear model calculations and basilar membrane vibration data. In: *Mechanics of Hearing*, E. de Boer and M.A. Viergever (eds), Martinus Nijhoff Publishers/Delft Univ. Press, 53-61.
55. Viergever, M.A. and Diependaal, R.J. (1985). Quantitative validation of cochlear models using the LG approximation. Internal Report, Delft Univ. of Technology.
56. Wever, E.G. (1970). *Theory of hearing*, Dover, New York. Republication of 1949.
57. Wilson, J.P. (1974). Basilar membrane vibration data and their relation to theories of frequency analysis. In: *Facts and models in hearing*, E. Zwicker and E. Terhardt (eds), Springer, Berlin, 56-64.
58. Wilson, J.P. and Johnstone, J.R. (1975). Basilar membrane and middle-ear vibration in guinea pig measured by capacitive probe, J. Acoust. Soc. Am. 57, 705-723.
59. Zweig, G., Lipes, R., and Pierce, J.R. (1976). The cochlear compromise, J. Acoust. Soc. Am. 59, 975-982.
60. Zwislocki-Mościcki, J. (1948). Theorie der Schneckenmechanik - qualitative und quantitative Analyse, Acta Otolaryng. suppl. 72.
61. Zwislocki, J. (1953). Review of recent mathematical theories of cochlear dynamics, J. Acoust. Soc. Am. 25, 743-751.

TRANSIENTS AND SPEECH PROCESSING IN A THREE-DIMENSIONAL MODEL OF THE HUMAN COCHLEA

Jeffrey G. Zais
Dept. of Aeronatics and Astronautics, Stanford University
Stanford, CA 94305

ABSTRACT

The three-dimensional mathematical model of the cochlea developed by Steele and Taber (1979) is used as the basis for calculations of the transient response of the cochlea. The fast Fourier transform is employed to perform the actual calculations. Due to space limitations, only five experimental results are simulated and summarized here: click responses, pure tone transients, single-period pitch perception, time-separation pitch perception, and the response of the basilar membrane to speech signals.

1. INTRODUCTION

Previous studies of the cochlea, both experimental and mathematical, have concentrated on the response to steady-state sinusoidal inputs. Békésy (1960) constructed an experimental model for the study of transients. He found that the place of maximum basilar membrane amplitude was developed in two cycles of excitation, and that even in this short time the amplitude distribution was close to the steady state solution. Dotson (1974) calculated the transient response of a 2-D model using the stationary phase technique. Other calculations of transients have been those modeling the basilar membrane as a set of idealized filters. In contrast, here the realistic 3-D cochlear model of Steele and Taber is used for all calculations. These steady-state solutions are used as normal mode shapes, and through the use of the Fast Fourier Transform (FFT) the response to any input can be determined. The chosen signal, of duration T, is assumed to repeat every T seconds. The amplitude of the signal is known at N discrete points evenly spaced through the interval. The frequency spectrum of the signal can be found using the FFT. To find the transient behavior, the response of the cochlea to the frequencies in the spectrum is found. At each station x/L, the frequency spectrum is multiplied by the corresponding amplitude and phase, and the time response of the basilar membrane is found by taking the inverse FFT of the result.

2. TRANSIENTS OF PURE TONES

The first example is the response of the basilar membrane to a 250 Hz sine-wave signal which is turned on at t=0, in addition to a steady-state 2560 Hz signal. Figure 1 presents a series of "snapshots" of the basilar membrane taken 1 msec apart. The times at the right of each curve denote the time in seconds since the 250 Hz tone was applied. The wave pattern of the

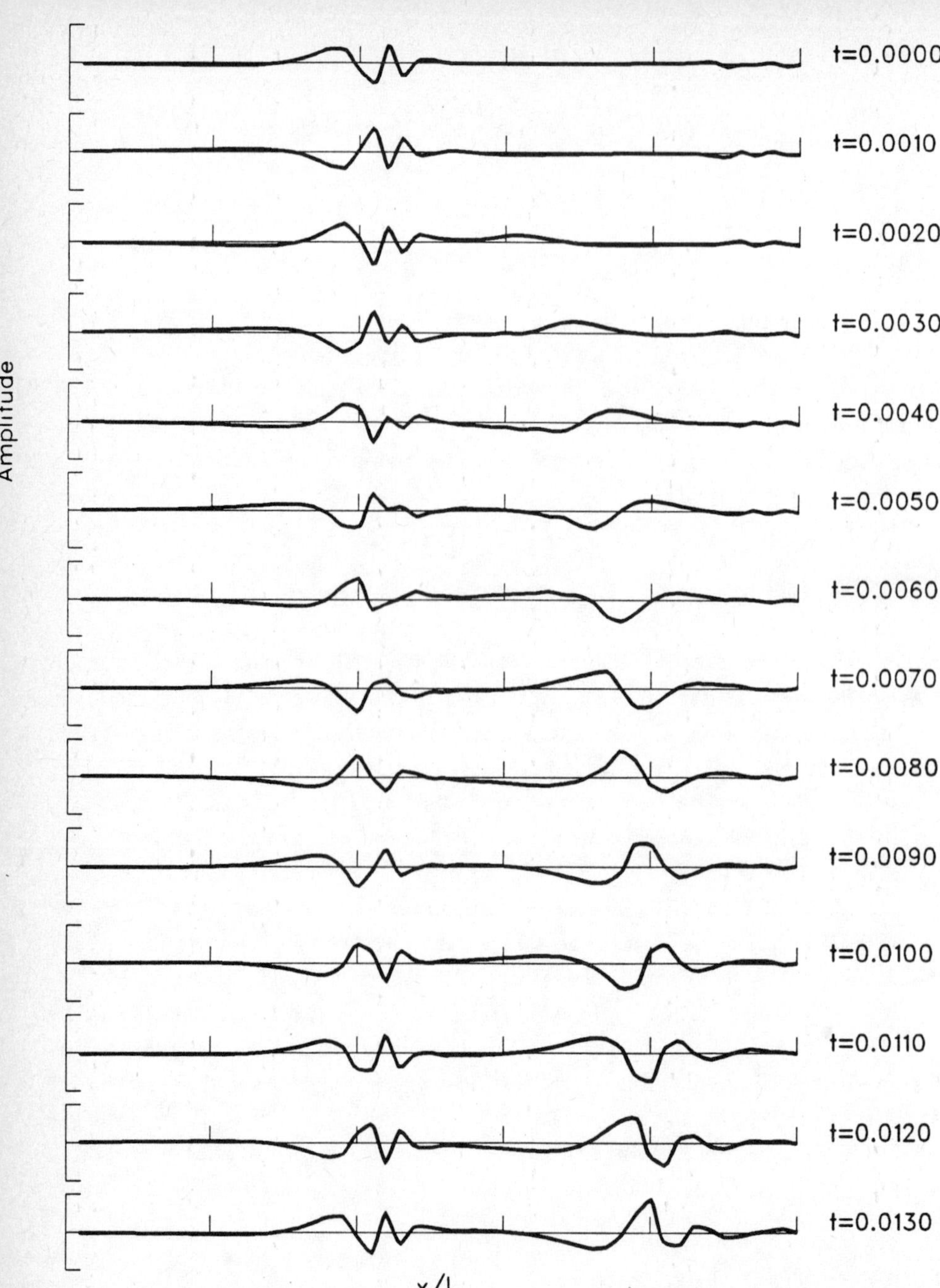

Figure 1. Series of "snapshots" of the basilar membrane amplitude (plotted on an arbitrary scale) in response to a 250 Hz sinusoidal tone added at t=0 to a steady tone of 2560 Hz. The times at the right of each snapshot indicate the time (in seconds) since the onset of the 250 Hz signal. The basilar membrane stiffness in this model of the human cochlea was choosen to agree with the frequency localization of Greenwood (1962).

lower tone passes through that of the higher tone, causing little disturbance to the latter's shape. After two cycles of the lower tone (0.008 sec) the wave pattern is nearly that of the steady-state case, except for the region near the helicotrema. The limitations of this method are evident in that the fundamental nonlinearities associated with pure-tone interaction (two-tone suppression, difference tones) can not be reproduced here.

3. PITCH PERCEPTION STUDIES

Pitch perception studies have been undertaken for over 50 years. Jenkins-Lee (1971) demonstrated the inadequacies of standard lab equipment in producing single-period sinusoids, and constucted a *Plasma-phone*, from which near-perfect input signals could be produced. These signals were then used in a study of the pitch perception of single sinusoids.

In Jenkins-Lee's experiments the listener was presented with a single sinusoid signal, and then, after a 500 msec pause, another single sinusoid of slightly longer or shorter period. As would be expected, the identification by the subject of which tone was higher increased as did the difference in frequency. The example presented (Figure 2) is for a frequency difference of 20% (*i.e.* 250 Hz and 300 Hz), for which the subjects gave correct responses over 80% of the time. In the plot of basilar membrane response, the signals can be clearly differentiated. For a small (5%) difference in freqeuncy (250 Hz and 262.5 Hz), for which the correct response was given only about 60% of the time, there is virtually no noticeable difference in response on this linear scale.

4. TIME SEPARATION PITCH

Another mechanism of pitch perception is "time-separation pitch." Here two tone bursts are presented at a time τ apart. The listener perceives a pitch associated not with the content of the tone burst itself, but with the spacing τ. Therefore, a certain pitch is perceived even though there is not necessarily any energy at that point in the spectrum–or on the basilar membrane.

A simulation of one of McClellan and Small's (1967) tests is found in Figure 3. In this set of experiments, a DC pulse pair (each pulse 0.23 msec in duration) was used, here modeled as two half-sine waves separated by $\tau = 10$ msec. The response (Figure 3-b) shows that the two pulses are clearly separated on the basilar membrane. In this case, the listeners could identify the pitch at $f=1/\tau=100$ Hz quite regularly. For a smaller spacing (τ=0.3 msec) the two waves on the basilar membrane are not distinct, since the "tip" of the second wave runs into the "tail" of the wave from the first pulse. For this separation, the listeners were not able to associate the clicks with the corresponding pitch as accurately as for the longer value of τ.

5. CLICK RESPONSES

There have been two studies of the response of *in vivo* animals to acoustic clicks. Robles *et al.* (1976) used the Mössbauer technique measuring the response of squirrel monkeys to 110 dB

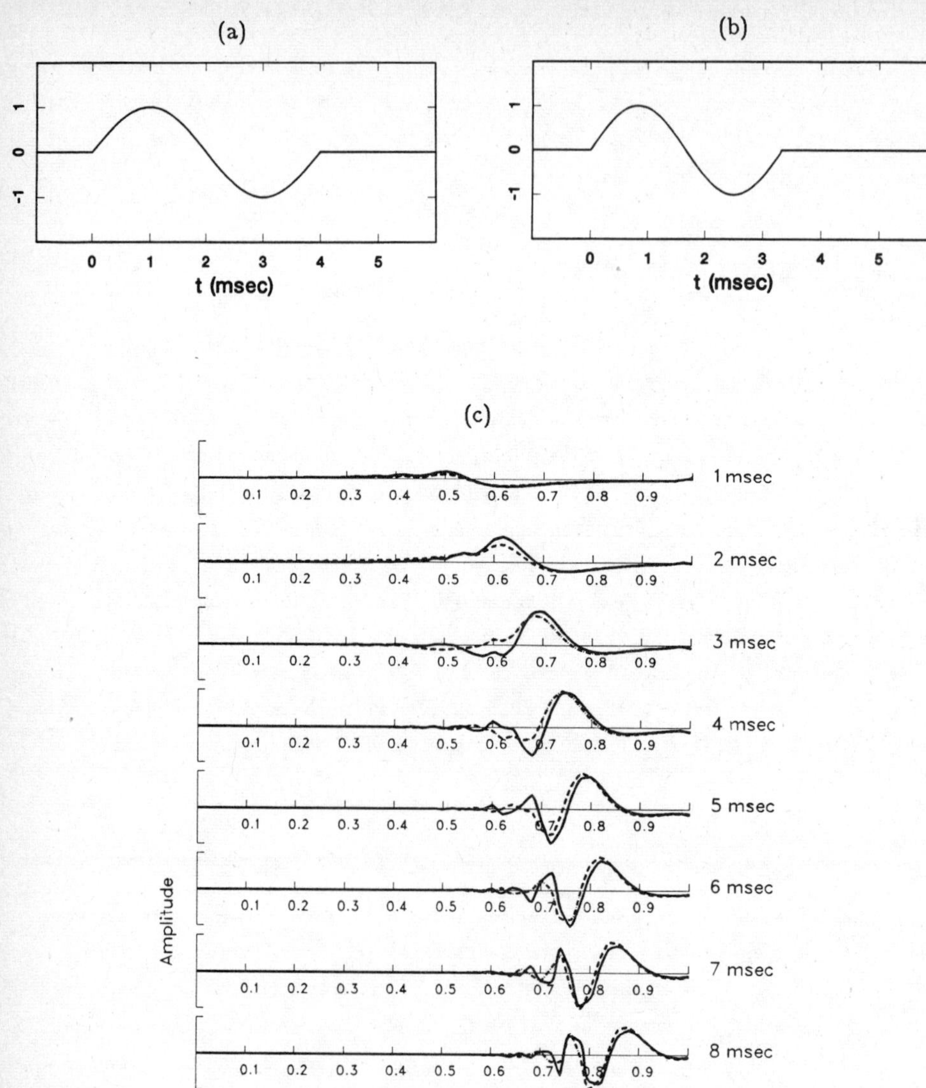

Figure 2. The idealized single-period sinusoids of (a) 250 Hz and (b) 300 Hz used by Jenkins-Lee (1971) in pitch perceptions studies. The ordinate is an arbitrary amplitude scale. (c) The response of the basilar membrane to (a) (solid line) and (b) (dashed line). For this separation in pitch, the listeners correctly identified the higher of the two pitches 80% of the time. For a steady-state signal, the peak for these frequencies occurs at $x/L = 0.78$ (300 Hz) and $x/L = 0.80$ (250 Hz).

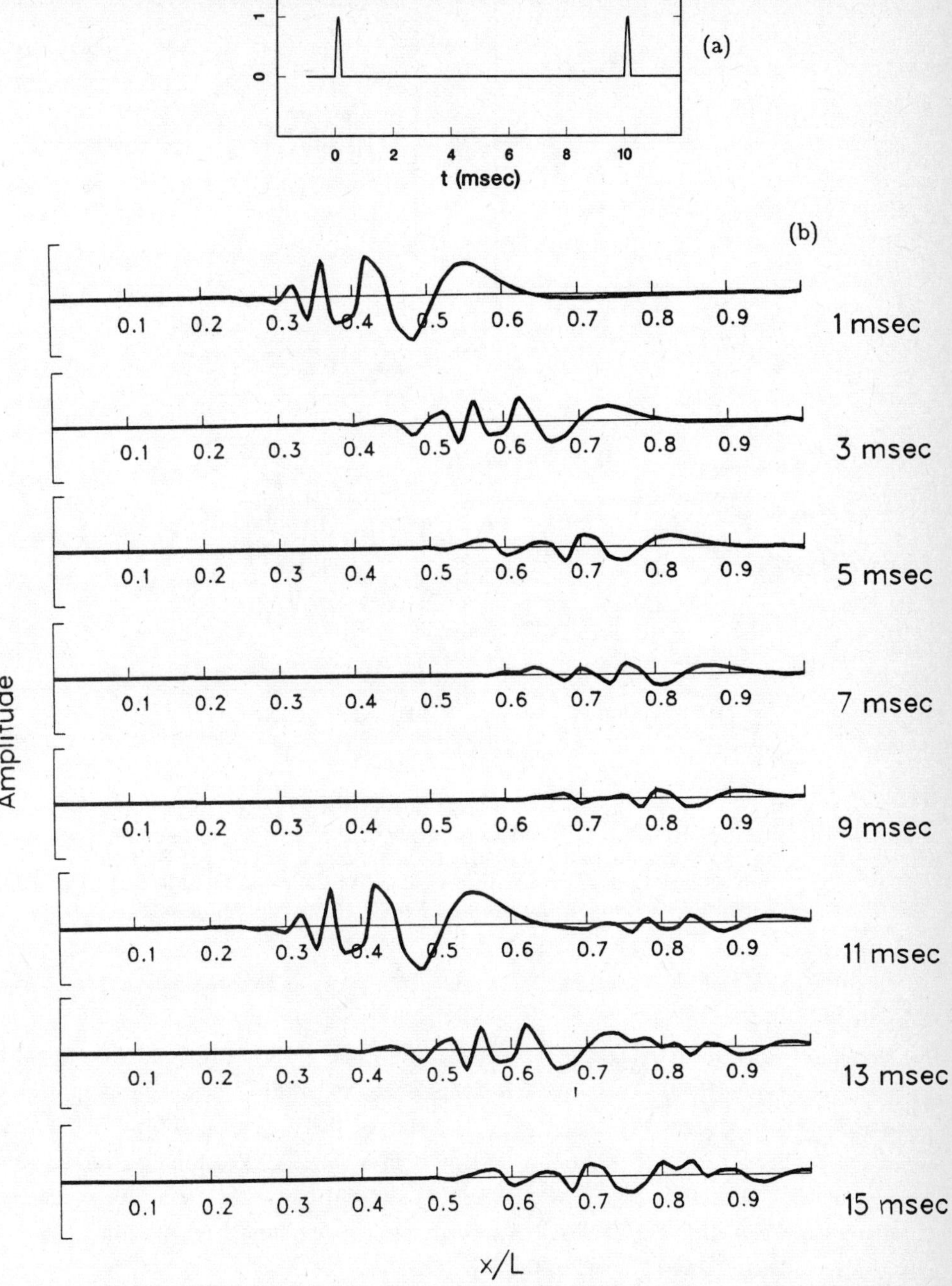

Figure 3. (a) Idealized signals used to model one of McClellan and Small's time-separation pitch studies. Here the spacing τ is 10 msec. (b) Basilar membrane respone to (a). For this value of τ, the responses to the two DC clicks can be separately identified.

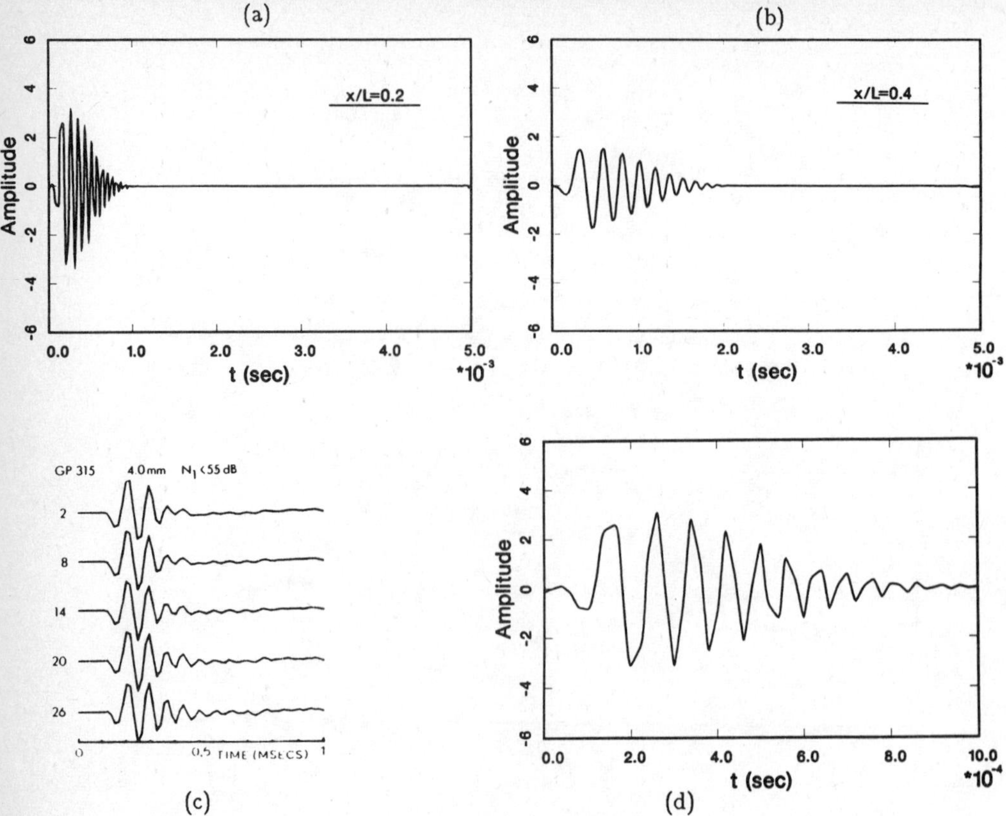

Figure 4. The computed response *vs.* time of the guinea pig model to a click, at (a) $x/L = 0.2$, and (b) $x/L = 0.4$. (c) Measurements of the click response of a guinea pig made by LePage and Johnstone (1980) at the 4.0 mm location. (d) Response at x/l=0.2 (3.6 mm) replotted so that the time scale has the same limits as (c). The model indicates lighter damping than (c), as was found by Robles, *et. al.* (1976).

clicks of 50-100 μsec , and found the same saturating nonlinearity as discovered in the steady-state case by Rhode (1971).

A similar experiment using the capacitance probe was performed on the guinea pig by LePage and Johnstone (1980). They used a 73 dB peak-equivalent SPL click of duration 15 μsec. They were able to reproduce the nonlinearity, and showed that it disappeared with the death of the animal. Their results show the oscillations from a single click continuing for up to 10 cycles. A model of the guinea pig (see Steele and Zais (1983) for details) was used to simulate this experiment. The click was modeled as a single peak in the input signal, with the results (Figure 4-d) similar to those obtained in the experiment.

6. SPEECH PROCESSING

The last example presented is the response of the basilar membrane to speech signals, in

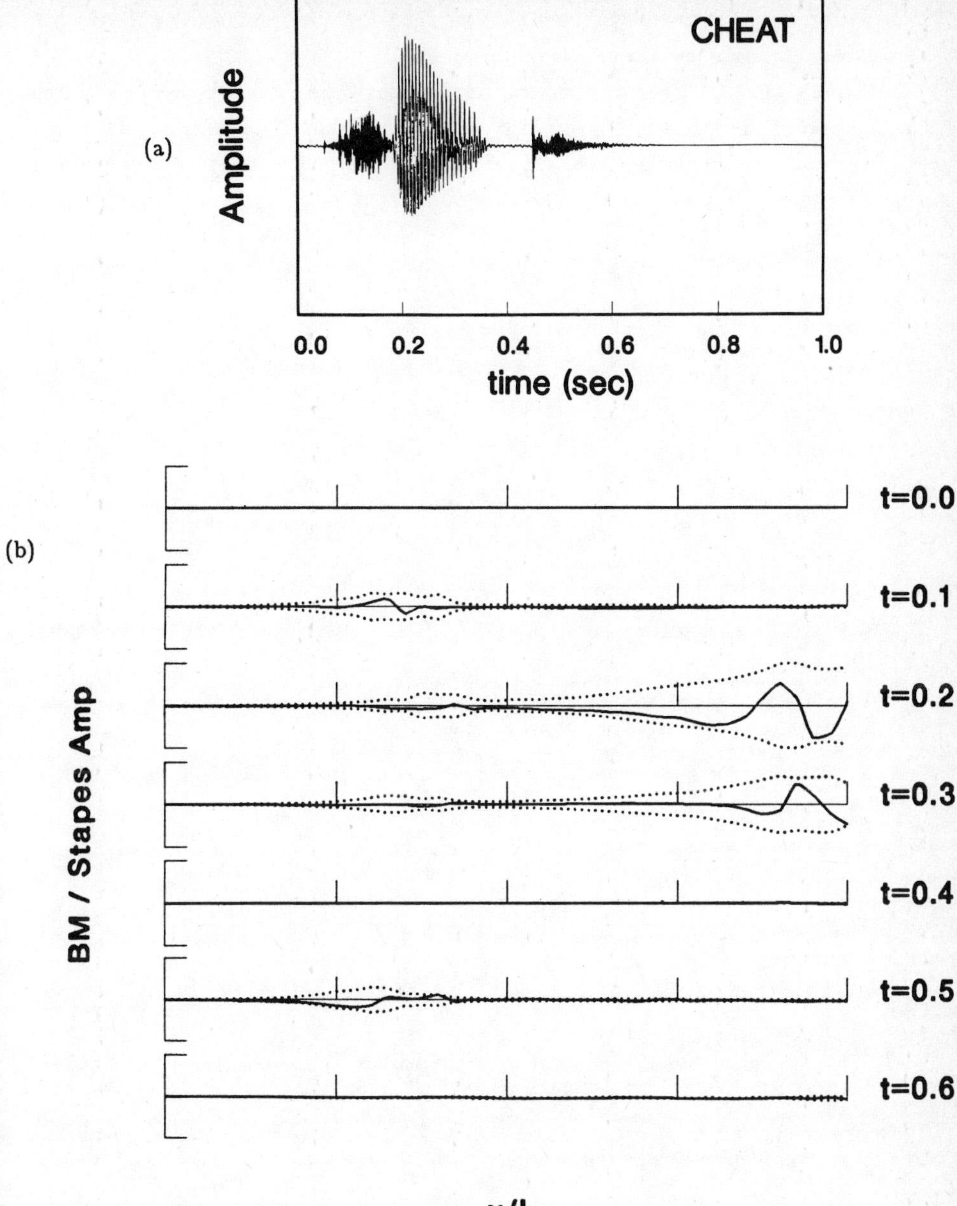

Figure 5. (a) Amplitude of the speech signal for the word "cheat." (b) Response of the basilar membrane to the signal of (a). The solid line is the basilar membrane, while the dotted lines indicate the approximate shape of the amplitude envelope. At right the the time in seconds is indicated.

particular the word "cheat." The signal (Figure 5-a) is 1 sec in length, and was sampled at a rate of 10 kHz. In the response (Figure 5-b), the peaks corresponding to two formants, f_1 and f_2, can be seen on the basilar membrane at t=0.2 and 0.3 sec. While the computing time for such a 1 second signal is prohibitive for more than general results, use of the FFT to compute the click response (such as in Figure 4), and then a standard convolution scheme (Oppenheim and Schafer, 1975) greatly decreases the computing time, allowing 3-D transient calculations to be carried out quite rapidly.

ACKNOWLEDGEMENTS

I would like to thank Prof. C.R. Steele and Dr. E.D. Schubert for their helpful suggestions. This work was funded by a grant from the National Institutes of Health.

REFERENCES

Békésy, G. von, (1960). *Experiments in Hearing.* (McGraw-Hill, New York.)

Dotson, R.D. (1974). *Transients in a Cochlear Model.* Ph.D. dissertation, Stanford University, Stanford, CA.

Greenwood, D.D. (1962). "Approximate calculations of the dimensions of traveling-wave envelopes in four species," *J. Acoust. Soc. Am. 34*, 1364–1369.

Jenkins-Lee, J.E. (1971). *Pitch Perception for Minimum Tone Bursts: Single Periods.* Ph.D. dissertation, Stanford University, Stanford, CA.

LePage, E.L. and Johnstone, B.M. (1980). "Nonlinear mechanical behaviour of the basilar membrane in the basal turn of the guinea pig cochlea," *Hearing Research 2*, 183–189.

McClellan, M.E. and Small, A.M. (1967). "Pitch perception of pulse pairs with random repetition rates," *J. Acoust. Soc. Am. 41*, 690–699.

Oppenheim, A.V. and Schafer, R.W. (1975). *Digital Signal Processing.* (Prentice-Hall, Englewood Cliffs, NJ.)

Rhode, W. (1971). "Observations of the vibration of the basilar membrane in squirrel monkeys using the Mössbauer technique," *J. Acoust. Soc. Am. 49*, 1218–1230.

Robles, L., Rhode, W.S., and Geisler, C.D. (1976). "Transient response of the basilar membrane measured in squirrel monkeys using the Mössbauer effect," *J. Acoust. Soc. Am. 59*, 926–939.

Steele, C.R. and Taber, L.A.(1979). "Comparison of WKB and experimental results for three-dimensional cochlear models," *J. Acoust. Soc. Am. 65*, 1007–1018.

Steele, C.R. and Zais, J.G. (1983). "Basilar Membrane Properties and Cochlear Response," in *Proc. Symposium on Mechanics of Hearing*, E. de Boer and M.A. Viergever, eds. (Delft University Press, 1983.)

THE MECHANICS OF THE BASILAR MEMBRANE AND MIDDLE EAR IN THE PIGEON

A.W. Gummer, J.W.Th. Smolders and R. Klinke
Zentrum der Physiologie
J.W. Goethe-Universität Frankfurt
Theodor-Stern-Kai 7
D-6000 Frankfurt 71, Federal Republic of Germany

INTRODUCTION

The frequency selectivity of primary auditory afferents has been allocated preferentially to different structures in various classes of tetrapod animals: to the basilar membrane (BM) in mammals (Sellick et al., 1982; Khanna and Leonard, 1982), to the hair cell in the turtle (Crawford and Fettiplace, 1980) and to hair cell stereocilia in the alligator lizard (Holton and Weiss, 1983). We have investigated the motion of the BM in the pigeon using the Mössbauer technique because the avian inner ear structures possess features which compare to those in other tetrapods. The avian BM is short by comparison to the mammalian BM (4 mm in pigeon) and is only slightly bent. There are numerous hair cells per radial section (14 - 54 in pigeon) which are tightly packed over the BM and neural limbus (Takasaka and Smith, 1971). The tallest stereocilia are firmly embedded in a porous tectorial membrane. There is a second class of cells, called hyaline cells, which rest on the BM, are not covered by the tectorial membrane and are densely innervated by efferent fibres. On a physiological level, the effect of temperature on single unit frequency threshold curves in the pigeon is different to that in the mammal (Gummer and Klinke, 1983) - there is a 1-octave reduction of best frequency per 10 ^{o}C reduction of cochlear temperature (Schermuly and Klinke, 1985). Von Békésy (1944) has shown that tonotopically mapped frequency tuning exists on the apical part of the chicken BM. It is not known, however, whether the avian BM supports travelling wave motion and whether it is capable of the frequency selectivity of primary auditory afferents.

METHODS

Experiments were performed on homing pigeons weighing 360 - 570 g, which were anaesthetized with pentobarbitone sodium (36 mg kg^{-1}) and artificially respired. Body temperature was controlled and the ECG was recorded. The head was warmed with an infra-red lamp. Animals were paralyzed with alcuronium chloride (10 mg kg^{-1}), approx. 1/2 hr before beginning the mechanical measurements.

Compound action potential (CAP) frequency threshold curves in response to tone pips (gaussian envelope, 4-ms duration, with carrier reversal) were used to provide an indication of the physiological condition of the cochlea. The skull remained open for the middle ear and BM recordings; the external auditory meatus on the contralateral side was closed.

For measurements from the BM the cochlea was exposed ventrally by removing the musculature between the carotid arteries and opening the skull ventro-laterally. Access to the BM is restricted to the first 1.4 mm from the basal end (35% BM length) because of the bend which is situated further apically.

The ventral region was isolated from the middle ear with bone wax. Scala tympani was then opened and the source placed on the BM after removal of perilymph. The source position relative to fine blood vessels in both limbi was noted. The scala was then allowed to refill; perilymph regeneration was often slow and sometimes incomplete. Nevertheless, an absorbant wick was positioned just outside the cochlea in order to remove excess perilymph which would have caused poorer counting statistics (up to 6 dB reduction of the signal-to-noise ratio). The surface of the counting tube was approximately parallel to the surface of the BM. CAP thresholds were sensitive to the presence of minute amounts of blood on the BM , the prolonged absence of perilymph and touching the BM with the probe used for manually placing the source, as reported by Sellick et al. (1982), and Khanna and Leonard (1982).

In later animals, measurements of the motion of the columella footplate (CFP) were made after completing the BM measurements (7 animals). The BM source was left in place and a larger source (x16) was used for the CFP measurement. Control experiments showed that the amplitude and phase responses of the CFP were unaffected by the presence of the BM source. The counting tube was oriented so that the angle between the measurement axis and the major axis of the columella was minimal; the tube position was then finely adjusted to optimize the count rate and the purity of the Mössbauer spectral line. This angle was 20 - 30°; the amplitude data were not corrected for this offset (0.5 - 1.2 dB).

The distance of the middle of the source from the basal end of the BM was measured at the end of the experiment, after further opening scala tympani.

The Mössbauer system and experimental paradigm are similar to that used by Johnstone and colleagues (Sellick et al., 1982), although the source activity was less. The threshold velocity of the detection system was about 0.08 mm s^{-1}. For measurement of BM motion data were collected at an iso-velocity of 0.08 mm s^{-1} and for the CFP motion at about 0.5 mm s^{-1} for frequencies above 0.7 kHz and 0.2 mm s^{-1} for lower frequencies.

Sound was presented closed field from a Brüel & Kjaer (4134) 1/2"-condensor microphone and the sound pressure was measured at about 1 mm from the tympanic membrane using a 1-mm probe tube microphone. The sound pressure was measured at each stage of the experiment. Between 0.125 and 11.314 kHz, the mean variation with frequency of the amplitude and phase was 21 dB and 0.85 cycles, respectively, for an open skull.

RESULTS AND DISCUSSION

Columella footplate response

Fig. 1 shows the CFP response; the displacement amplitudes are corrected for a sound pressure of 100 dB SPL at the tympanic membrane. Measurements were made at 1/4-octave intervals for 0.125 - 11.314 kHz, together with intermediate frequencies above 4.757 kHz, in order to ensure a more detailed sampling of the high frequency region of the phase response which, according to convention, is plotted on a linear frequency axis. The mean amplitude response can be divided into four main sections; namely, a low frequency region with mean amplitude of 0.3 μm for 0.125 - 0.707 kHz, followed by slopes of -8 dB oct^{-1} for 0.841 - 2.828 kHz, -16 dB oct^{-1} for 2.828 - 5.657 kHz and -8 dB oct^{-1} for 5.657 - 11.314 kHz. The individual responses contained irregularities above 2.828 kHz, which in some cases appeared as sharp amplitude notches - their frequencies and depths varied across animals and, to a lesser extent, with time for a given animal. The irregularities were manifested in the phase response as an excessively large phase lag (tending to 2 cycles at high frequencies) or a large phase reversal (negative group delay). The phase responses were consistent with a minimum phase function calculated from straight-line segment approximations to the amplitude response. The irregularities were independent of sound level. It is almost certain that the irregularities are due to some frequency dependent, transverse motion of the columella because, as shown in Fig. 1, when the measurement axis was shifted dorsally by 33° (animal 96), by changing only the orientation of the counting tube, the phase lag was reduced by about 1 cycle at high frequencies. The effect was reversible.

Irregularities were not reported for the motion of the CFP by Saunders and Johnstone (1972), presumably because the low activity of the source which was available at that time precluded measurement at finer frequency intervals. However, irregularities in this frequency range were reported for the incus response in the guinea pig (Wilson and Johnstone, 1975). They pose a particular problem for the case of the pigeon because the total phase excursion of the BM, relative to sound pressure, will be seen to be no more than 2 cycles, so that the BM displacement could appear to lead the CFP displacement in the relevant frequency range. An apparent phase lead, additional to the asymptotic value of 0.25 cycles at low frequencies, is also strikingly evident for frequencies up to 8 kHz in the guinea pig (Wilson and Johnstone, 1975; Figs. 14 & 15). However, in that case the effect was not so troublesome because the

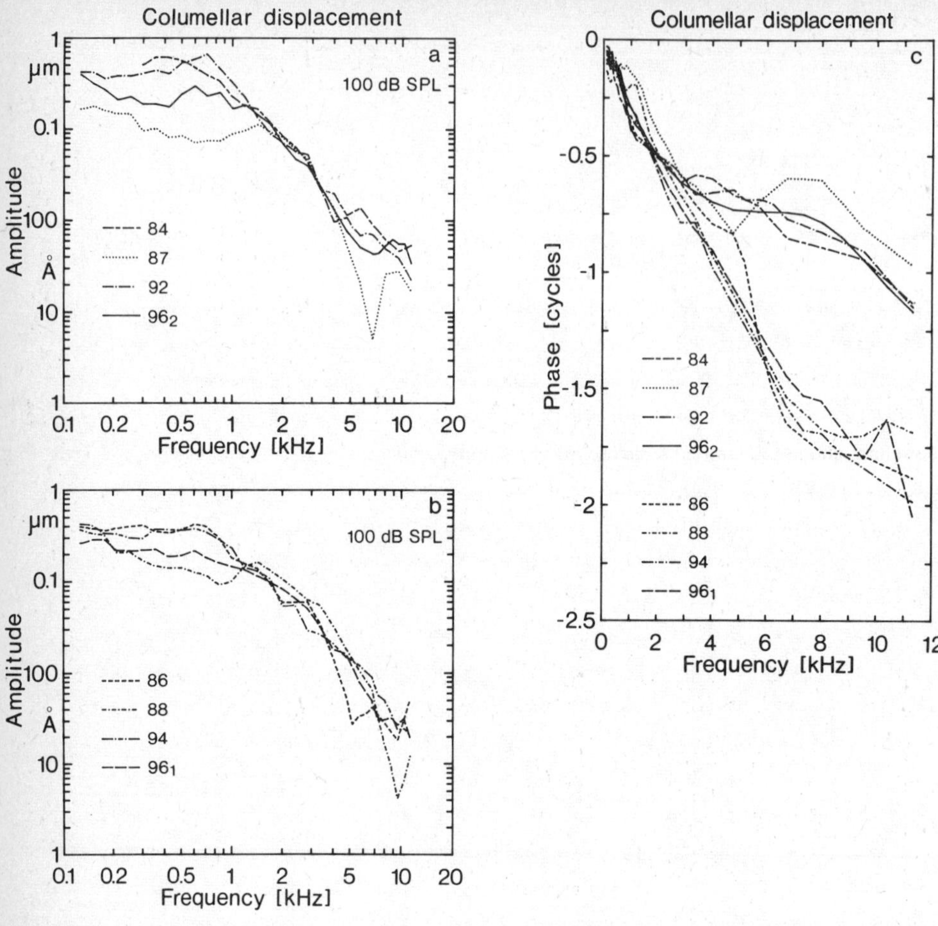

Figure 1.(a), (b) Amplitude and (c) phase of the CFP displacement relative to sound pressure at the tympanic membrane at 100 dB SPL as a function of frequency, with an open skull, for 7 pigeons identified by numbers. The cochlea was closed and intact for animals 84, 87, 88, 96. For the other animals the cochlea was open, scala tympani was full with perilymph and there was a Mössbauer source on the BM. For animal 96 the counting tube was shifted from its usual orientation (curve 96_1) to a location 33° more dorsally (curve 96_2). The amplitude responses are grouped in two figures according to the two types of phase responses. In this and all following figures the plotting scales for amplitude and phase are 12 dB oct^{-1} and 0.125 ms, respectively.

most apical recording was situated at the 16-kHz point. Since the irregularities appear to be due to a transverse motion of the middle ear structures, it is unlikely that they represent the effective input to the cochlea. Therefore, in this report, the BM motion will not be referenced to some averaged CFP motion.

Basilar membrane response

Figs. 2 and 3 show the BM response measured at different locations. The amplitude data are presented as iso-velocity curves because it has been shown in the mammal that, under ideal physiological conditions, the tuned region of BM iso-velocity curves at threshold matches that of single unit threshold tuning curves, (Sellick et al., 1982; Khanna and Leonard, 1982). The iso-velocity curves have been divided into two groups according to sensitivity, as measured by the sound pressure at the characteristic frequency (CF) of the BM. Because the curves can have ill-defined "tips", we have defined the BM CF as that frequency above which the required sound pressure began to rise above the minimum region, by analogy to Wilson and Johnstone (1975). It is seen that the BM response is tuned and tonotopically organized so that the BM CF shifts to lower frequencies with distance from the base of the BM. For a given frequency in the tuned region, the phase lag increases with distance from the base, thus providing direct evidence for a travelling wave. The change of phase with distance is larger in the more sensitive group. For high frequency stimulation the slope of the phase curve is considerably reduced and tends to the slope of those CFP phase responses that have smaller phase accumulation (upper group in Fig. 1 c), indicating standing wave motion in this frequency range.

The BM iso-velocity tuning curves measured in our experiments do not match the single unit threshold tuning curves reported for the pigeon by Sachs et al. (1974). Ninety percent of fibres with CFs in the range 1.4 - 4 kHz had CF-thresholds of 0 - 20 dB SPL and $Q_{10\ dB}$ values situated in a band defined by 2 - 5 at 1.4 kHz and 3.5 - 6.5 at 4 kHz. In our experiments, the sound pressure required to produce a velocity of 0.08 mm s^{-1} at BM CF ranged from 39 - 75 dB SPL and $Q_{10\ dB}$ from 0.5 - 2.

The sensitivity of the pigeon BM is dependent on the physiological condition of the basilar papilla. CAP thresholds between 1 - 2 kHz proved to be the best monitor of cochlear condition. No correlation was found between the CAP threshold at BM CF and the SPL required to produce 0.08 mm s^{-1} at BM CF. Fig. 4 shows the effect of damage on the BM response after a 24-dB loss of CAP threshold at 1.4 kHz. The loss was probably due to sound overexposure resulting primarily from the high sound pressures used to produce the upper SPL region of the intensity functions. The BM response shows a loss of 7 - 16 dB for 1 - 4 kHz, a reduction of BM CF (about 0.5 oct) and of $Q_{10\ dB}$ (1.2 to 0.5), and a well-defined high-frequency amplitude plateau; there was a concomitant reduction of phase lag.

The effect of physiological condition on the tonotopic organization of the BM is illustrated in Fig. 5. The BM CF-distance data have been divided into two groups according to the SPL at BM CF. It is evident that for the sensitive group there is a strong correlation between logCF and distance (r=0.95), whereas for the other group the correlation is weak (r=0.73). The regression for the sensitive group pre-

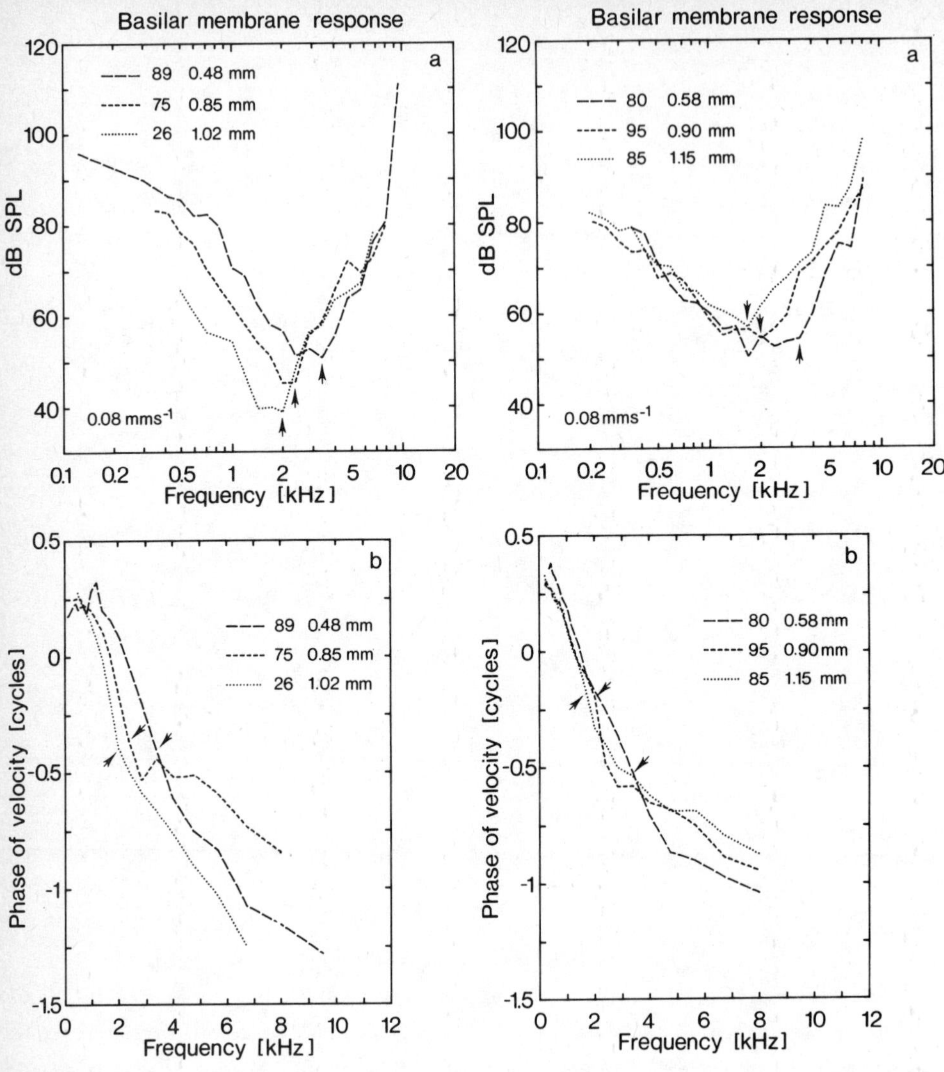

Figure 2. (a) Sound pressure at the tympanic membrane required to produce an iso-velocity of 0.08 mm s^{-1} on the BM, and (b) phase of BM velocity relative to sound pressure, as a function of frequency, for positions of 0.48 - 1.02 mm from the basal end, for 3 pigeons in which the required SPL at the characteristic frequency (CF) of the BM was 39 - 51 dB SPL. Arrows indicate the BM CF.

Figure 3. As in Fig. 2, but for positions of 0.58 - 1.15 mm, with SPL at BM CF of 54 - 57 dB SPL.

dicts an upper BM CF of 6 kHz for the pigeon, which equals the maximum CF of primary auditory afferents (Sachs et al., 1974). The mapping constant of 0.63 mm oct^{-1} compares to 0.56 mm oct^{-1} inferred from sound over-exposure experiments in the chick (Ryals and Rubel, 1982). For the other group the maximum BM CF is 0.45 oct below that for the sensitive group, agreeing with reports for the effects of cochlear de-

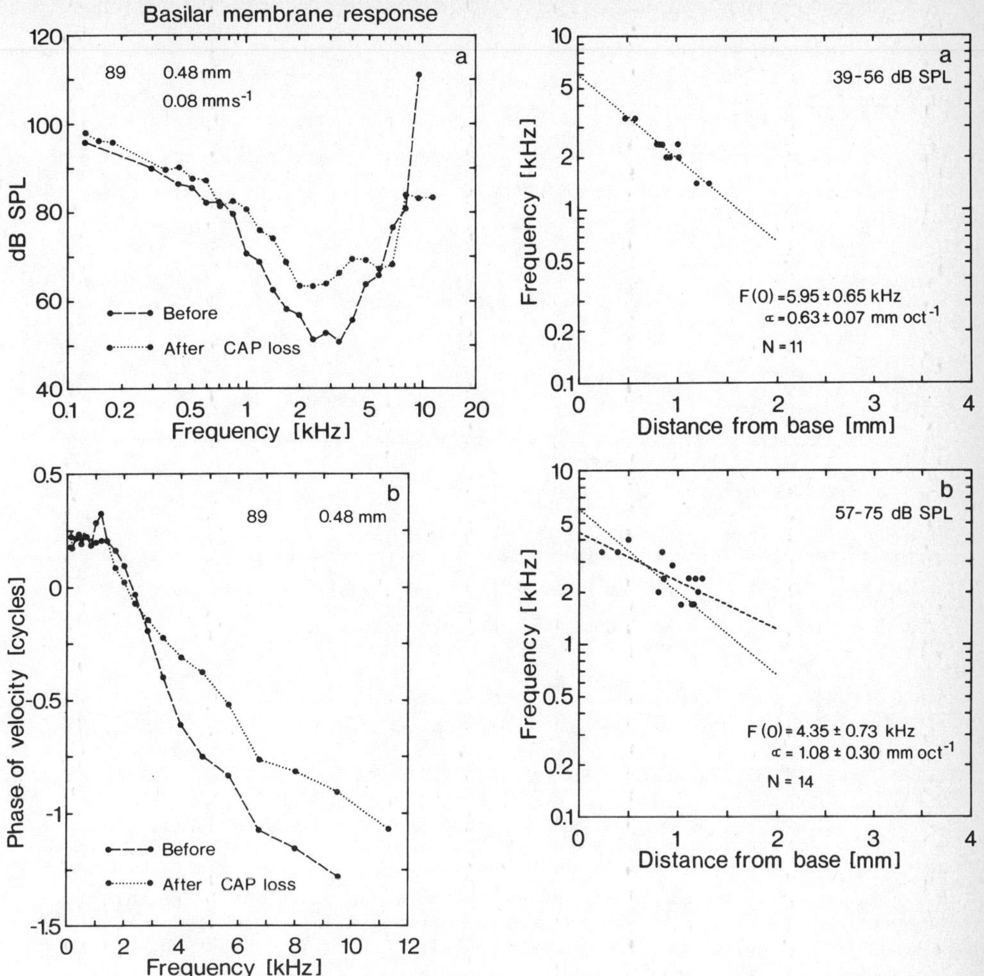

Figure 4. (a) BM iso-velocity curve (0.08 mm s^{-1}) and (b) phase of BM velocity, relative to sound pressure, for one pigeon (89), measured before and after a 24-dB loss of CAP threshold at 1.414 kHz. Distance from basal end: 0.48 mm. In detail, the CAP thresholds at 1.414 kHz were: 44 dB SPL before cochlear opening; 57 dB SPL immediately prior to collection of Mössbauer data (16 min after source placement); 65 dB SPL after iso-velocity measurements for 1 - 11.314 kHz (48 min); 77 dB SPL after measurement of intensity functions at 2.378, 4.757 and 1.189 kHz (69 min); 81 dB SPL after iso-velocity measurements for 0.125 - 1 kHz (17 min), beginning at the highest frequency. Collection of data for the second BM response began after completing the low-frequency tail of the first BM tuning curve and measuring the CAP-frequency threshold curve.

Figure 5. Characteristic frequency (CF) of the BM as a function of distance from the basal end of the pigeon BM, for 25 source placements in 24 animals. The data are grouped according to the SPL required to produce an iso-velocity of 0.08 mm s^{-1} at BM CF: (a) 39 - 56 dB SPL and (b) 57 - 75 dB SPL. The dotted line in (a) and the dashed line in (b) are the regression lines for the two groups; for comparison, the line from (a) has been replotted in (b), also as a dotted line. The estimated frequency axis intercept, F(0), and mapping constant, α , are given on the figure.

terioration on BM CF in the basal turn of the mammal (Sellick et al., 1982; Khanna and Leonard, 1982). However, for positions apical to the 0.7-mm point on the pigeon BM, the BM CF tends to increase with cochlear damage.

Intensity functions for most animals were linear over the entire frequency range; the exceptions resembled the functions reported by Sellick et al. (1982; Fig. 7) for recordings made at the end of the series when the source had lost more than half of its activity. The 'in situ' count rate in our experiments was about four times less than that for Sellick et al. (1982) because the source was not as active and the recording configuration through a deep hole in the neck was prohibitive. Thus, it is possible that the mismatch between the BM and single unit responses in the pigeon is due, in the case of our sensitive group of recordings, to sound over-exposure.

ACKNOWLEDGEMENTS

The authors are indebted to Prof. B.M. Johnstone for his invaluable expertise given at the beginning of the experimental series. This investigation was supported by the Deutsche Forschungsgemeinschaft (SFB 45).

REFERENCES

Békésy, G.von, "Über die mechanische Frequenzanalyse in der Schnecke verschiedener Tiere." Akust. Zeitschrift 9, pp. 3-11, 1944.

Crawford, A.C. and Fettiplace, R., "The Frequency Selectivity of Auditory Nerve Fibres and Hair Cells in the Cochlea of the Turtle." J. Physiol. 306, pp. 79-125, 1980.

Gummer, A.W. and Klinke, R., "Influence of Temperature on Tuning of Primary-like Units in the Guinea Pig Cochlear Nucleus." Hearing Research 12, pp.367-380, 1983.

Holton, T. and Weiss, T.F., "Receptor Potentials of Lizard Cochlear Hair Cells with Free-standing stereocilia in response to tones." J. Physiol. 345, pp. 205-240, 1983.

Khanna, S.M. and Leonard, D.G.B., "Basilar Membrane Tuning in the Cat Cochlea." Science 215, pp. 305-306, 1982.

Ryals, B.M. and Rubel, E.W., "Patterns of Hair Cell Loss in Chick Basilar Papilla after Intense Auditory Stimulation. Frequency Organization." Acta Otolaryngol. 93, pp. 205-210, 1982.

Sachs, M.B., Young, E.D. and Lewis, R.H., "Discharge Patterns of Single Fibres in the Pigeon Auditory Nerve." Brain Res. 70, pp. 431-447, 1974.

Saunders, J.C. and Johnstone, B.M., "A Comparative Analysis of Middle-Ear Function in Non-mammalian Vertebrates." Acta Otolaryngol. 73, pp. 353-361, 1972.

Schermuly, L. and Klinke, R., "Change of Characteristic Frequency of Pigeon Primary Auditory Afferents with Temperature." J. Comp. Physiol.A, 156, pp.209-211, 1985.

Sellick, P.M., Patuzzi, R. and Johnstone, B.M., "Measurement of Basilar Membrane Motion in the Guinea Pig using the Mössbauer Technique." J. Acoust. Soc. Am. 72, pp. 131-141, 1982.

Takasaka, T. and Smith, C.A., "The Structure and Innervation of the Pigeon's Basilar Papilla." J. Ultrastructure Research 35 , pp. 20-65, 1971.

Wilson, J.P. and Johnstone, B.M., "Basilar Membrane and Middle-Ear Vibration in Guinea Pig Measured by Capacitive Probe." J. Acoust.Soc.Am. 57,pp.705-723, 1975.

ON THE MECHANICS OF THE HORSESHOE BAT COCHLEA

H. Duifhuis[1,2] and M. Vater[3]
[1]Biophysics Dept., Rijksuniversiteit Groningen, the Netherlands
[2]also: Institute for Perception Research, Eindhoven, the Netherlands
[3]Zoology Dept., Universität München, Federal Republic of Germany

ABSTRACT

The basilar membrane (BM) in the greater horseshoe bat has a peculiar thickness and width profile, which suggests that BM-stiffness in the basal half turn is more than one order of magnitude greater than in the second half turn and more apically. The transition is quite abrupt.

Motivated by new data on the cochlear frequency map we analyse the possibility that the stiffer basal half turn acts as an acoustic interference filter. Its increased stiffness and impedance transitions at the stapes and at the transition mentioned, make this a feasible mechanism. We propose that 2¼ wavelengths of the reflected call frequency of the echolocating bat match the length of the interference filter. Using a scaled version of the middle ear description given by Matthews (1980) and linear, passive, long-wave, 1-dimensional cochlear mechanics we analyse the effect of the interference filter and compute the tympanic membrane input impedance. Experimental data on the latter (Wilson and Bruns, 1983a) are interpreted in terms of the proposed mechanism: fine structure in the input impedance appears to reflect the interference filter passbands.

1. INTRODUCTION

The auditory system of horseshoe bats appears to be well designed for processing the constant frequency component of the echolocation signal. Within a narrow frequency band around 83 kHz thresholds are low and tuning is extremely sharp. These properties, together with the over-representation of this frequency range, have been traced back to morphological specializations of the cochlea. The most salient feature is the thickness and width profile of the basilar membrane. Two regions of constant BM-parameters can be distinguished, with the basally located region (~4.4 mm) having an extraordinary thickness (review: Neuweiler et al, 1980).

Bruns (1976b) established a cochlear frequency map based on the swelling of outer hair cell nuclei in response to high intensity tones. According to this map, the bat's resting frequency (f_r: constant frequency component emitted by non-flying bat) is located at the abrupt discontinuity in basilar membrane parameters at 4.4 mm. The frequency range in which the relevant echos occur, from f_r to f_r+3 kHz, would be mapped onto 3.15 mm of the thickened BM just basalward of this point.

Using the HRP-technique, Vater et al (1985) have shown that this frequency range is mapped apically to the transition point onto the 2nd half turn, where innervation density is maximal. These results necessitate a new interpretation of the function of the thickened basilar membrane region. In this paper we propose that this basal part of the cochlea is specialized to act as an acoustic interference filter. Consequently, its functional rôle is seen as a local peripheral mechanical element, enhancing sensitivity and tuning properties, rather than in direct mechano-electrical transduction and relay of information to the central auditory system.

2. NEW WORKING HYPOTHESIS

We propose that the basal 4.4 mm of the horseshoe bat's cochlea functions as a - virtually passive - interference filter. The interference filter has a multimodal bandpass characteristic (combfilter). One of these bandpass modes is tuned to the bat's f_r to f_r+3 kHz range. The mode frequencies are determined by the length of the basal part and its BM-stiffness, but they are (almost) independent of its BM-mass, which in combination with the stiffness determines the BM-resonance frequency of the basal part.

In order for such a mechanism to be feasible, the basal part of the cochlea (part 1) should have:

1- appropriate acoustic impedance (mis-)matches at its two interfaces, viz. the middle ear-part 1 transition, and the part 1-part 2 transition;

2- adequate transmission properties in the relevant frequency range.

To test the hypothesis we analyse the transmission properties of a model peripheral ear of the horseshoe bat.

3. MODEL OF THE PERIPHERAL EAR OF THE HORSESHOE BAT

Conceptually we divide the peripheral ear into external ear, middle ear, and cochlea, with the cochlea subdivided into part 1 (basal half turn), part 2 (2nd half turn) and the apical part (remainder). The elements are discussed in the subsequent subsections.

Acoustic properties are formulated in terms of pressure, p, volume velocity, U, and acoustic impedance, Z_a. Quantities are given in SI-units.

3.1 *The external ear*

We assume that the primary purpose of the external ear (apart from its directivity) is to lower the cut-off frequency, as in an exponential horn. Above this cut-off frequency the acoustic impedance at the eardrum, $Z_{a.t}$, equals the ratio of the specific acoustic impedance of air and the eardrum area, S_t. Thus we assume that for the frequencies of interest

$$Z_{a.t} = Z_s(\text{air})/S_t = 415/2.5\cdot10^{-6} = 1.66\cdot10^8 \text{ Pa}\cdot\text{s/m}^3. \qquad (1)$$

Since this value is real, we use the shorthand R for $Z_{a.t}$.

3.2 *The middle ear*

The middle ear can be represented by the four pole K_{me}, which relates pressure and volume velocity at the eardrum to those at the stapes. A more physical middle ear model can be formulated at the expense of more parameters. Following Matthews (1980) we take the equivalent network shown in Fig. 1. Here Z_{d1} models shunt leakage through the eardrum, Z_{mec} the middle ear cavity, Z_{d2} eardrum and malleus properties, T the pressure and volume velocity transformer ratio, Z_j the stiffness in the ossicle joints, and Z_2 further ossicle and cochlea interface properties. Of course, the elements of the matrix K_{me} can be expressed in terms of these parameters.

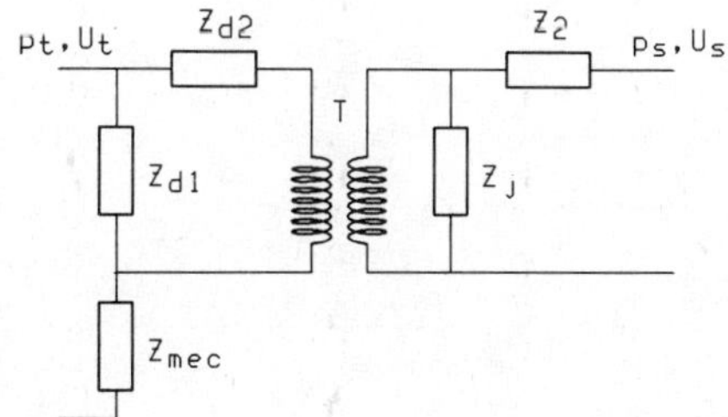

Figure 1. Middle car model after Matthews (1980). Parameter values used for bat: Z_{mec} RC-parallel, $R=2.10^{12}$, $C=5.10^{-14}$; Z_{d1} LCR series, $L=10^3$, $R=1.7\cdot10^9$, $C=1.5\cdot10^{-15}$; Z_{d2} LCR series, $L=1.25\cdot10^3$, $R=4\cdot10^8$, $C=6.5\cdot10^{-15}$; Z_j single C, $C=10^{-20}$; Z_2 LCR series, $L=10^7$, $R=2\cdot10^{12}$, $C=10^{-18}$; all SI acoustic impedance units, and T=85.

3.3 *The cochlea*

The first 4.4 mm of the cochlea, designated as part 1, has relatively constant BM-width and BM-height. Hence it can be modelled as a transmission line with constant parameters. At 4.4 mm it is loaded with the input impedance of part 2. Since BM-parameters are also relatively constant over some 3 mm of part 2 (cf Bruns 1976a) and since changes thereafter are gradual, we take this load impedance equal to the characteristic impedance of part 2. In the present study we concentrate on transmission properties up to and just beyond the transition point.

Formulating the transmission line model in terms of acoustic impedances, except for the BM-impedance which is expressed as a specific acoustic impedance, one has (cf, e.g., Zwislocki, 1953,1983):

$$Z'_x = -\frac{1}{U_x}\frac{\partial p}{\partial x} = \frac{\rho}{S_{sc}v}\frac{\partial v}{\partial t} = \frac{i\omega\rho}{S_{sc}} \tag{2a}$$

$$Y'_y = -\frac{1}{p}\frac{\partial U_x}{\partial x} = \frac{S_{sc}wb}{pS_{sc}} = \frac{2b}{Z_{BM}} \tag{2b}$$

where Z'_x is the longitudinal acoustic impedance per meter; Y'_y the shunt admittance per meter; S_{sc} the scala area (width b × height h); w the average 'vertical' BM-velocity; p the scala pressure (trans membrane pressure 2p); Z_{BM} the basilar membrane impedance 2p/w; v and U_x the particle and volume velocity, respectively, in the x-direction; ω the angular frequency; ρ the fluid density; and $i = \sqrt{-1}$.

Characteristic impedance Z_o and wave number k follow directly:

$$Z_o = \{Z'_x/Y'_y\}^{\frac{1}{2}} = \frac{1}{S_{sc}}\{\tfrac{1}{2}i\omega\rho h Z_{BM}\}^{\frac{1}{2}} \tag{3}$$

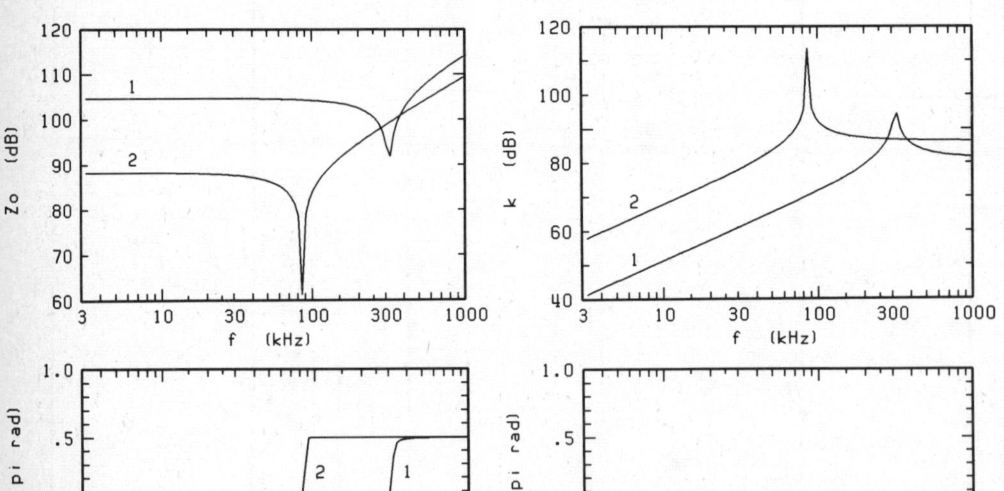

Figure 2. Characteristic impedances $Z_{o1,2}$ and wave numbers $k_{1,2}$ of parts 1 and 2 of the horseshoe bat cochlea. Parameter values: $\rho = 10^3$, $S_{sc} = .3\cdot10^{-6}$, $h = 10^{-3}$, $m_1 = 0.01485$, $s_1 = 59.4\cdot10^9$; $\delta_1 = 0.04$; $m_2 = .004585$ $s_2 = 1.339\cdot10^9$; $\delta_2 = 0.002$.

and

$$k = \{-Z'_x Y'_y\}^{\frac{1}{2}} = \{-\frac{2i\omega\rho}{hZ_{BM}}\}^{\frac{1}{2}} = \frac{\rho\omega}{Z_o S_{sc}} \quad (4)$$

where

$$Z_{BM} = i\omega m + d + s/i\omega = \frac{s}{i\omega}\left[1 - (\frac{\omega}{\omega_o})^2 + i\delta(\frac{\omega}{\omega_o})\right] = \frac{s}{i\omega} H(\omega) \quad (5)$$

with $\omega_o = \sqrt{s/m}$, $\delta = d/\sqrt{sm}$, where m, d, and s are BM-mass, damping, and stiffness per unit area. Transmission through part 1 can be described by a four pole L, the elements of which depend on Z_{o1}, k_1 and on the length of 1, l = 4.4 mm. Magnitudes and phases of Z_{o1}, Z_{o2}, k_1, and k_2 are given in Fig. 2 for the parameter values mentioned. They are given in m, kg, and s units; the values are discussed in Sec. 5.

4. THE BENEFIT OF THE INTERFERENCE FILTER

The assumption that the basalmost half turn of the cochlea functions as an acoustic interference filter implies that it operates as a multimodal bandpass filter. The passbands occur at those frequencies for which n±¼ wavelengths match its length (n an integer). A passband tuned to the relevant frequency range, i.e. just above f_r, causes both enhanced sensitivity and enhanced sharpness of tuning in this range.

The assessment of the benefit is not trivial. We have chosen to compare the ear with cochlea part 1 and a reasonably efficient middle ear (case a) to the ear lacking part 1 but with the same middle ear (case b). Assuming the same incident sound field we compute p_2 in both cases, just beyond the transition point, and define the gain factor

$$g = |p_2(a)/p_2(b)| \, , \quad (6)$$

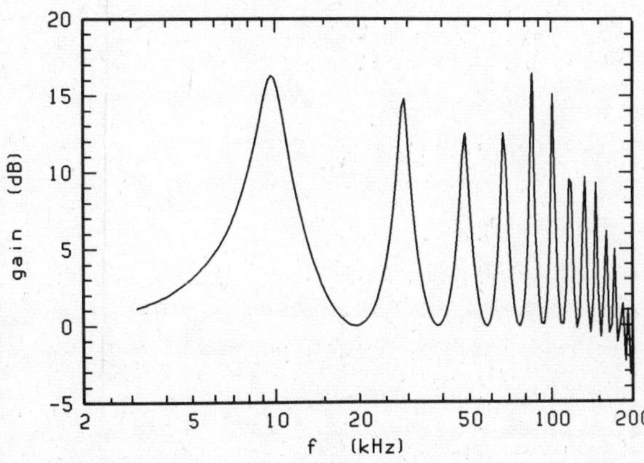

Figure 3. The gain factor produced by the first half turn of the cochlea acting as an acoustic interference filter, the 5th passband mode of which is tuned to the important echolocation frequency range. Parameters as before.

which, obviously, depends on K_{me}, L, and Z_{o2}. We remark that g is not necessarily maximum when the stapes impedance match is optimum (i.e., no reflections in case a). For the parameter values mentioned before g is given in Fig.3.

5. DISCUSSION

5.1 *Cochlea parameters*

The basic assumption of this paper proposes that one of the pass-band modes of the interference filter (cochlea part 1) is tuned to the echolocation signal frequency range. Travelling wave data from Wilson and Bruns (1983b) suggest that in this range $k_1 \cdot l = \frac{9}{4} \cdot 2\pi$. This fixes k_1. The value of s_1 follows if scala height, area, and damping coefficient δ_1 are known. The thickness ratio then gives s_2, and m_2 follows from f_{o2}. We are less certain about m_1 and f_{o1}. One might propose that the ratio of m_1 and m_2 is equal to the thickness ratio. In fact the ratio may be somewhat smaller because the effective BM-mass includes some 'attached' fluid mass. Geometrical parameters were derived from Bruns (1976b). For the thickness ratio it gives ~3.5; we assume that stiffness is proportional to the cubed thickness. Parameter values are listed in the Fig. 2 caption. As to be expected from the difference in relevant frequency range, stiffness and in particular mass differ from values used to describe 'normal' mammalian cochleae.

5.2 *Middle ear parameters*

The middle ear model in Fig. 1 has many parameters which are, unfortunately, not assessable on the basis of available data. However, it is possible to formulate certain constraints.

Z_{mec} is primarily determined by the middle ear cavity volume and the eardrum area, which yields the stiffness. A guess about the time-constant gives a value for the leak resistance.

The transformer ratio T has been estimated to be 30 on the basis of geometrical data by Wilson and Bruns (1983a). They note that their transmission data show a lever ratio of about 5.6 rather than 2, as assumed in the earlier estimate. Hence we take T = 85.

Bearing this in mind we started from the parameter values used by Matthews, modelling the cat middle ear, and scaled these upward by a reasonable factor. The values were modified somewhat to yield a reasonable approximation of the input impedance at the eardrum, $Z_{in.t}$, as measured by Wilson and Bruns (1983a). The values used are given in the Fig. 1 caption. Figure 4 gives the computed input impedance together

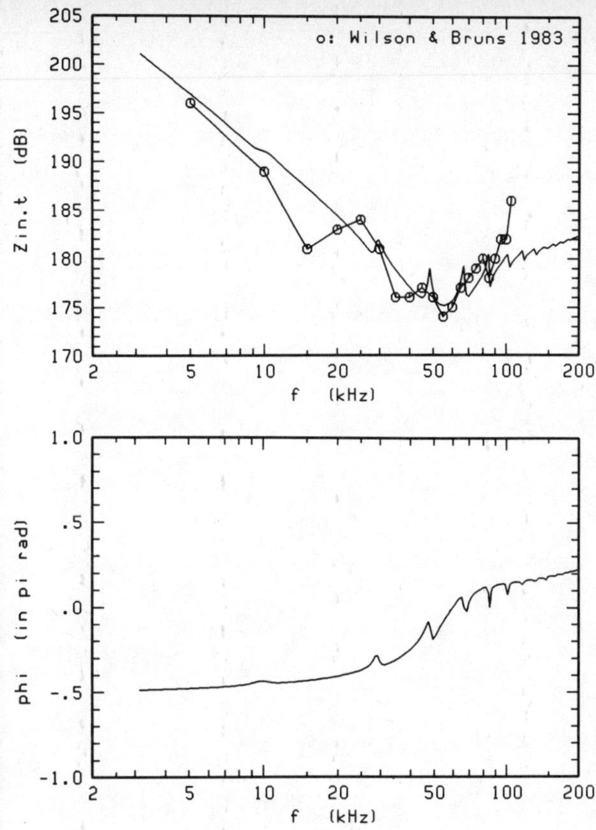

Figure 4. Magnitude and phase of the eardrum input impedance. The data points are from Wilson and Bruns 1983a, their Fig. 7. The magnitude is in dB re 1 Pa·s/m³.

with data points from Wilson and Bruns, which were converted to acoustic impedance taking $S_t=2.5\cdot10^{-6}$ m^2, and allowing 4 dB difference between umbo velocity and average eardrum velocity. Their phase data is limited to 3 points and confounded by a delay, and therefore not reproduced. The amplitude data match the computed curve satisfactorily. On the one hand that is no surprise given the abundance of parameters, on the other hand, several physical constraints have been taken into account. Three out of the four subharmonics at which the interference filter is supposed to resonate are reflected in the data near the computed notches. Data are available at 5 kHz intervals, thereby missing details of the fine structure.

5.3 *Evaluation of the gain factor*

The bandpass peak near f_r as shown in Fig. 3 is quite sharp, and therefore useful. Its precise form is rather sensitive to the exact match of the relevant wavelength to the length of part 1 and to the damping coefficient δ_2. Resonance frequency and damping in part 1 play a secondary rôle. The finding that the 5th mode of the filter appears to be used has the following implications. First, it produces also enhancement in the low-frequency range. Secondly, given the mode frequency, the modes are the narrower the higher the mode number. This is of advantage as long as adjacent modes are sufficiently far apart.

5.4 *Conclusion*

We have not yet fully exploited middle ear parameter optimization, an undertaking which should be constrained by more detailed data. The cochlea part also needs further experimental and theoretical study. On the basis of the results obtained so far, we propose that the interference mechanism plays a significant rôle in the normal horseshoe bat cochlea.

ACKOWLEDGEMENTS

This project started in the summer of 1984, when HD was invited by Professor E.Zwicker to visit his Electro-Acoustics Institute in Munich. It is supported by Sonderforschungsbereich 204.

REFERENCES

Bruns, V., "Peripheral Auditory Tuning for Fine Frequency Analysis by the CF-FM Bat *Rhinolophus ferrumequinum*. I. Mechanical specialization of the cochlea," J.comp.Physiol. 106, pp. 77-86, 1976a.

-II. Frequency mapping in the cochlea," J.comp.Physiol. 106, pp. 87-97, 1976b.

Matthews, J.W. Mechanical modeling of nonlinear phenomena observed in the peripheral auditory system. Doct.diss. Washington Univ.,St Louis MO, 1980.

Neuweiler, G., Bruns, V. and Schuller, G., "Ears adapted for the detection of motion, or how echolocating bats have exploited the capacities of the mammalian auditory system," J.Acoust.Soc.Am. 68, pp. 741-753, 1980.

Vater, M., Feng, A.S. and Betz, M., "An HRP-study of the frequency map of the horseshoe bat cochlea: Morphological correlates of the sharp tuning to a narrow frequency band," J.comp.Physiol. in press 1985.

Wilson, J.P. and Bruns, V., "Middle-ear mechanics in the CF-bat *Rhinolophus ferrumequinum*," Hearing Research 10, pp. 1-13, 1983a.

Wilson, J.P. and Bruns, V., "Basilar membrane tuning properties in the specialised cochlea of the CF-bat *Rhinolophus ferrumequinum*," Hearing Research 10, pp. 15-35, 1983b.

Zwislocki, J., "Wave motion in the cochlea caused by bone conduction," J.Acoust.Soc.Am. 25, pp. 986-989, 1953.

Zwislocki, J.J., "Sharp vibration maximum in the cochlea without wave reflection," Hearing Research 9, pp. 103-111, 1983.

RESONANCE AND REFLECTION IN THE COCHLEA: THE CASE OF THE CF-FM BAT, *RHINOLOPHUS FERRUMEQUINUM*

Christine E. Miller
Department of Biomedical Engineering, Duke University
Durham, NC 27706

ABSTRACT

A two-dimensional mathematical model of the cochlea of the greater horseshoe bat *Rhinolophus ferrumequinum* is constructed. From 0-4.5 mm, the partition has large, constant mass and stiffness with *in vacuo* resonance of 83 kHz, the bat's emitted resting frequency. From 4.5-16 mm, partition mass is small and stiffness decreases exponentially with distance from the stapes. Solution is by VLFEM (Very Large Finite Element Method), an analytical/numerical hybrid technique designed for cochlear models with discontinuities.

Basilar membrane amplitude is calculated for twenty input frequencies as a function of distance from the stapes. The specialized basal region attenuates amplitude at all frequencies, with a sharp minimum at 83 kHz. Reflection is significant for frequencies below 83 kHz. The results agree qualitatively with experimental measurements, indicating maximal neural stimulation at minimal basilar membrane motion.

1. INTRODUCTION

Fine tuning in the fibers of the cochlear nerve challenges current cochlear models for explanation. On a relative tuning scale, the biggest challenge is perhaps the greater horseshoe bat, *Rhinolophus ferrumequinum (Rf)*. Suga *et al.* (1976) found tuning curves from single auditory nerve fibers and cochlear nuclear neurons in this bat with high- and low-frequency slopes as steep as 3,500 and 2,000 dB/octave, respectively, and Q-10dB's as high as 400. In addition, the anatomy of the *Rf* cochlea is unique, with a sharp discontinuity in basilar membrane (BM) structure where the neurons with largest Q's are located. A relationship between the mechanical discontinuity and the neural tuning is naturally suspected. This paper examines that relationship with a mathematical model of the *Rf* cochlea.

Rf is a CF-FM echolocating bat, emitting a long (up to 50 ms) constant frequency (CF) sound followed by a brief (2-3 ms) frequency mod-

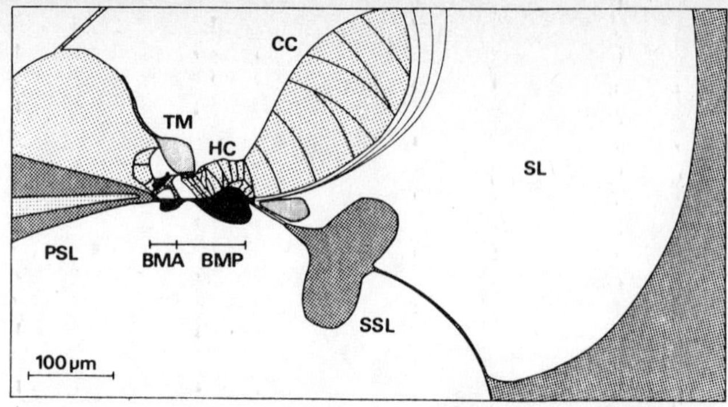

Figure 1. Cross section of the specialized basal region of the cochlea of the greater horseshoe bat, *Rhinolophus ferrumequinum*. PSL: primary spiral lamina; BMA: basilar membrane arcuate zone; BMP: basilar membrane pectinate zone; SSL: secondary spiral lamina; SL: spiral ligament; CC: cells of Claudius; HC: cells of Hensen; TM: tectorial membrane. (Illustration from Bruns, 1979).

ulated (FM) sweep (Bruns, 1976a). The stationary bat emits a CF of approximately 83 kHz, the resting frequency (F_R). The moving bat lowers the emitted sound to keep the echoed frequency or reference frequency constant at 50-200 Hz higher than F_R.

Bruns (1976a, 1980) describes the unusual basal region in the cochlea as follows (see Fig. 1):

(1) The BM is 35 µm thick, primarily due to the large pectinate zone.

(2) The secondary spiral lamina (SSL) is greatly hypertrophied and connected to the wall by a thin curved shelf of bone. The outer edge of the BM connects to the secondary spiral lamina by filaments from the inner margin of the spiral ligament.

(3) The BM abruptly decreases in thickness to 10 µm between 4.3 and 4.6 mm. The SSL maintains its hypertrophied form to 7.8 mm, where it very gradually shrinks.

By studying the swelling of outer hair cells, Bruns (1976b) found that F_R coincides with the 4.5 mm mechanical discontinuity. $F_R/2$ is located at 7.8 mm, while the frequencies within 3 kHz above F_R use the 3.15 mm basal to 4.5 mm. Q-10dB's for neurons at F_R are largest (average 140), dropping on either side of F_R below 20 for frequencies below 70 and above 90 kHz (Suga *et al.*, 1976).

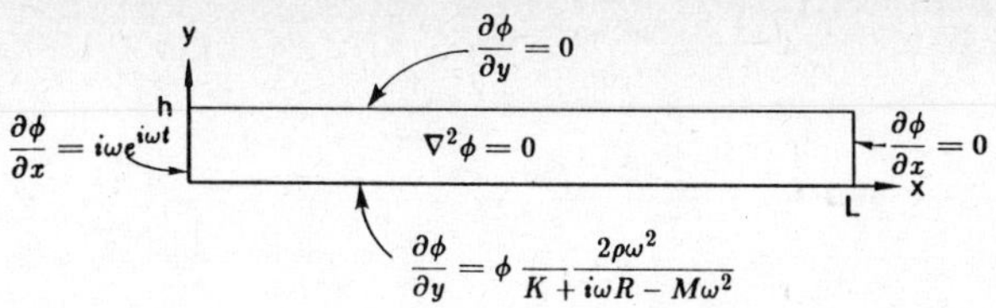

Figure 2. Boundary value problem for the two-dimensional cochlear model. The domain is fluid-filled with ϕ the fluid velocity potential. The basilar membrane is at y = 0 with stiffness K(x), mass M(x), and damping R(x).

VLFEM was designed (Miller, 1985) to have the accuracy of finite element or finite difference techniques and some of the speed of asymptotic techniques. Since it maintains the exact form of the solution, it is used here to examine reflections due to the discontinuity.

2. MODEL

A two-dimensional box model of the cochlea is appropriate for a first approach to this problem. Scala tympani and vestibuli have equal, constant cross sections and rigid wialls and are filled with a linear, incompressible, inviscid fluid. All structures of scala media are ignored except the BM, represented by a point impedance function $Z(x;i\omega) = R(x) + i[\omega M(x) - K(x)/\omega]$. R(x), M(x), and K(x) are damping, mass, and stiffness factors per unit area, and ω is the circular frequency of stapes vibration. Fig. 2 illustrates the complete boundary value problem.

Two models are studied here. Model B incorporates the properties of the specialized basal region of the *Rf* cochlea from 0-4.5 mm. Mass and stiffness are calculated from the cross section of Fig. 1 with SSL included, since it is relatively free to vibrate with the BM. The *in vacuo* resonant frequency of the BM in this region is $\sqrt{K/M}/2\pi = 83$ kHz, assumed to be our model bat's resting frequency. Lower mass and an exponential stiffness variation are used from 4.5-16 mm. Model A, used for comparison, has uniformly varying properties identical to those of Model B from 4.5-16 mm (see Table 1).

	MODEL A	MODEL B	
	0-16 mm	0-4.5 mm	4.5-16 mm
h (mm)	1.0	1.0	1.0
ρ (g/mm^3)	1.0×10^3	1.0×10^3	1.0×10^3
M(x) (g/mm^2)	1.5×10^{-6}	8.46×10^{-3}	1.5×10^{-6}
K(x) (μN/mm^3)	$2.8 \times 10^8 e^{-.32x}$	2.3×10^9	$2.8 \times 10^8 e^{-.32x}$
R(x) (μNs/mm^3)	6.0	8.0	6.0

Table 1. Model parameters.

3. METHODS

The method is described fully in Miller (1985); only a brief outline is given here. Lines of constant x divide the domain illustrated in Fig. 2 into elements over which BM properties are constant. The velocity potential in each element is

$$\phi = \sum_{n=0}^{\infty} \left[A_n e^{-ik_n x} + \tilde{A}_n e^{ik_n x} \right] \cosh k_n (y - h)\, e^{i\omega t} \tag{1}$$

where A_n and $\tilde{A}_n$ are complex constants. The complex wavenumbers k_n are the solutions to

$$k_n h \tanh k_n h = \frac{2\rho\omega^2 h}{K + i\omega R - M\omega^2} \tag{2}$$

where ρ is the fluid density and h is the scala height. The change in k_n is constant between elements, thus determining element size. The eigenfunctions $\cosh k_n(y-h)$ in each element are expressed in terms of a Fourier cosine series, so that nodal pressures and velocities for each mode can be equated directly between elements. Only a few terms or modes of the series are needed since it converges rapidly for small $k_n h$. Element stiffness matrices relating nodal pressures to velocities are calculated and assembled into a global matrix, from which velocities are found and used to calculate the A_n and $\tilde{A}_n$ of (1).

4. RESULTS

Figs. 3a and 3c show BM response curves for Model A. The low mass creates broad amplitude peaks, with maximum deflection for 83 kHz (F_R)

at 4.5 mm, for $F_R/2$ at 7.8 mm, and for $F_R/4$ at 11.0 mm. Thus 3.2 mm of BM resolves each octave of frequency.

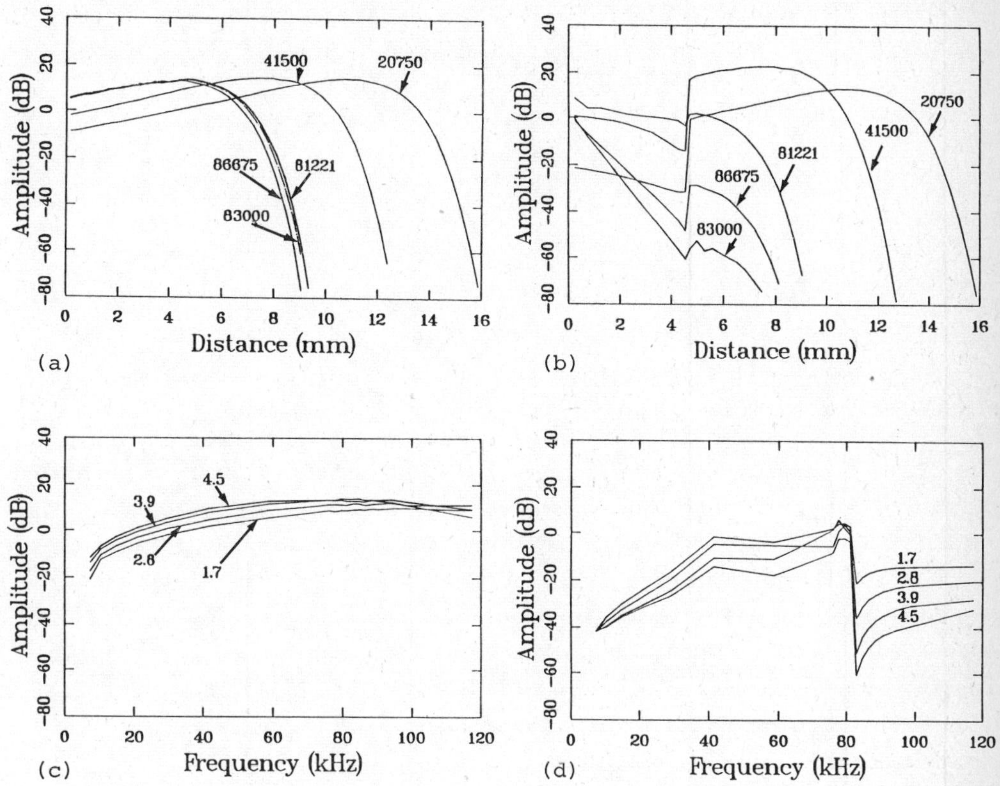

Figure 3. Ratio of basilar membrane/stapes displacement amplitude calculated for the two-dimensional model. (a) Amplitude for Model A as a function of distance from the stapes for five frequencies. Curve labels give frequency in Hz. (b) Amplitude for Model B as a function of distance from the stapes. (c) Amplitude for Model A as a function of frequency for four locations along the basilar membrane. Labels give distance from the stapes in mm. (d) Amplitude for Model B as a function of frequency at four locations.

Figs. 3b and 3d show BM response curves for the *Rf* model, Model B. The high mass, high stiffness region from 0-4.5 mm reduces the amplitude of waves of all frequencies. Beyond this region, the response for frequencies below 83 kHz is similar to that of Model A. Response at 83 kHz is affected most, with amplitude reaching a local minimum of -60 dB at 4.5 mm. This is a 74 dB drop from the maximum at 4.5 mm in Model A. The amplitude curves for frequencies both above and below 83 kHz lie above it; the drop is rapid from the low frequency side and more gradual on the high frequency side. Fig. 4 demonstrates this by plotting amplitude versus both frequency and distance from the stapes; 83 kHz is located at the sharp valley of the surface.

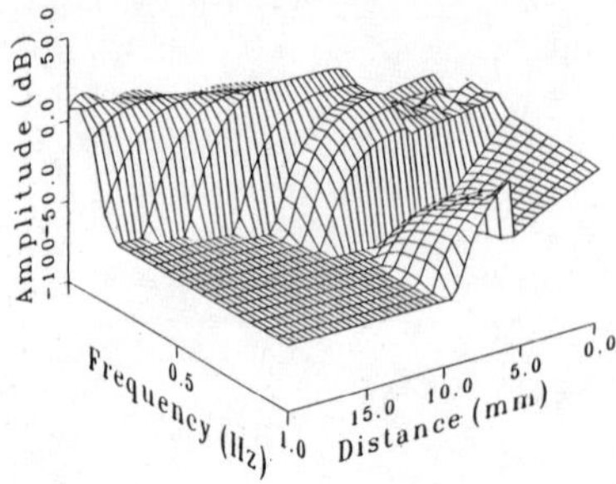

Figure 4. Ratio of basilar membrane/stapes amplitude for Model B as a function of both frequency and distance. The frequency axis is non-dimensional and non-linear; 83 kHz lies in the valley at 0.44.

Fig. 5 shows the energy of the incident (right-traveling) and reflected (left-traveling) waves at frequencies above and below 83 kHz. Figs. 5a and 5c show for Model A at 78 and 87 kHz that approximately 1/1000 of the incident energy is reflected. An even smaller percentage of total energy is reflected in Model B for frequencies above 83 kHz (Fig. 5d). Both incident and reflected waves are damped in the basal region. For frequencies below 83 kHz in Model B, however, a significant proportion of total energy is reflected in the basal region to show almost standing-wave characteristics.

5. DISCUSSION

We have seen in a two-dimensional cochlear model that replacing a 4.5-mm piece of the BM with "normal", exponentially varying stiffness by a section of constant stiffness and large mass causes a reduction in the amplitude of BM motion. Reduction is maximal when the input frequency is the *in vacuo* resonant frequency of the BM. The condition at 83 kHz can probably be more accurately defined as antiresonance, which like resonance occurs when reactance (the imaginary part of impedance) vanishes, but unlike resonance occurs when input resistance is large. The model behavior is opposed to what has commonly been regarded as desirable in a model, that is, sharply peaked maximums. Although no experimental evidence exists to confirm vanishing reactance for the BM at 83 kHz, the model has some properties in common with experimental observations. The small frequency range near 83 kHz, eliciting almost

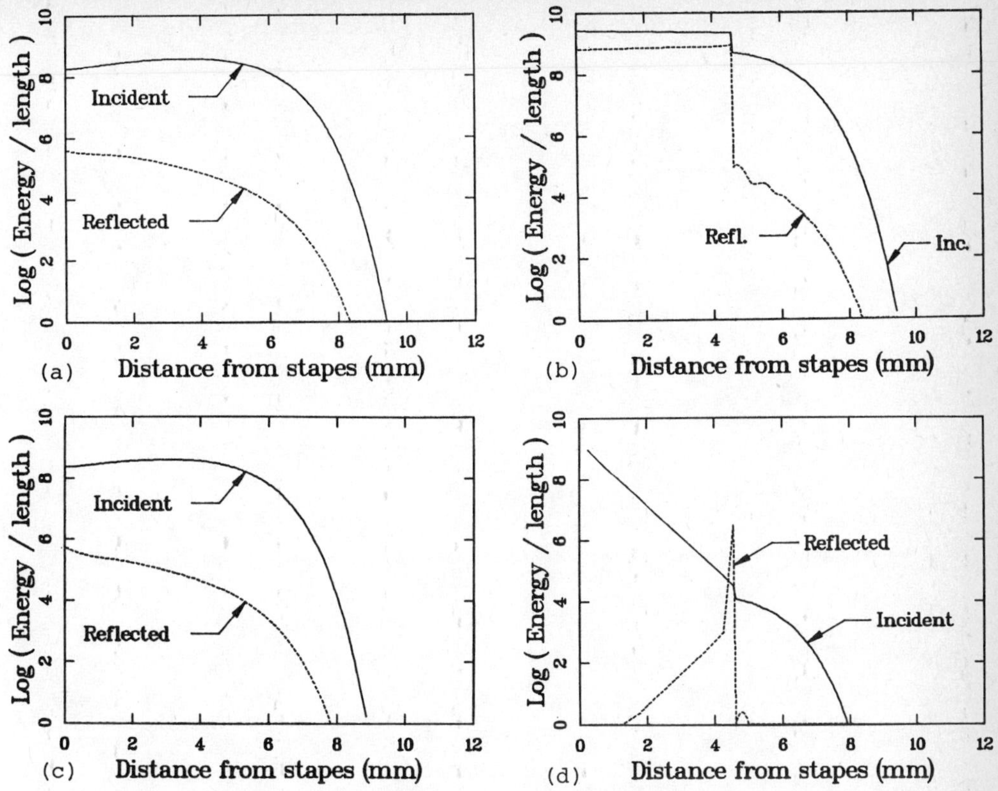

Figure 5. Energy density in basilar membrane motion due to incident and reflected waves. Energy units are μN-mm (10^{-9}J) per mm^2 of membrane. (a) Model A, 78 kHz; (b) Model B, 78 kHz; (c) Model A, 87 kHz; (d) Model B, 87 kHz.

equal BM response without the specialized region, causes widely different BM response in the model with specialized resonant region. This suggests Bruns' (1979b) observation of greatly expanded frequency mapping close to 83 kHz in the *Rf* cochlea. Also, Wilson and Bruns (1983) found sharp minimums in measured BM tuning curves in the basal region of the *Rf* cochlea at 83 kHz. Their tuning curves for three locations are shown in Fig. 6. The fact that maximal neural stimulation can occur near minimal BM motion is thus already established, and mechanical antiresonance may play a role in the phenomenon.

The sudden change in BM impedance at 4.5 mm suggests reflection at the interface. This occurs for frequencies below 83 kHz and may explain the decreased amplitude from 0-4.5 mm for these below-resonant frequencies. Reflection from the discontinuity is not as evident at frequencies above 83 kHz, probably because vibration amplitudes are already

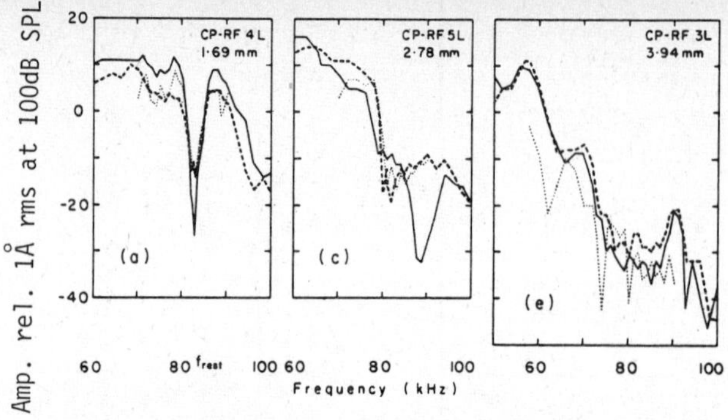

Figure 6. Basilar membrane amplitude measured *in vivo* in the greater horseshoe bat by Wilson and Bruns. Solid line: pectinate zone; Dashed line: arcuate zone; Dotted line: difference. (From Wilson and Bruns, 1983).

low due to the mass-loading effect. When reactance vanishes at 83 kHz, damping predominates to greatly attenuate vibration.

ACKNOWLEDGEMENTS

Computer calculations were done at the National Biomedical Simulation Resource at Duke University, funded by a grant from the Division of Research Resources, National Institutes of Health.

REFERENCES

Bruns, V., "Peripheral auditory tuning for fine frequency analysis by the CF-FM bat, *Rhinolophus ferrumequinum*. I. Mechanical specializations of the cochlea." J. Comp. Physiol. 106, pp. 77-86, 1976a.

Bruns, V., "Peripheral auditory tuning for fine frequency analysis by the CF-FM bat, *Rhinolophus ferrumequinum*. II. Frequency mapping in the cochlea." J. Comp. Physiol. 106, pp. 87-97, 1976b.

Bruns, V., "Functional anatomy as an approach to frequency analysis in the mammalian cochlea." Verh. Dtsch. Zool. Ges., pp. 141-154, 1979.

Bruns, V., "Basilar membrane and its anchoring system in the cochlea of the greater horseshoe bat." Anat. Embryol. 161, pp. 29-50, 1980.

Miller, C. E., "VLFEM analysis of a two-dimensional cochlear model." J. Appl. Mechanics, to appear, 1985.

Suga, N., Neuweiler, G., and Moller, J., "Peripheral auditory tuning for fine frequency analysis by the CF-FM bat, *Rhinolophus ferrumequinum*. IV. Properties of peripheral auditory neurons." J. Comp. Physiol. 106, pp. 111-125, 1976.

Wilson, J.P. and Bruns, V., "Basilar membrane tuning properties in the specialized cochlea of the CF-Bat, *Rhinolophus ferrumequinum*." Hear. Res. 10, pp. 15-35, 1983.

THE COMPLETE SOLUTION OF THE BASILAR MEMBRANE CONDITION IN TWO DIMENSIONAL MODELS OF THE COCHLEA

J.S.C. van Dijk
Institute of Phonetic Sciences, University of Amsterdam
Herengracht 338 1016 CG Amsterdam The Netherlands

ABSTRACT

In two-dimensional models of the cochlea, basilar membrane behaviour has to satisfy a mixed type boundary condition. In this paper this condition will be studied using the complex plane. Its solution is given in terms of the most general solution of Laplace's equation for pressure. The result shows complex membrane pressure as a linear combination of two independent functions. Taking advantage of the well-known frequency to place mapping, the solution is written as a frequency invariant closed shape. This representation is useful in giving a physical interpretation of membrane motion in relation to fluid flow in cochlear models. Due to the frequency invariance, no time consuming numerical methods are necessary to find response curves for membrane pressure or velocity.

1. INTRODUCTION

During the last two decades the fiber structure of the basilar membrane of a few members of the animal kingdom has been thoroughly described by several investigators (for example Spoendlin, 1968, 1970; Voldřich and Úlehlová, 1981). This can be conceived as a corroboration of the original observations of Hensen (1862), who proposed to Helmholtz - be it from a dynamical point of view - to observe this membrane as a typical example of an anisotropic medium. As a model maker Helmholtz (1877) replaced the membrane with a row of resonators. Conceptually he tuned a single resonator in harmony with the first resonance frequency of a (bundle of) fiber(s). In this manner the basis of most one- and two-dimensional models was established. In most models the membrane is surrounded by an ideal fluid. Hard walls and prescribed piston-like motions at the windows limit the fluid's extent. This spatial limitation of the fluid often leads to the formulation of sophisticated boundary value problems in a quite natural way. In these problems a mixed type boundary condition gives a comprehensive description of physical laws which both membrane and fluid motion have to fulfill. However, from these models it is rather difficult to derive an explicit expression for the basilar membrane motion which leads to insight in the combined action of membrane and fluid motion. Therefore we shall reinvestigate this almost classical problem and try to understand its meaning in cochlear modelling.

2. STATEMENT OF THE PROBLEM

Consider a row of damped harmonic oscillators. For the sake of convenience that system will be called a membrane. It is assumed that the membrane is part of the axis $y = 0$ of the complex $z = x + iy$ plane (Fig. 1). The membrane is surrounded by an ideal fluid. There is no direct coupling between adjacent oscillators. When the membrane and the fluid are in motion, the dynamical equilibrium between membrane and fluid pressure is expressed by

$$m(x)\,\frac{d^2y(x,t)}{dt^2} + r(x)\,\frac{dy(x,t)}{dt} + s(x)y(x,t) = -\,2\,p(x,0,t). \tag{2.1}$$

$y(x,t)$ is the membrane deflection at the point x of the real axis. $m(x)$, $r(x)$ and $s(x)$ respectively are mass, resistance and stiffness of the oscillator at x. $-2\,p(x,0,t)$ represents the hydrodynamic pressure between the upper and lower side of the membrane at x. The hydrodynamic pressure $p = p(x,y,t)$ obeys Laplace's equation

$$\frac{\partial^2 p}{\partial x^2} + \frac{\partial^2 p}{\partial y^2} = 0 \quad , \tag{2.2}$$

which follows from the condition of incompressibility and the linear Euler equations. One of the last equations, namely

$$\rho\,\frac{\partial v}{\partial t} = -\,\frac{\partial p}{\partial y} \quad , \tag{2.3}$$

is of special importance. In (2.3) ρ is the density of the fluid and $v = v(x,y,t)$ is the component of the fluid velocity parallel to the axis $x = 0$. At the membrane this equation can be written as

$$\rho\,\frac{d^2y(x,t)}{dt^2} = -\,\frac{\partial p(x,0,t)}{\partial y} \quad , \tag{2.4}$$

because we assume that the membrane always follows the motion of the fluid. In fact, the indirect coupling between the membrane oscillators is expressed in (2.4). In case of harmonic fluid oscillations a general shape of the membrane pressure is given by

$$p(x,0,t) = \hat{p}(x)\cos(\omega t + \phi(x)) \quad . \tag{2.5}$$

$\hat{p}(x)$ is the real amplitude of the pressure oscillation at x, $\phi(x)$ is the phase at that point and ω the circular frequency. Accepting the restriction to study only forced membrane oscillations the problem reads: at the membrane $\hat{p}(x)$ and $\phi(x)$ have to be found so that not only (2.1) and (2.4) have been satisfied, but Laplace's equation (2.2) has been fulfilled as well.

3. PARAMETERS

When studying linear oscillators, usually resistance and stiffness are made relative to the mass of an oscillator. In order to simplify matters the mass is taken independent of place on the membrane. Consequently parameter functions $\beta(x)$ and $\omega_0(x)$ are introduced according to

$$2m\beta(x) = r(x) \qquad \text{and} \qquad m\omega_0{}^2(x) = s(x) \quad , \tag{3.1}$$

where m is a constant. $\beta(x)$ controls the rate of decay of free vibrations of the oscillator at x. A well-known measure for this decay is the logarithmic decrement. This quantity is taken to be constant as well. It is easy to show that this holds true when $\beta(x)$ is proportional to $\omega_0(x)$. In practice, one often deals with small values of the damping. That means, the constant of proportionality must be very small. Therefore we put

$$\beta(x) = \sin\varepsilon\ \omega_0(x) \quad , \tag{3.2}$$

where ε is a small positive number. For mathematical reasons $\omega_0(x)$ is called the resonance frequency at x. Concerning the relation between place and resonance frequency, Békésy's (1960) view is adopted. This well-known logarithmic relation can be written as

$$\omega_0(x) = Q\exp(-\frac{ax}{2}) \quad , \tag{3.3}$$

in which Q and a are positive numbers. Typical values of parameters are given in section 7.

4. THE NORMALIZED MEMBRANE EQUATION

Forced membrane oscillations follow from (2.1) in which the pressure is defined according to (2.5). Write (2.5) in its complex counterpart. Then it is found from (2.1) and (2.4) that the complex amplitude $\overset{+}{p}(x)$ of oscillations proportional to $\exp(-i\omega t)$ has to satisfy the membrane condition

$$B(x)\,\frac{\partial p^{+}(x)}{\partial y} + 2\rho\omega^{2}p^{+}(x) = 0 \quad , \tag{4.1}$$

in which

$$B(x) = \{ (\frac{\omega_0(x)}{\omega})^{2} - 2i\frac{\beta(x)}{\omega} - 1 \}\, m\omega^{2} \quad . \tag{4.2}$$

Resonance takes place if $\mathrm{Re}\,B(x) = 0$. Inserting (3.2) and (3.3) in (4.3), it is readily shown that resonance occurs if x and ω satisfy $0 = \frac{a}{2}x + \ln\frac{\omega}{Q}$. However, this condition forms part of the elementary conformal mapping

$$\xi = \frac{a}{2}x + \ln\frac{\omega}{Q} \;,\; \eta = \frac{a}{2}y \quad , \tag{4.3}$$

from which follows that resonance takes place at the origin of the complex $\zeta = \xi + i\eta$ plane (Fig.1).

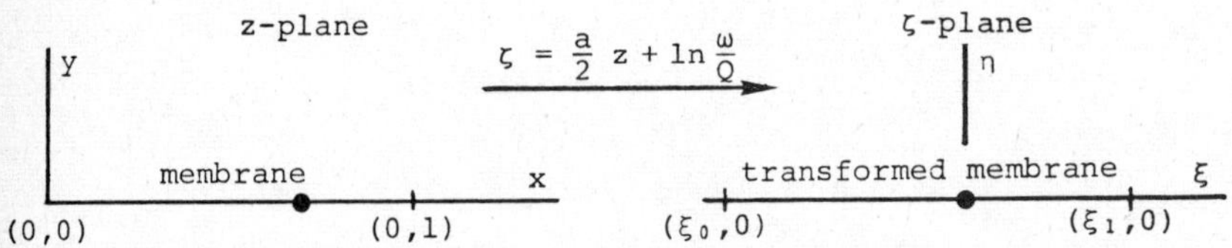

Fig.1. Along the real axis of the $z = x + iy$ plane the membrane stretches from $x = 0$ to $x = 1$. In this interval resonance takes place (•). The mapping (4.3) maps the point of resonance onto the origin of the $\zeta = \xi + i\eta$ plane. At the real axis of this plane the membrane condition takes the frequency invariant shape (4.4).

Using (3.2), (3.3) and (4.3) at the axis $\eta = 0$ of the ζ-plane the membrane condition takes the frequency invariant shape

$$\mu A(\xi)\frac{\partial p^{+}(\xi)}{\partial \eta} + p^{+}(\xi) = 0 \qquad \mu = \frac{ma}{4\rho} \quad , \tag{4.4}$$

where

$$A(\xi) = (e^{-\xi} - e^{+i\varepsilon})(e^{-\xi} + e^{-i\varepsilon}) \quad . \tag{4.5}$$

5. THE COMPLETE SOLUTION OF THE MEMBRANE CONDITION

The mapping (4.3) is conformal. Therefore p^{+} has to satisfy Laplace's equation in the ζ-plane as well. The most general solution of this equation can be written as $p^{+}(\zeta,\bar{\zeta}) = F(\zeta) + G(\bar{\zeta})$. F and G respectively are arbitrary analytical functions of the complex conjungate coordinates ζ and $\bar{\zeta}$. Subjecting this solution to the membrane condition (4.4), it appears that the main contribution to p^+ can be written as

$$p^+(\zeta,\bar{\zeta}) \approx C_1\{(1-e^{\zeta+i\varepsilon})(1+e^{\zeta-i\varepsilon})\}^{-i\nu} + \\ + C_2\{(1-e^{\bar{\zeta}+i\varepsilon})(1+e^{\bar{\zeta}-i\varepsilon})\}^{+i\nu} , \qquad \nu = \frac{1}{2\mu} . \tag{5.1}$$

The constants C_1 and C_2 are still indeterminate. ν follows from (4.4). For practical purposes it is convenient to introduce curvilinear coordinates near resonance. These are defined by

$$r(\xi,\eta)\exp(i\upsilon(\xi,\eta)) = (1-e^{\zeta+i\varepsilon})(1+e^{\zeta-i\varepsilon}) . \tag{5.2}$$

As follows from (5.1) and (5.2) at the membrane -i.e. η=0- the pressure takes the shape

$$p^+(\xi,0) = C_1\exp\{\nu(\upsilon(\xi,0) - i\ln r(\xi,0))\} + \\ + C_2\exp\{-\nu(\upsilon(\xi,0) - i\ln r(\xi,0))\} , \tag{5.3}$$

where $r(\xi,0)$ and $\upsilon(\xi,0)$ are given by

$$r(\xi,0) = 2e^{\xi}\{\sinh^2\xi + \sin^2\varepsilon\}^{\frac{1}{2}} \tag{5.4}$$

$$\upsilon(\xi,0) = \begin{cases} \pi + \arctan\dfrac{\sin\varepsilon}{\sinh\xi} , & \xi < 0 . \\ \arctan\dfrac{\sin\varepsilon}{\sinh\xi} , & \xi > 0 . \end{cases} \tag{5.5}$$

6. BASILAR MEMBRANE MODELS

Now boundary conditions at the ends of the transformed membrane (Fig.1) will be defined. At $\xi = \xi_0$ the system is set into motion. For the present this condition is taken as simple as possible: $p^+(\xi_0,0) = 1$. Concerning $\xi = \xi_1$ we will consider two different conditions. The first one is the well-known hard-wall condition, whereas the second one models a helicotrema-like condition. Thus, conditions to which (5.3) is subjected are:

$$\text{model 1}: \quad p^+(\xi_0,0) = 1 ; \quad \frac{\partial p^+(\xi_1,0)}{\partial\xi} = 0 ; \tag{6.1}$$

$$\text{model 2}: \quad p^+(\xi_0,0) = 1 ; \quad p^+(\xi_1,0) = 0 . \tag{6.2}$$

The solution of model 1 is

$$p^{+}(\xi,0) = \frac{\cosh\{\nu(\upsilon(\xi,0) - \upsilon(\xi_1,0)) + i\nu \frac{\ln r(\xi_1,0)}{\ln r(\xi,0)}\}}{\cosh\{\nu(\upsilon(\xi_0,0) - \upsilon(\xi_1,0)) + i\nu \frac{\ln r(\xi_1,0)}{\ln r(\xi_0,0)}\}} . \tag{6.3}$$

The solution of model 2 is easily found by replacing both in numerator and in denominator of (6.3) the hyperbolic cosine with the hyperbolic sine. If the constant ν is sufficiently large, left of resonance (6.3) can be approximated very well. Then, for $\xi_0 < \xi < 0$

$$p^{+}(\xi,0) \approx \exp\{\nu(\upsilon(\xi,0) - \upsilon(\xi_0,0)) + i\nu \ln \frac{r(\xi_0,0)}{r(\xi,0)}\} , \tag{6.4}$$

which is the well-known phase intregral (Steele and Taber, 1979) subjected to the condition $p^{+}(\xi_0,0) = 1$ only. In the next section some plots based on (6.4) are given. At the right of resonance the approximation (6.4) fails.

7. SOME RESULTS

As a typical example of the present analysis some results are shown which we derived from (6.4). These have to be considered as main constituents of - for example - the solution of model 1 (6.3). Values of parameters are: $m = 0.05$ g/cm^2, $\rho = 1$ g/cm^3, $a = 3$ cm^{-1}. The length of the membrane is 3.5 cm. Resonance occurs at $x = 1.75$ cm for $f = 1000$ Hz. ε, the damping controlling parameter, varies from 0 to 0.1.

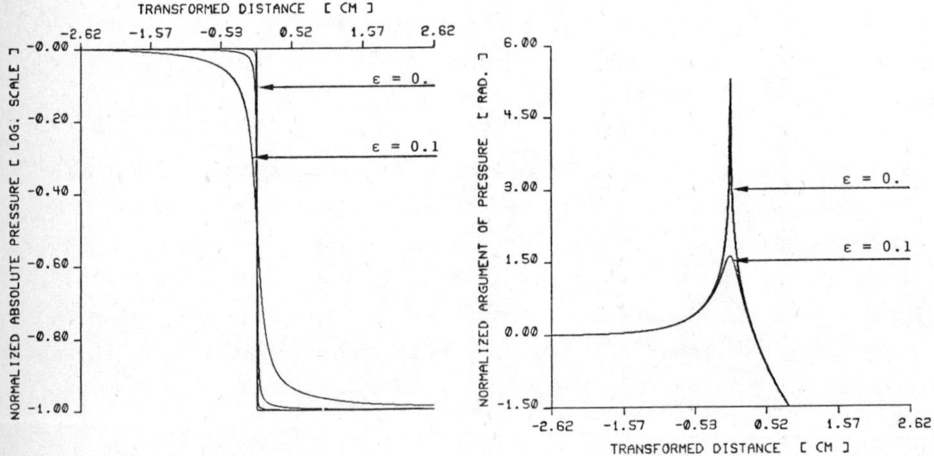

Fig.2. Normalized amplitude and phase of the pressure (6.4) for different values of the damping parameter ε. $f = 1000$ Hz. The abcissa is the transformed distance ξ (4.3). In the first figure the ordinate is a ln-scale normalized to $\nu\pi$. In the second figure radians have been nor-

malized to ν. ν is dimensionless and follows from (5.1) and (4.4).

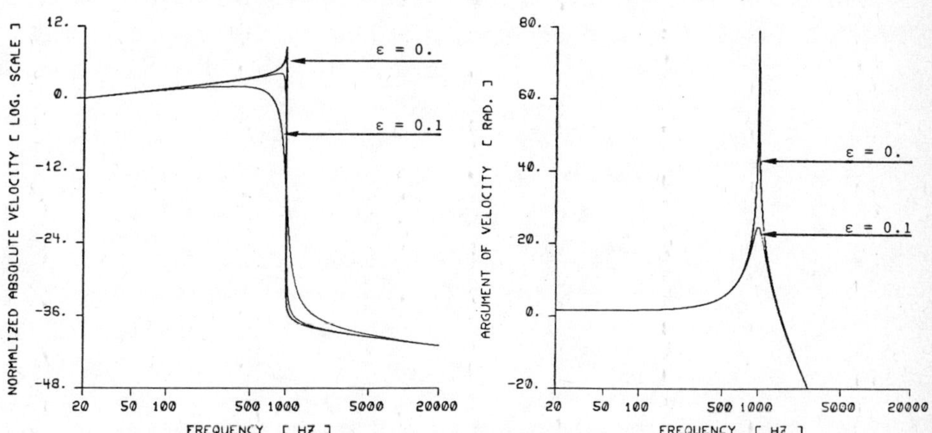

Fig.3. Velocity response curves which follow from (4.3), (4.4) and (6.4). For every frequency the velocity is relative with respect to the fluid velocity parallel to the membrane at the point of propulsion. Ordinates of the amplitude scale are normalized ln-units. Ordinates of the phase scale are radians. At $\xi = \xi_0$, $\upsilon(\xi,0)$ (5.5) has been approximated with π.

8. CONCLUSIONS AND DISCUSSION

Application of the frequency to place mapping (4.3) leads to a frequency invariant shape of the membrane condition. When the general solution of Laplace's equation for pressure is subjected to this condition, the result shows the complex membrane pressure as a superposition of two phase integrals which straightforwardly leads to an explicit expression for the membrane pressure (5.3). The solution still contains two indeterminate constants. In this way it is needless to search for a special second order equation (De Boer, 1983, 1984) to which the membrane pressure has to satisfy. Moreover, the original proposal of Steele and Taber (1979) has been established and improved. The constants follow from appropriate boundary conditions at the ends of the membrane. If these conditions meet with conditions at the windows and near the helicotrema in boundary value problems for one of the cochlear scalae, our solution forms part of these models. In this manner questions concerning short or long waves have become inadequate. Between resonance and helicotrema the pressure is negligible with respect to the pressure at the other side of resonance. For example, in (6.3) - with parameters as in section 7 - this level difference is about -360 dB. Thus, after resonance the pressure difference at the membrane is approximately zero. In consequence of this, the main contribution to the motion in that region is the additional fluid necessary to satisfy the condition of compatibility (law of conservation of mass) for each scala separately. This explains the differences between numerical model studies (Allen, 1977; Neely, 1981; Viergever, 1980) and phase integral solutions in two dimensional cochlear mechanics.

ACKNOWLEDGMENTS

I thank H.A. Lauwerier for helpful and stimulating comments, M. Nieberg and Th. Philips for valuable discussions and assistance in figure preparation, J. Bossema for typing the manuscript, H. Deighton and L. Landy for improving the text.

REFERENCES

Allen, J.B., "Two Dimensional Cochlear Fluid Model: New Results." J. Acoust. Soc. Am. 61, pp. 110-119, 1977.

von Békésy, G., "Experiments in Hearing." , McGraw-Hill, New York, 1960.

de Boer, E., "Wave Reflection in Passive and Active Cochlea Models." In Mechanics of Hearing, Ed. by E. de Boer and M.A. Viergever, Martinus Nijhoff Publishers, Delft University Press, 135-142, 1983.

de Boer, E., "Auditory Physics. Physical Principles in Hearing Theory II.", Physics Reports, 105, no 3, pp. 143-226, 1984.

Helmholtz, H.L.F., "On the Sensations of Tone as a Physiological Basis for the Theory of Music." Second English Edition, pp. 406-411, Dover Publications, New York, 1954.

Hensen, V., In Helmholtz, H.L.F., "On the Sensations of Tone", pp. 145-146, Dover Publications, New York, 1954.

Neely, S.T., "Finite Difference Solution of a Two Dimensional Mathematical Model of the Cochlea." J. Acoust. Soc. Am. 69, pp. 1386-1393, 1981.

Spoendlin, H., "Ultrastructure and Peripheral Innervation Pattern of the Receptor in Relation to the First Coding of the Acoustic Message." In Hearing Mechanisms in Vertebrates, pp. 89-125, London, 1968.

Spoendlin, H., "Structural Basis of Peripheral Frequency Analysis." In Frequency Analysis and Periodicity Detection in Hearing. Ed. by R. Plomp and G.F. Smoorenburg, Sythoff, Leiden, 1970.

Steele, C.R., and Taber, L.A., "Comparison of WKB and Finite Difference Calculations for a Two-dimensional Cochlear Model." J. Acoust. Soc. Am. 65, pp. 1001-1006, 1979.

Viergever, M.A., "Mechanics of the Inner-ear." Academic Dissertation, Delft University Press, 1980.

Voldřich, L. and Úlelová, L. "Cochlear Mechanics." Krátký Film, Praha, 1981.

LONGITUDINAL STIFFNESS COUPLING IN A 1-DIMENSIONAL MODEL OF THE PERIPHERAL EAR

Robert E. Wickesberg[1] and C. Daniel Geisler[1,2]
[1]Department of Neurophysiology
[2]Department of Electrical and Computer Engineering
University of Wisconsin - Madison
Madison, Wisconsin 53706

ABSTRACT

A one-dimensional computer model of the peripheral ear was explored using pure-tone inputs. This model was adopted after frequency-domain modeling showed that the differences between one- and two-dimensional models were small. The inner ear representation was the classical mass-spring-damper transmission line. In the time-domain simulations, this classical model was modified to have adjacent elements of the cochlear partition model lightly coupled to each other via springs.

Simulation of the entire peripheral ear demonstrated that the parameters used in Neely's (1981), Allen's (1977) and our model were fairly well chosen. All three representations yielded cochlear input impedances that accurate approximated Lynch et al.'s (1982) results at higher frequencies, and they yielded sound pressure levels at the partition that approximated Nedzelnitsky's (1980) measurements at low frequencies. In addition, very low values of cochlear partition damping resulted in highly peaked displacement curves that were approximately correct in amplitude (near 1 A at 0 dB SPL for 1 kHz).

The effect of moderate longitudinal stiffness coupling was to broaden significantly the width of the resonant peak in the lightly damped case, at the cost of only slightly reducing the peak's magnitude. Proper values of this longitudinal coupling resulted in tuning-curve shapes and values of Q_{10} that were quite realistic.

1. INTRODUCTION

Several recent cochlear models have achieved realistically sharp frequency tuning (Q_{10}'s) by adding tuned micro-mechanical mechanisms, either passive (Allen, 1980) or active (Neely and Kim, 1983), to the classical resonant-circuit basilar-membrane representation (Geisler, 1976). We report here that realistically sharp frequency tuning can be obtained with the classical one-dimensional, transmission-line model simply by introducing a longitudinal stiffness coupling between neighboring elements of the basilar membrane model. In addition, we also report on several interesting input characteristics of linear, transmission-line cochlear models.

2. MODEL

The basic differential equations for the classical linear, one-dimensional model (shown in Figure 1) describe isolated basilar-membrane strips vibrating in an inviscid, incompressible fluid. The one-dimensional (1-D) formulation is known to be somewhat inaccurate (e.g., Steele and Taber, 1979), particularly near the place of maximum displacement, but its relative simplicity, compared to a two-dimensional (2-D) model, makes it a very attractive investigative tool. Accordingly, we have principally used 1-D simulations in this study, although we have also done some 2-D simulations for comparison purposes. When convenient, we solved the equations in the frequency domain, but a time-domain solution involving centered second-differences was used for the more complex solutions. The frequency domain models of the cochlear partition used at least 2800 sections, and the height of the cochlea was represented by 8 points in the 2-D models. Such fine resolution was needed to prevent instabilities (e.g., oscillations in the displacement envelope such as are seen to a very limited extent in the left-most curve of Figure 2). All the data presented here were calculated using a DEC VAX/750 computer.

The parameters of our basilar membrane representation were chosen to achieve the cat's spatial tonotopic organization as determined by

$$\frac{\partial P_d}{\partial x} = -2\rho a_s \qquad P_d = 0$$

$$\frac{\partial^2 P_d}{\partial x^2} = \frac{2\beta\rho}{A} a_{bm}$$

Figure 1. The basic differential equations, and boundary conditions, describing the classical one-dimensional model of basilar-membrane motion, where P_d is the pressure difference across the membrane, A is the cross-sectional area of the cochlear duct, ρ is the fluid density, β is the width of the basilar membrane, and a_{bm} and a_s represent the accelerations of the basilar membrane and the stapes, respectively.

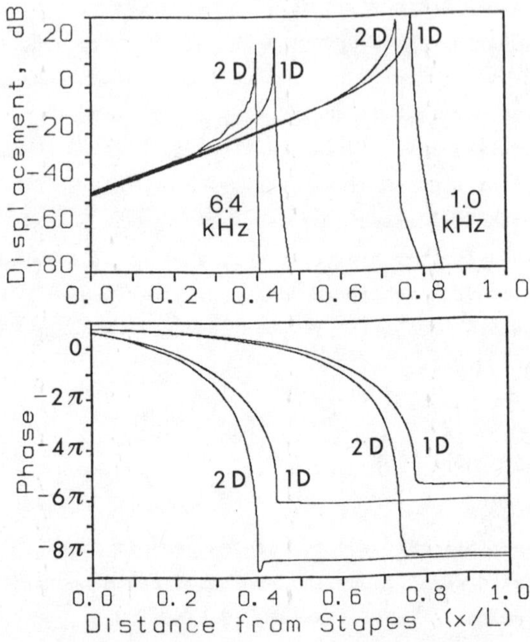

Figure 2. Comparison of the response amplitudes and phases produced by 1-D and 2-D linear models of the same cochlea for 6.4 kHz and 1.0 kHz inputs. The mass was a constant 0.05 gm/cm^2; damping was a constant 5 dyne-s/cm^3.

Liberman (1982). Basilar membrane mass was kept constant at 0.05 gm/cm^2, while the membrane stiffness ranged from $6.4x10^9$ $dynes/cm^3$ at the base to $1.6x10^4$ $dynes/cm^3$ at the apex. Membrane damping was fixed at one value for all points during any one simulation. The input to the basilar membrane representation was the acceleration of the stapes, which was determined by convolving the stimuli with a middle-ear transfer function derived from the data of Guinan and Peake (1967). The apical end of the membrane was terminated with a pressure short-circuit.

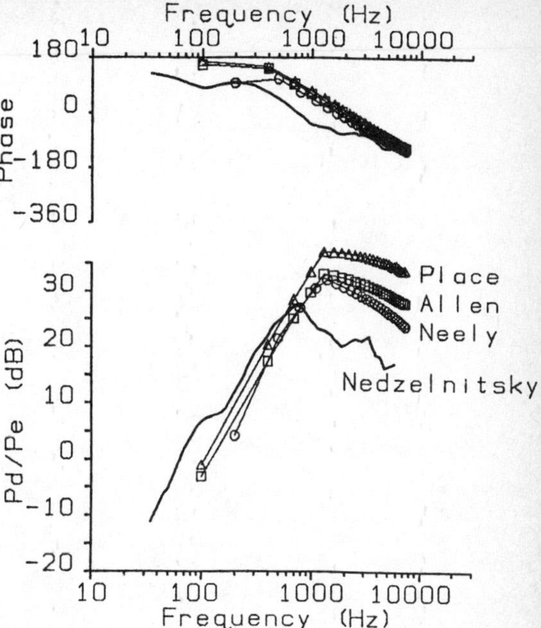

Figure 3. Comparisons between Nedzelnitsky's (1980) measurement of the amplitude and phase of the pressure difference, P_d, at the basal end of the cochlea, relative to the sound pressure at the eardrum, P_e, with the results from 1-D models using Allen's (1977), Neely's (1981) and our ("place") parameters.

3. COMPARISON OF ONE- AND TWO-DIMENSIONAL MODELS

Figure 2 shows the displacement envelopes and phase curves from linear 1-D and 2-D simulations using identical basilar membrane mass, damping and stiffness parameters. These results do not differ very much in shape, but there are some obvious differences. Peak displacements in the 2-D simulation occur more basally and are slightly less peaked than those of the 1-D model (this difference would be more pronounced, if higher values of damping were used). This basal shift of the peak displacement when going from 1-D to 2-D formulations has also been reported by Steele and Taber (1979). In order to approximate the 2-D solutions with a 1-D model, we had to decrease the mass, which results in a slight rounding of the tip, increases the phase roll-off, and shifts the peak apically, and then make a corresponding decrease in stiffness to shift the peak basal-ward. The 2-D representation also produces appreciably more phase lag than the 1-D model (a minumum-phase system). This implies that minimum-phase models of cochlear function (e.g. Geisler and Sinex, 1983) are somewhat inaccurate and must be used with caution.

In view of the rather small qualitative differences in the outputs of the 1-D and 2-D models, we decided to use the 1-D representation for the remainder of the work presented here. The 2-D model was used occasionally to insure that the conclusions reached with the 1-D model were valid.

4. INPUT CHARACTERISTICS

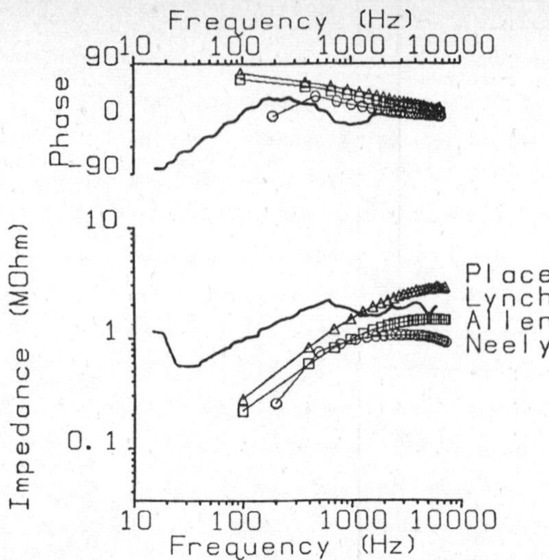

Figure 4. Comparisons between Lynch et al.'s (1984) measurement of the amplitude (in cgs acoustic MOhms) and phase of the cochlear input impedance with the input impedances across the cochlear partition for 1-D models using Allen's (1977), Neely's (1981) and our ("place") parameters.

Some of the input characteristics of the cat cochlea are known (Nedzelnitsky, 1980; Lynch et al., 1982). To help judge the adequacy of various basilar membrane representations, designed to produce desired deflections at various places along the membrane, it is of interest to see if they also display realistic input characteristics without further modifications. Figure 3 shows the pressure transfer functions obtained from our model (labeled "place" to reflect the matching of Liberman's frequency-to-place map of the cochlea), as well as those of Allen (1977) and Neely (1981). Notice that in spite of their differences, all three models match Nedzelnitsky's amplitude data (1980) fairly well in shape, but are too large in amplitude at frequencies above 1 kHz. Since the three models have roughly equivalent values of basal basilar membrane stiffness (the over-riding element at such frequencies), this agreement is not too surprising. The slightly higher amplitudes obtained using Allen's and our parameters are due to the higher frequencies represented (up to 57 kHz in our model). The phases of the pressure transfer functions are shown in the upper part of this figure. Above 2 kHz the phases obtained from all three models are in very good agreement with Nedzelnitsky's results. Below 2 Khz the phases of the models tend toward 180 degrees, while Nedzelnitsky's measurements tend toward 90 degrees. We must also note here that at frequencies below 400 Hz, Neely's parameters produced large reflections from the helicotrema, and hence the point for his model at 200 Hz should be viewed with caution, although it does agree with the measured phase.

In Figure 4, the input impedances of the three models are compared with experimental results from Lynch et al. (1982). The model input impedance was defined as basal pressure difference divided by stapes velocity (the input impedance across the cochlear partition). The experimental input impedance, on the other hand, was calculated using scala vestibuli pressure only. But, as Lynch et al. (1982) point out, this is a valid comparison above 100 Hz. Notice that above 1 kHz, all model parameters yield input impedances of approximately 1 cgs acoustic megohm. The deviations of the models from realistic behavior at low frequencies are probably due to effects of the helicotrema and the round window, which have not been taken into account accurately. The phase comparisons are shown in the upper part of Figure 4. Again, above 2 kHz all the models are in very good agreement with the measured results, but below 2 kHz an almost 90 degree difference develops. Lynch et al. (1982) also estimated the input impedances across the cochlear partition from Nedzelnitsky's (1980) and Guinan and Peake's (1967) data, and they show the phase tending toward 90 degrees only for frequencies below 100 Hz.

5. DISPLACEMENT OF THE BASILAR MEMBRANE

We have calculated the amplitude and phase for our 1-D model as functions of input frequency for the 8.6 kHz point along the basilar membrane. Curves for two different values of basilar membrane damping (200 and 5 dyne-s/cm^3) are shown in Figure 5 along with actual basilar membrane displacement data from Rhode (1978). The amplitude curve obtained with the higher damping value matches Rhode's displacement data very well and the corresponding phase curve is very similar in shape to the measured one. At the higher frequencies Rhode's phase data roll off faster than the model's, and they plateau at about -26 radians in comparison to about -18 radians for the model. However, this 8-radian difference is nearly the same as that seen between the 1-D and 2-D curves in Figure 2.

Figure 5 also shows the dramatic differences that changes in damping bring. The peak for the lightly damped model is almost 40 dB greater than that generated with the highly damped model. However, the low-frequency tails and high-frequency slopes of the two curves are identical (cf. Hubbard and Geisler, 1972).

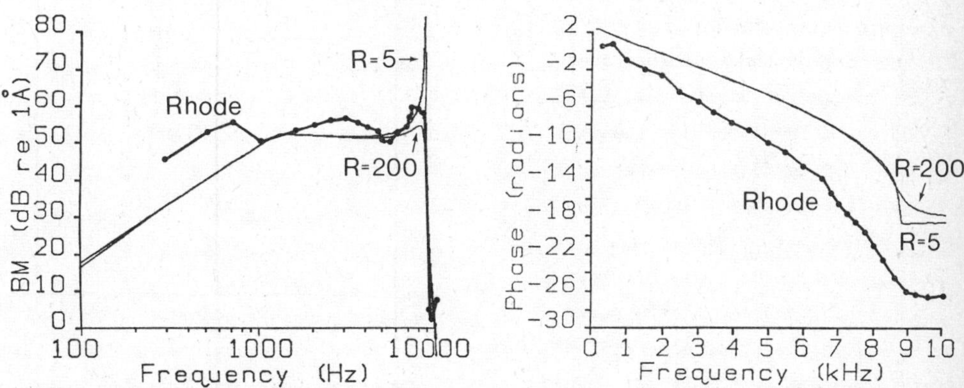

Figure 5. Comparison of amplitude and phase data for the linear 1-D model of the peripheral ear with Rhode's (1978) displacement and phase data. Data for both a highly damped (200 dyne-s/cm^3) and a lightly damped simulation (5 dyne-s/cm^3) are shown. The model results have been determined at the point along the basilar membrane resonating at 8.6 kHz for sinusoidal inputs from 100 Hz to 10 kHz in 100 Hz increments. The displacements have been normalized for a 100 dB SPL input, and phase is relative to stapes motion.

Next notice the overall magnitude of the displacement. The model curves, like Rhode's data, have been normalized for a 100 dB SPL input at the tympanic membrane. The tip of the highly damped curve is about 50 dB below the level expected of a linear system, if a 0 dB SPL input were to produce a threshold displacement of 1 Å (Rhode, 1978). Thus, highly damped models, using parameters that yield reasonable input characteristics (Figures 3 and 4), do not yield realistic values of peak membrane displacement near threshold. The addition of "spectral zeroes" to further attenuate, at some frequencies, the output of those models, in order to achieve better approximations of neural tuning curves (Allen, 1980), thus leads to unrealistically low amplitudes. On the other hand, the lightly-damping simulation very nicely approximates Rhode's amplitude results in the areas away from the resonant point, and around resonance has a tip-to-tail ratio of almost 40 dB. Moreover, the peak is only about 10 dB below the linearly expected level (again assuming a threshold displacement of 1 Å). This amplitude difference is due to the attenuation of the middle ear. Since the 1 kHz, 30 dB SPL input produces a displacement of 30 dB re 1 Å (Figure 2), the damping basal to the 1 kHz place must be slowly decreasing to maintain a constant threshold displacement.

6. LATERAL STIFFNESS COUPLING

The major problem with the lightly damped model is the very high Q_{10} values it produces. For example, the peak shown in Figure 5 has a Q_{10} value of 25. However, if each segment of the cochlear model is not independent of its neighbors, but connected by a spring, then the equation for the displacement, y, of a strip with width dx becomes:

$$(1)\quad P_d(x) = m\ddot{y}(x) + r\dot{y}(x) + k(x)y(x) - k_c(y(x-dx) - 2y(x) + y(x+dx))$$

where the dot denotes a time derivative; P_d is the pressure difference; m, r, and k are the mass, damping and stiffness of the strip, and k_c is the lateral coupling.

The results of the lateral stiffness coupling are shown in Figure 6. As can be seen, there are three principal effects. First, the

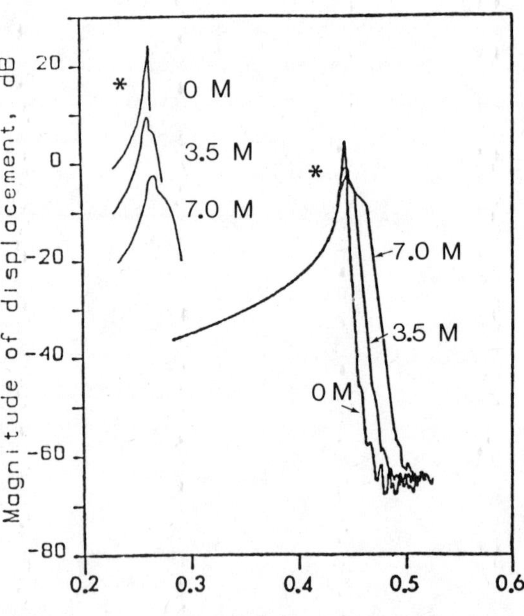

Figure 6. Effect of varying the longitudinal stiffness coupling, k_c, on the shape of the displacement curve. To better visualize the effects on the tip, the peaks have been redrawn in the upper left hand corner of the figure.

width of the peak is increased (most clearly seen in the insert), since the large-amplitude excursions of the resonant strip tend to pull up the strips on either side. Secondly, the amplitude of the peak is decreased as the lesser amplitudes of the strips on either side tend to hold down the resonant strip. Thirdly, the apical shoulder shifts toward the apex, but the slope of that shoulder does not change nor does the basal response. The net effect of the coupling is to lower the Q_{10} of the vibration patterns at the resonant point (from 25 down to 8.5 for the 3.5×10^6 coupling used in this example), at the cost of just a few dB in peak amplitude. Using this same level of coupling throughout the cochlea, displacement curves have been generated, which are quite realistic in general shape, in the absolute levels of displacement amplitudes, and in equivalent Q_{10} values.

The inclusion of longitudinal stiffness in cochlear models was investigated previously by Allen and Sondhi (1979). They found that even small amounts of longitudinal coupling were undesirable. The difference may be due to our use of a much lower damping and to many more spatial points, as with fewer sections we produced effects on the high-frequency slope similar to those reported by Allen and Sondhi.

7. DISCUSSION

The results presented here are useful in several respects. First, a 1-D transmission-line model with lateral coupling has been shown to produce very realistic displacement curves. The tails of these curves match Rhode's Mössbauer data very well, and a low damping value yields 40 to 50 dB tip-to-tail ratios. The low-damping model also has a low-frequency displacement of 1 Å at 0 dB SPL. This value corresponds well with Rhode's estimate of the neural threshold in the squirrel monkey and is only slightly below the estimate of 3.5 Å given by Sellick et al. (1982) for the guinea pig. The good agreement between Rhode's middle-intensity data and the more highly damped amplitude and phase curves implies that large damping values are needed. Yet the necessity of matching the highly-tuned displacements obtained by other workers (e.g., Khanna and Leonard, 1982) requires a lightly damped model. A variable damping which increases with displacement, such as is incorporated into several models (e.g., Hall, 1974) is clearly supported.

Secondly, the comparison of the 1-D model outputs using Allen's (1977), Neely's (1981) and our parameters with the experimental results of Nedzelnitsky (1982) and Lynch et al. (1984) indicates that the mass and stiffness values for all three models are fairly accurately chosen. However, there are significant amplitude and phase discrepancies between the experimental and model input impedance at low frequencies. This indicates the necessity for formulations of the models which more accurately reflect the contributions of the helicotrema and round window.

ACKNOWLEDGEMENTS This work was supported by NIH grants NS-12732 and NS-07026.

REFERENCES

Allen, J. B., "Two-dimensional cochlear fluid model: New results," J. Acoust. Soc. Am. 61, pp. 110-119, 1977.

Allen, J. B., "Cochlear micromechanics - A physical model of transduction," J. Acoust. Soc. Am. 68, pp. 1660-1670, 1980.

Allen, J. B. and Sondhi, M. M., "Cochlear macromechanics: Time domain solutions," J. Acoust. Soc. Am. 66, pp. 123-132, 1979.

Geisler, C. D., "Mathematical models of the mechanics of the inner ear," in Handbook of Sensory Physiology, Auditory System, Ed's. W. D. Keidel and W. D. Neff, Springer-Verlag, New York, pp. 391-415, 1976.

Geisler, C. D. and Sinex, D. G., "Comparison of click responses of primary auditory fibers with minimum-phase predictions," J. Acoust. Soc. Am. 73, pp. 1671-1675, 1983.

Guinan, J. J., Jr. and Peake, W. T., "Middle-ear characteristics of anesthetized cats," J. Acoust. Soc. Am. 41, pp. 1237-1261, 1967.

Hall, J. L., Two-tone distortion products in a nonlinear model of the basilar membrane," J. Acoust. Soc. Am. 56, pp. 1818-1828, 1974.

Hubbard, A. E. and Geisler, C. D., "A hybrid-computer model of the cochlear partition," J. Acoust. Soc. Am. 51, pp. 1895-1903, 1972.

Khanna, S. M. and Leonard, D. G., "Basilar membrane tuning in the cat cochlea," Science 215, pp. 305-306, 1982.

Liberman, M. C., "The cochlear frequency map for the cat: Labeling auditory-nerve fibers of known characteristic frequency," J. Acoust. Soc. Am. 72, pp. 1441-1449, 1982.

Lynch, T. J., III, Nedzelnitsky, V., and Peake, W. T., "Input impedance of the cochlea in cat," J. Acoust. Soc. Am. 72, 108-130, 1982.

Nedzelnitsky, V., "Sound pressures in the basal turn of the cat cochlea," J. Acoust. Soc. Am. 68, pp. 1676-1689, 1980.

Neely, S. T., "Finite difference solution of a two-dimensional mathematical model of the cochlea," J. Acoust. Soc. Am. 69, pp. 1386-1393, 1981.

Neely, S. T. and Kim, D. O., "An active cochlear model shows sharp tuning and high sensitivity'" Hearing Res. 9, pp. 123-130, 1983.

Rhode, W. R., "Some observations on cochlear mechanics'" J. Acoust. Soc. Am. 64, pp. 158-176, 1978.

Sellick, P. M., Patuzzi, R. and Johnstone, B. M., "Measurement of basilar membrane motion in the guinea pig using the Mossbauer technique," J. Acoust. Soc. Am. 72, pp. 131-141, 1982.

Steele, C. R. and Taber, L. A., "Comparison of WKB calculations and experimental models results for three-dimensional cochlear models," J. Acoust. Soc. Am. 65, pp. 1007-1018, 1979.

MOSSBAUER MEASUREMENTS OF THE MECHANICAL RESPONSE TO SINGLE-TONE AND TWO-TONE STIMULI AT THE BASE OF THE CHINCHILLA COCHLEA

Luis Robles, Mario A. Ruggero and Nola C. Rich
Departamento de Fisiología y Biofísica, Facultad de Medicina, Universidad de Chile, Santiago, Chile and
Department of Otolaryngology, University of Minnesota, Research East, Minneapolis, MN 55414

ABSTRACT

Basilar membrane (BM) motion was measured at a site 3.5 mm from the basal end of the chinchilla cochlea using the Mössbauer technique. In preparations with little surgical damage, mechanical responses were as sharply tuned as auditory nerve fibers with the same characteristic frequency (CF, about 8.4 kHz). High-frequency plateaus were observed in both isovelocity tuning curves and phase-frequency curves. Input-output functions at frequencies around CF were strongly nonlinear. Another type of nonlinearity, two-tone suppression, was also demonstrated in several cochleas, with suppression effects as large as 28 dB.

INTRODUCTION

The application of the Mössbauer technique to the measurement of basilar membrane (BM) motion by Johnstone and Boyle (1967) made it possible to study *in vivo* cochlear vibrations in the guinea pig and to demonstrate frequency tuning sharper than previously reported by von Békésy (1960). Using the same technique, Rhode (1971) found even sharper frequency tuning curves and discovered nonlinear vibrations in the squirrel monkey BM at its most sensitive frequencies. However, the results of these early applications of the Mössbauer technique could not completely account for the sharpness of tuning of auditory nerve fibers. The discrepancy between neural and mechanical data, particularly evident in comparisons involving capacitive probe measurements in the guinea pig (Wilson and Johnstone, 1975 and Evans and Wilson, 1975), prompted some investigators to postulate the existence of a "second filter" in the cochlea (e.g., Evans and Wilson, 1973). The need for such a "second filter", however, has been made questionable by recent demonstrations in the guinea pig (Sellick *et al.*, 1982) and cat (Khanna and Leonard, 1982) of BM responses with sharpness of tuning comparable to that observed in auditory nerve fibers. In addition, Sellick *et al.* (1982) reported a strong nonlinearity, similar to that described by Rhode (1971) and by Rhode and Robles (1974). In the present series of experiments we have examined BM responses at the base of the chinchilla cochlea. We have confirmed the existence of a nonlinearity at frequencies around the tip of the tuning curve and have measured mechanical tuning curves almost identical to auditory-nerve frequency-threshold tuning curves.

METHODS

Chinchillas weighing about 500 g were anesthetized with sodium pentobarbital (65 mg/kg i.p.). The left pinna was resected and the lateral wall of the bony ear canal was chipped away to allow insertion of the earphone assembly. The middle ear cavity was widely opened and the tensor tympani was sectioned. A silver-wire electrode was placed on the round window to record the compound action potential (AP) in response to tone bursts (Santi _et al_., 1982). The basal turn of the cochlea was opened by first thinning the bone with a fine burr and then removing bone fragments with a metal pick. The radioactive source was an 80 um square of rhodium foil doped with Co-57; the absorber was Fe-57-enriched palladium foil. The source-absorber combination has an isomer shift of 0.07 mm/s. Perilymph was drained by suction and the source gently placed on the BM using a glass pipette. The Mössbauer radiation was detected using a proportional counter and a single-channel analyzer. Counts were binned into a 32-bin period histogram locked to the sinusoidal stimulus. Velocity and phase of the BM motion were estimated using a least-squares fit program. The acoustic stimulation system and the methods used for single-unit recordings have been described in detail (Ruggero and Rich, 1983).

RESULTS

The results presented here were obtained in 27 chinchillas. Due to anatomical constraints the source was always positioned on the BM within a narrow range of distances from the basal end (2.9-4.1 mm, with a mean of 3.5 mm or 20 % of the total length). The first measurements were always obtained at frequencies around the characteristic frequency (CF), 8-9 kHz. At each

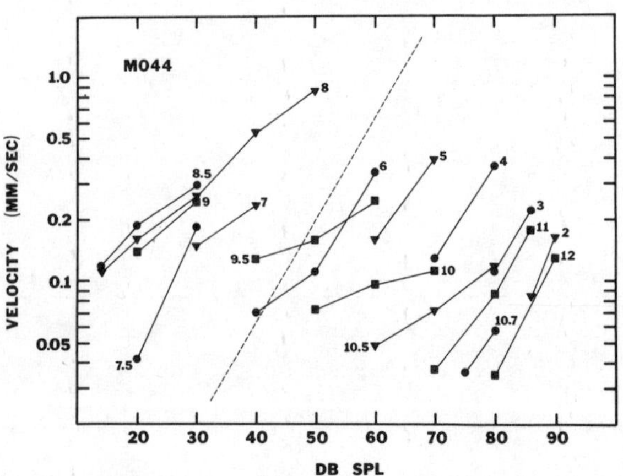

Figure 1. BM input-output functions.

frequency, data were collected for 100 s at SPLs yielding peak velocities in the range of 0.03 to 1 mm/s. Figure 1 shows BM intensity functions obtained in chinchilla M044. These functions, as all those obtained in animals that had sharp tuning and high sensitivity, are strongly nonlinear for frequencies around CF, but linear for frequencies lower than 7 kHz and higher than 10.5 kHz. The nonlinearity is of a compressive type: the intensity functions have a slope of less than unity (dashed line).

There was large variability of tuning and sensitivity at the tip of the velocity responses obtained in different animals, and this variability usually correlated well with AP thresholds. Figure 2 shows isovelocity curves derived from the intensity functions by interpolating the SPLs required to obtain a BM velocity of 0.1 mm/s. These tuning curves, measured in four of the five cochleas with the most sensitive and sharply tuned responses (SPL minima at 20 dB or less), illustrate the similarity in the shape of the tuning curves from the most sensitive preparations (namely, those with AP threshold elevations of 15 dB or less). They have CFs at 8 to 8.75 kHz, minima at 13 to 20 dB SPL, Q10 values of 5.2 to 6.1, tip-to-tail (at 1 kHz) ratios of 48 to 77 dB and high-frequency slopes on the order of 300 dB/oct. Eight animals had less sensitive responses than those shown in Fig. 2, with minima between 21 and 40 dB SPL. Their isovelocity tuning curves had lower CFs, Q10 values, and tip-to-tail ratios than more sensitive preparations. In most animals (including those with substantial sensitivity loss at CF) there was a remarkable constancy in the shape and absolute level of the tail of the tuning curves.

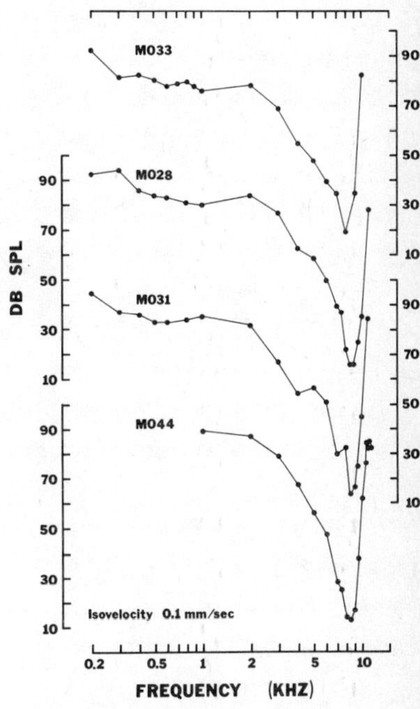

Figure 2. Isovelocity frequency tuning curves for four sensitive cochleas.

Some of the isovelocity tuning curves (e.g., M044 in Fig. 2) have a high-frequency plateau at 80-90 dB SPL, which was observed whenever we extended our measurements to frequencies well above CF. In some experiments, measurements around CF obtained both before and after measuring the plateau showed only moderate decreases of sensitivity.

There was always a reduction of BM velocity with time for the frequencies at the tip of the tuning curve. Meanwhile there was almost no change in the response at the tail and at the high-frequency amplitude plateau. Figure 3 shows intensity curves obtained at several frequencies and various times in one of the best experiments. Over time there is an almost parallel shift of the intensity curves to higher SPLs for the frequencies around CF (8, 8.5 and 9 kHz), while the curves for 6 kHz and lower frequencies show little or no change in 8 hours. Corresponding isovelocity tuning curves for the same data (not shown) reveal a loss in sharpness of tuning, an increase in the minimum SPL and a shift of the CF to lower values. The deterioration of mechanical sensitivity is often, but not always, closely reflected in elevations of AP threshold at CF.

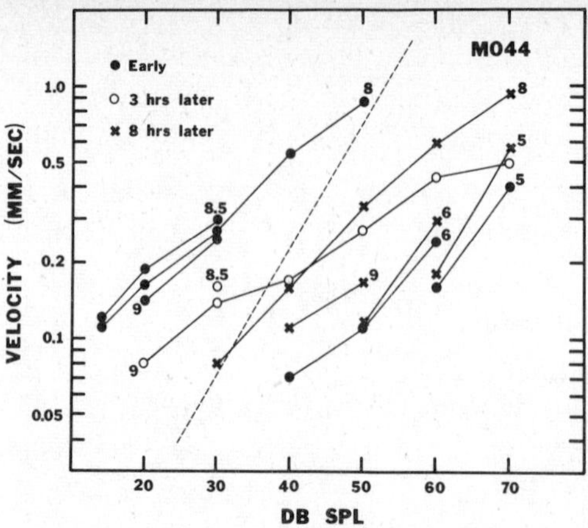

Figure 3. BM input-output curves showing deterioration of response sensitivity with time.

Basilar membrane response phases for the preparations for which isovelocity tuning curves are shown in Fig. 2 are plotted in Fig. 4 as scala tympani displacement referred to rarefaction in the ear canal close to the eardrum. The curves show an increasing phase lag with frequency, varying slowly from about 400 Hz to 4 kHz, having rapid phase changes around CF (indicated by arrows) and reaching a plateau at 7-7.5 π or 9-9.5 π radians at higher frequencies. The slopes near CF average -4.9 radians/kHz, which corresponds to a group delay of 0.8 ms.

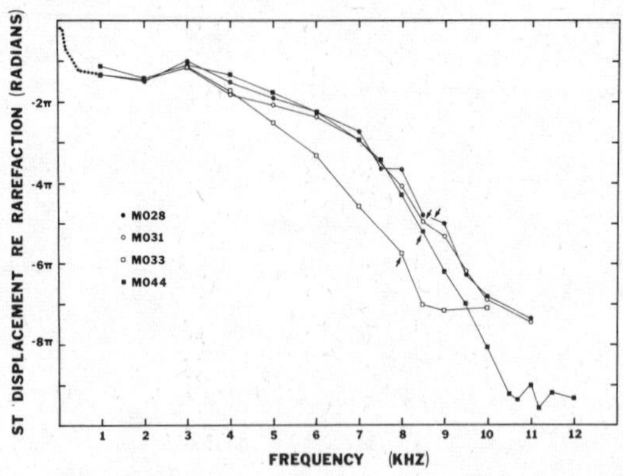

Figure 4. BM response phases as a function of stimulus frequency.

In Fig. 5 the frequency tuning of the mechanical responses is compared to a "synthetic" frequency-threshold curve derived from responses of chinchilla auditory-nerve fibers. For the mechanical data we have plotted the mean value of the isovelocity curves for the five best experiments, of which four are shown in Fig. 2, after shifting the curves to a mean CF of 8.35 kHz (open circles). From the isovelocity curve we have derived an isoamplitude curve (closed circles) for 19.1 A, which is the amplitude corresponding to 0.1 mm/s at 8.35 kHz. For the neural data we have plotted the mean value of the frequency-threshold curves of afferent fibers with CFs between 7 and 10 kHz (triangles), once they have been shifted to correspond to the mechanical mean CF. At the tip, the mean neural curve is almost identical to either mechanical curve; at the tail, it lies between isovelocity and isoamplitude, but it is closer to the isoamplitude curve.

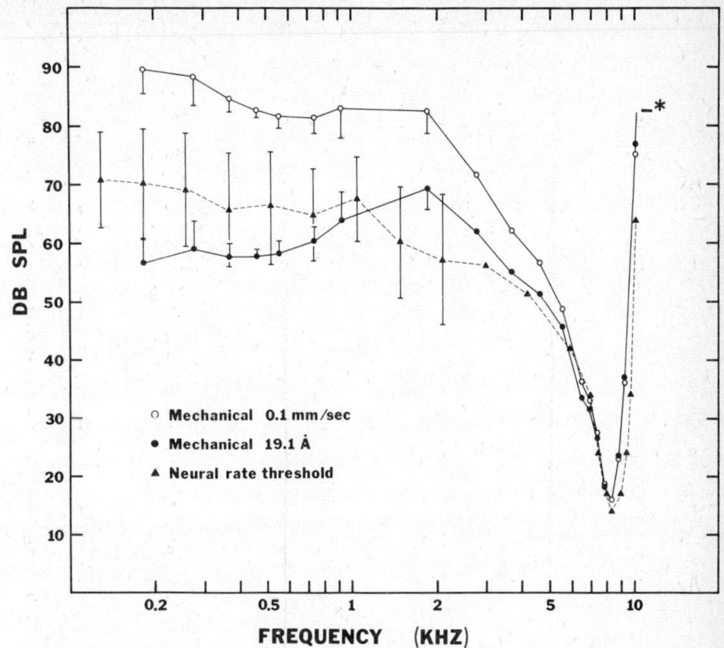

Figure 5. A comparison of the frequency tuning of the 3.5-mm region of the BM and of afferent fibers with corresponding CF.

The BM response to two-tone stimuli was studied in six animals after measuring the responses to single tones. A probe tone was set at a frequency near or equal to CF and at an intensity that produced a clear response when presented alone (30 to 60 dB SPL). A second tone (suppressor) was set at a frequency higher than CF and at an intensity that did not produce a measurable response when presented alone (at or lower than 80 dB SPL). Two-tone suppression was demonstrated in five of the six animals studied. Figure 6 shows the change in the intensity functions produced by a suppressor tone at two frequencies higher than CF and at several intensities. The solid symbols display the intensity functions for the probe tone alone, obtained immediately before each of the two-tone measurements, after correction for the progressive loss of sensitivity with time. The open symbols show intensity functions for the probe tone when presented simultaneously with a suppressor tone. There is a clear reduction of the velocity response of the probe tone for most of the two-tone conditions. Because the magnitude of the two-tone suppression tended to decrease with time, we have indicated in the figure the time (referred to placement of the source) at which each curve was obtained. Suppression magnitudes were estimated by measuring the horizontal shift of the intensity function for the probe tone, caused by the addition of the suppressor tone (Javel, 1981). The largest suppression obtained in this experiment (for the 10 kHz, 60 dB suppressor) varied from 15 to 28 dB. The effect of the suppressors on the intensity function of the probe is not merely to reduce its sensitivity, but also to increase its slope, making it more linear.

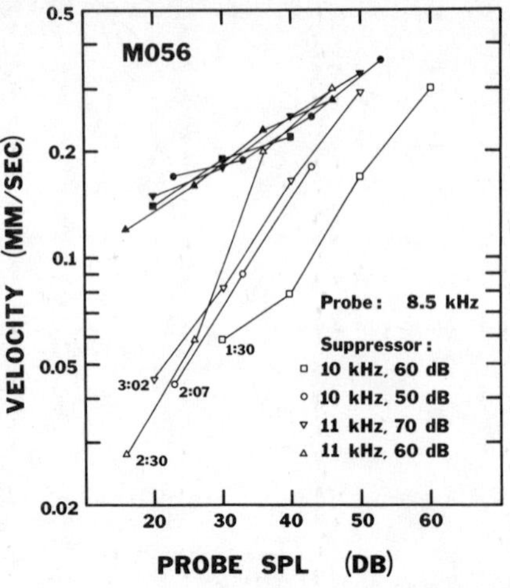

Figure 6. BM input-output curves demonstrating two-tone suppression. Curves for responses to 8.5 kHz tones presented alone (filled symbols) are compared to responses in the presence of suppressors at 10 or 11 kHz (open symbols).

DISCUSSION

Our measurements show that the mechanical response of the BM is very labile and that in order to obtain sensitive and sharply tuned responses extreme care is required in the surgical manipulation of the cochlea. The results obtained in some of our experiments demonstrate that it is possible to obtain sharp mechanical tuning curves with low SPL minima and with a loss of sensitivity around CF slow enough to allow collection of reasonably complete data for the tip of the tuning curve. The Q10 values of 5.2 to 6.1 measured in our best experiments are similar to the highest values reported for the BM of guinea pigs (Sellick _et al_., 1982 and 1983) and cats (Khanna and Leonard, 1982). They are also higher than the mean Q10 values obtained for chinchilla auditory nerve fibers of similar CF in our laboratory (see Fig. 5) and elsewhere (Dallos and Harris, 1978; Salvi _et al_., 1983).

We have found a compressive nonlinearity in the BM input-output functions for single tones, confirming the reports by Rhode (1971) in the squirrel monkey and Sellick _et al_. (1982) in the guinea pig. We have also measured two-tone suppression of magnitude greater than that reported by Rhode (1977) in the squirrel monkey, for suppressors at intensities 30-50 dB lower than the ones used in his study. Our values of suppression, however, appear to be lower than those reported for auditory nerve fibers in the cat (Javel, 1981). A valid quantitative comparison between suppression at the BM and at the auditory nerve must await more complete mechanical, as well as neural, data obtained for similar CFs in the same species.

High-frequency plateaus were observed in several isovelocity tuning curves and in all phase-frequency curves, similar to those reported by Rhode (1971) in squirrel monkey, Wilson and Johnstone (1975) in guinea pig and Wilson and Evans (1983) in cat. As far as we have been able to ascertain, these plateaus (or changes of slope in the isovelocity curves) seem to be a normal feature of the mechanical responses of undamaged preparations.

Our observations on the lability of the preparations, and especially on the progressive loss of sensitivity and tuning with time, support the view that the normal response of the BM depends on nonlinear feedback from the outer hair cells that boosts its low-level sensitivity and sharpness of tuning and that is extremely susceptible to injury.

ACKNOWLEDGEMENTS

Dr. John A. Costalupes helped to edit the manuscript. This work was supported by NSF Grant BNS-8304587 and NINCDS Grant NS12125.

REFERENCES

Békésy, G. von, Experiments in Hearing, ed. by E. G. Wever, Mc Graw Hill, New York, 1960.

Dallos, P. and Harris, D., "Properties of auditory nerve responses in absence of outer hair cells." J. Neurophysiol. 41, 365-383, 1978.

Evans, E. F. and Wilson, J. P., "The frequency selectivity of the cochlea." In Basic Mechanisms in Hearing, ed. by A. Møller, Academic Press, New York, pp. 519-554, 1973.

Evans, E. F. and Wilson, J. P., "Cochlear tuning properties: concurrent basilar membrane and single nerve fiber measurements." Science 190, 1218-1221, 1975.

Javel, E., "Suppression of auditory nerve responses I: Temporal analysis, intensity effects and suppression contours." J. Acoust. Soc. Am. 69, 1735-1745, 1981.

Johnstone, B. M. and Boyle, A. J. F., "Basilar membrane vibration examined with the Mössbauer technique." Science 158, 389-390, 1967.

Khanna, S. M. and Leonard, D. G. B., "Basilar membrane tuning in the cat cochlea." Science 215, 305-306, 1982

Rhode, W. S., "Observations of the vibration of the basilar membrane in squirrel monkeys using the Mössbauer technique." J. Acoust. Soc. Am. 49, 1218-1231, 1971.

Rhode, W. S., "Some observations on two-tone interaction measured with the Mössbauer effect." In Psychophysics and Physiology of Hearing, ed. by E. F. Evans and J. P. Wilson, Academic Press, London, pp. 27-38, 1977.

Rhode, W. S. and Robles, L., "Evidence from Mössbauer experiments for nonlinear vibration in the cochlea." J. Acoust. Soc. Am. 55, 588-596, 1974.

Ruggero, M. A. and Rich, N. C., "Chinchilla auditory-nerve responses to low frequency tones." J. Acoust. Soc. Am. 73, 2096-2108, 1983.

Salvi, R., Hamernik, R. P. and Henderson, D., "Response patterns of auditory nerve fibers during temporary threshold shift." Hearing Res. 10, 37-67, 1983.

Santi, P. A., Ruggero, M. A., Nelson, D. A. and Turner, C. W., "Kanamycin and bumetanide ototoxicity: anatomical, physiological and behavioral correlates." Hearing Res. 7, 261-279, 1982.

Sellick, P. M., Patuzzi, R. and Johnstone, B. M., "Measurement of basilar membrane motion in the guinea pig using the Mössbauer technique." J. Acoust. Soc. Am. 72, 131-141, 1982.

Sellick, P. M., Patuzzi, R. and Johnstone, B. M., "Comparison between the tuning properties of inner hair cells and basilar membrane motion." Hearing Res. 10, 93-100, 1983.

Wilson, J. P. and Evans, E. F., "Some observations on the 'passive' mechanics of the cat basilar membrane." In Mechanisms of Hearing, ed. by W. R. Webster and L. M. Aitkin, Monash University Press, Clayton, Victoria, Australia, pp. 30-35, 1983.

Wilson, J. P. and Johnstone, J. R., "Basilar membrane and middle-ear vibration in guinea pig measured by capacitive probe." J. Acoust. Soc. Am. 57, 705-723, 1975.

PARAMETER SENSITIVITY
IN A MATHEMATICAL MODEL OF BASILAR MEMBRANE MECHANICS

Kathleen A. Morrish
Richard S. Chadwick+
Shihab A. Shamma*
John Rinzel

Mathematical Research Branch, NIADDK
NIH, Bethesda, MD 20205

+Biomedical Engineering and Instrumentation Branch
DRS, NIH, Bethesda, MD 20205

*Department of Electrical Engineering
University of Maryland, College Park, MD 20742

ABSTRACT

A mathematical model of cochlear processing is developed to describe the transformation from acoustic stimulus to intracellular hair cell potential. It incorporates a linear formulation of three-dimensional basilar membrane mechanics, subtectorial fluid-cilia displacement coupling, and a simplified description of the inner hair cell nonlinear transduction process. When the model parameters of the basilar membrane stage are set to values characteristic of the guinea pig, good agreement with experiment is obtained for single tone responses. When the parameters are varied, the basilar membrane tuning can change dramatically.

1. INTRODUCTION

The motivation behind this effort is to formulate a composite model of the various cochlear stages which is founded upon biophysical principles and experimental data. This report concentrates on the basilar membrane stage. A linear, three-dimensional formulation of basilar membrane mechanics which is dependent upon explicit physiologically identifiable parameters is employed. The model is outlined in Section 2. A discussion of the subtectorial fluid-cilia displacement coupling and the inner hair cell nonlinear transduction process can be found in Shamma et al. (1985).

In Section 3, theoretical tuning curves are presented for parameters set appropriately for the guinea pig. The sensitivity of the tuning curves to various parameter changes is described in Section 4. It is shown that appropriate choice of system constants and functions is essential for reasonable behavior of the results. This is discussed in light of available experimental data in Section 5.

2. THE BASILAR MEMBRANE MODEL

For purposes of the basilar membrane model, the cochlea consists of a straight, rigid tube with square cross-section filled with a viscous, incompressible fluid. The tube is divided into two equal chambers, the scala vestibuli and scala tympani, by an interior surface, part of which is rigid and part viscoelastic (the basilar membrane). Both parts are assumed to be the same length as the cochlea itself. The viscoelastic part is a highly anisotropic plate with variable width and thickness, and a damping per unit area proportional to basilar membrane velocity. It is simply supported at stationary side boundaries. Displacement of the stapes at the basal end ($s = 0$) causes a pressure difference in the two chambers, which drives the basilar membrane motion. Pressure between the chambers is equalized at the apical end of the cochlea ($s = 1$) by the helicotrema.

Amplitudes are assumed to be small enough that the linearized system is a good approximation to the full equations. At low frequency (< 1500 Hz for the parameters in Table 1), specialized asymptotics simplify the calculations. The result is a very efficient, three-dimensional model with forward and backward running waves and explicit formulae. The current model can be shown to be a low frequency limit of the model of Holmes and Cole (1984). The details have been discussed elsewhere (Chadwick, 1985).

3. RESULTS WITH GUINEA PIG PARAMETERS

There is little data on basilar membrane response and cochlear geometry at the apical end of the cochlea. By using information in a paper by Fernandez (1952), shape parameters and functions can be obtained for the guinea pig. Tuning curves generated by the model at $s > 0.5$ can be related to experimental results from von Bekesy (1960). Lack of experimental data on phase at low frequency precludes an evaluation of the model in terms of the phase.

By adjusting membrane damping and stiffness, it is possible to produce results from the model (Fig. 1) which agree with experiment in several respects. First, the cochlear map produced by the model agrees with that obtained by von Bekesy over the range from $s = 0.86$ to $s = 0.56$. This includes frequencies from 300 Hz to 1400 Hz. Second, the high frequency slopes conform to experiment. To elaborate on this, it is necessary to discuss the notion of high frequency slope.

The high frequency slope is defined here as the slope obtained with a linear regression on all computed points from 6 dB below the peak to 12 dB below the peak. This definition was chosen to take advantage of the well-defined peak of the tuning curve while utilizing frequencies within the range of interest for this model. Note that the definition may yield values which are smaller than those reported elsewhere: experimental high frequency slopes are usually calculated on the basis of a few points obtained at the highest frequencies measured. However, it has not yet been shown either experimentally or theoretically that the slope of the tuning curve in the high frequency limit should approach a constant. In particular, the current model and von Bekesy's data do not exhibit such behavior over the range of frequencies examined. Slopes produced by the model and measured as above range from 22 to 29 dB/octave between s = 0.926 and s = 0.676. Von Bekesy obtained corresponding slopes between 23 and 25 dB/octave over the same s-values.

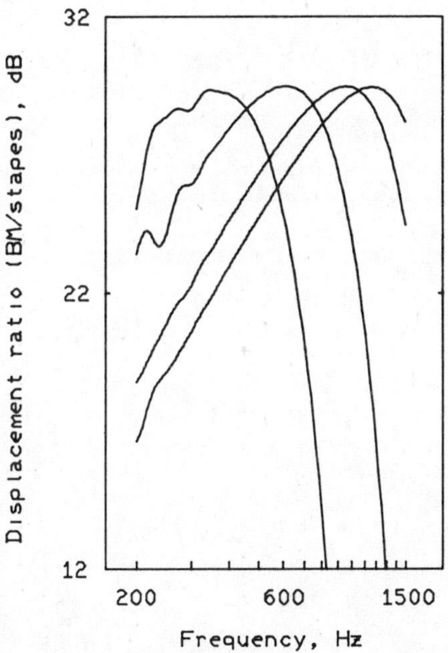

Figure 1. Tuning curves generated by the model with parameters from Table 1. The curves, peaking from left to right, are at s = 0.800, 0.700, 0.622, and 0.590.

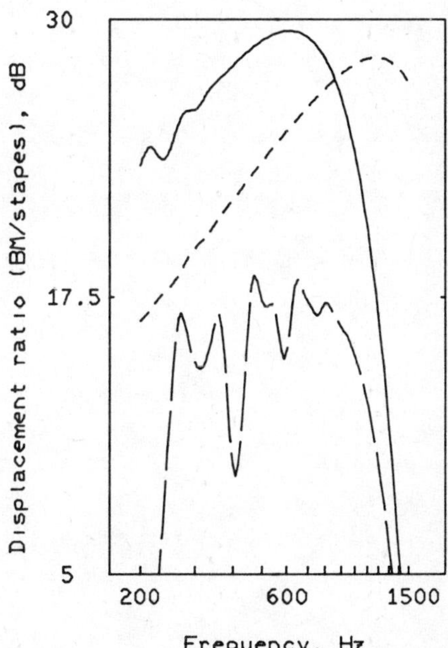

Figure 2. Tuning curves at s = 0.700 for several basilar membrane thickness functions: the long dashed line is a constant approximation (h = 0.35), the short dashed line is a linear approximation (h = 1 - 0.9s), and the solid line represents a piecewise exponential (see Table 1).

The results differ from experiments in certain respects. The low frequency slope obtained by von Bekesy is larger than that predicted by the model. It is not surprising, therefore, that the theoretical curves exhibit oscillations on the low frequency side not seen in von Bekesy's published experimental results. These oscillations are due to the wave reflected at $s = 1$.

4. PARAMETER VARIATIONS

To illustrate the sensitivity of the mechanical basilar membrane model, the set of parameters for the guinea pig is varied systematically. The mechanical parameters explored are membrane damping, fluid viscosity and membrane stiffness. The shape parameters investigated are basilar membrane length and thickness.

MECHANICAL PARAMETERS. The peak of the tuning curve moves to lower frequencies and decreases in height as damping is increased. An increase in viscosity or a decrease in characteristic stiffness (D_0 in Table 1) affect the model similarly. Higher viscosity (or greater damping, or lower stiffness) causes the energy of the input to dissipate more quickly, so that the peak is decreased in height. It also dissipates the high frequency responses more effectively than the low frequency responses, shifting the peak to lower frequencies. The viscosity-dependent shift has been observed experimentally (Wilson and Evans, 1983).

Increased damping also causes broadening of the wave envelope on the basilar membrane. As the wave develops a significant height at the apical end, reflections become more important. This results in low frequency oscillations in the tuning curve.

SHAPE PARAMETERS. When the length of the basilar membrane is decreased, the tuning curve shifts to higher frequencies and increases in height. This is reasonable, because a decrease in length results in a decrease in both mass of the basilar membrane and surface area of the cochlea with which to dissipate the energy of the input tone. If basilar membrane characteristic thickness (h_0 in Table 1) is varied without changing its shape (h in Table 1), only minor changes occur. However, the shape significantly affects tuning. A linear approximation to the membrane thickness caused a shift of the tuning toward higher frequencies. A constant thickness caused all tuning to be lost (Fig. 2). The decrease in peak height observed for the constant thickness is due to increased dissipation of the input signal near $s = 0$. The irregular oscillations stem from the reduced stiffness gradient. As the stiffness gradient decreases, the basilar membrane tuning broadens, which may lead to a significant reflected wave.

TABLE 1. Guinea Pig Parameters for the Cochlea and Basilar Membrane (BM)

maximum width of BM, B_0	0.025 cm	
maximum thickness of BM, h_0	0.0007 cm	
length of BM	1.85 cm	
membrane damping constant	115. dyne/cm^3	
maximum membrane stiffness, D_0	0.094 dyne cm	
density of fluid	1.0 g/cm^3	
viscosity of fluid	0.008 cm^2/s	
width of BM	$B_0(8 + 5s)/13$	
height of scala tympani	$B_0(6.5 - 17s)$,	$s < .176$
	$B_0(3.8 - 1.7s)$,	$s > .176$
thickness of BM, h	h_0,	$s < .2$
	$h_0[0.1 + 0.9\exp(0.7 - 3.5s)]$,	$s > .2$
membrane stiffness, D	D_0h^3	

5. DISCUSSION

Exploring parameter sensitivity in the basilar membrane model can demonstrate that the accuracy of the low frequency approximation is affected by parameter values. Certain modifications of the quantities in Table 1 produce an anomalous resonance peak in the tuning curve, which appears when a correction term becomes large. The location of the peak can be predicted analytically. By monitoring the size of the correction terms, poor approximation can be avoided.

The low frequency oscillations seen in Fig. 1 are not unique to this model. The theory of Holmes and Cole also exhibits them (Holmes, 1985). In fact, it has been shown (Holmes, 1979) that any hydroelastic system with finite boundaries has the potential for reflections. By adjusting the viscosity, membrane damping, or boundary condition at the apical end, or by removing the backward travelling wave from the model, reflections can be controlled or eliminated. If reflections contribute to the cochlear response, they may be masked by noise and difficult to measure. It has been suggested that certain phase phenomena observed experimentally may be due to reflections (Allen, 1983).

It is the difficulty in measuring phenomena in the inner ear that makes mathematical models of hearing so valuable. Because this model depends mainly upon explicit, physiologically identifiable parameters, the effects of varying these parameters can be approximated. As additional data is accumulated, the potential for predicting quantities such as membrane stiffness may be realized.

Fig. 2 illustrates that certain approximations to physiological quantities in the system can yield poor results. This suggests that the cochlea may be optimally designed in some ways, and that overlooking or oversimplifying certain design parameters in mathematical models may lead to inappropriate conclusions. The delicate process of choosing appropriate design parameters is aided greatly by experimental data. Unfortunately, little is known about responses at the apical end of the cochlea. It is hoped that more information will soon be available so that physiological and anatomical parameters of the cochlea will be known with greater certainty.

REFERENCES

Allen, J. B., "Magnitude and Phase-Frequency Response to Single Tones in the Auditory Nerve". J. Acoust. Soc. Am. 73, pp. 2071-2092, 1983.

von Bekesy, G., Experiments in Hearing. McGraw-Hill, New York, 1960.

Chadwick, R. S., "Three Dimensional Effects on Low Frequency Cochlear Mechanics". Mech. Res. Communications (in press).

Fernandez, C., "Dimensions of the Cochlea (Guinea Pig)". J. Acoust. Soc. Am. 24, pp. 519-523, 1952.

Holmes, M. H. and Cole, J. D., "Cochlear Mechanics: Analysis for a Pure Tone". J. Acoust. Soc. Am. 76, pp. 767-778, 1984.

Holmes, M. H., "Frequency Discrimination in the Mammalian Cochlea". R. P. I. Math. Report No. 150, 1985.

Holmes, M. H., "A Spectral Problem in Hydroelasticity". J. Diff. Equations 32, pp. 388-397, 1979.

Shamma, S. A., Chadwick, R. S., Wilbur, W. J., Rinzel, J. and Morrish, K. A., "A Biophysical Model of Cochlear Processing: Intensity Dependence of Pure Tone Responses". J. Acoust. Soc. Am. (submitted for publication).

Wilson, J. P. and Evans, E. F., "Some Observations on the "Passive" Mechanics of Cat Basilar Membrane". In Mechanisms of Hearing, Eds. W. Webster and L. Aitkin. Monash University Press: Australia, 1983.

COCHLEAR MICROMECHANICS

MICROMECHANICS OF THE COCHLEAR PARTITION

Stephen T. Neely
Boys Town National Institute
555 North 30th Street
Omaha, Nebraska 68131

ABSTRACT

Cochlear micromechanics describes the radial displacement of hair bundles of the inner and outer hair cells in response to forces generated by fluid pressure gradients near the cochlear partition. A simple micromechanical model of a radial cross-section of the cochlear partition is a lumped mass, stiffness, and damping. This simple representation is adequate to simulate "traveling waves" and a frequency-to-place correspondence in a cochlear model, but inadequate to simulate "neural-like" tuning in cochlear mechanics. A second, coupled resonant element appears to be necessary to explain the sharp tuning which is typical of the cochlea. This second resonant element may be passive or may be part of a mechanical "cochlear amplifier" which contributes mechanical energy (at the expense of electrochemical energy) to provide high sensitivity at the threshold of hearing.

1. INTRODUCTION

Cochlear micromechanics describes the mechanics of the cochlear partition consisting of the organ of Corti (including the basilar membrane) and the tectorial membrane (see Fig. 1). The micromechanics of the cochlear partition are closely coupled to the macromechanics of the cochlear fluid. The micromechanics are also influenced by bidirectional transduction at the outer hair cells. Thus, the micromechanics of the cochlear partition must be considered within the context of bidirectional coupling to both cochlear macromechanics and cochlear transduction.

The lower surface of the organ of Corti (see Fig 1) is the basilar membrane (BM) which separates scala media and scala tympani. Within the organ of Corti are the sensory hair cells. There are three rows of outer hair cells and one row of inner hair cells. The inner hair cells (IHC) receive most of the afferent innervation and are

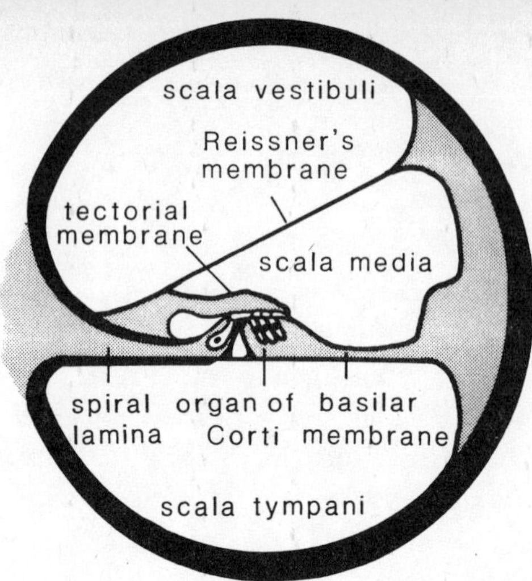

Figure 1. Schematic illustration of a radial cross-section through the cochlea.

responsible for sending information to the central auditory system (Spoendlin, 1979). The outer hair cells (OHC) receive most of the efferent innervation and are capable of modifying cochlear mechanics through bidirectional transduction (Weiss, 1982). The upper surface of the organ of Corti is called the reticular lamina (RL). The hair bundles (HB) of the inner and outer hair cells extend up above the RL toward the TM. The TM covers the RL with a gap of a few microns between their two adjacent surfaces. The tallest hairs (stereocilia) of the OHC (and perhaps also the IHC) are embedded in the underside of the TM (Lim, 1980).

The cochlear partition has a tonotopic organization with high frequencies being represented at the base near the vestibule and low frequencies being represented at the apex near the helicotrema. The frequency to place map generally has a constant number of millimeters per octave over most of the cochlear partition (Greenwood, 1960; Liberman, 1982). This means that a linear distance along the partition corresponds with a logarithmic change in frequency representation.

When von Békésy (1960) observed traveling waves on the cochlear partition in response to pure tones, the spatial distribution of the response was too broad to account for all of the ear's fine frequency discrimination. At that time, the interpretation of this observation was that the cochlea did not provide any sharply tuned filters and that fine frequency discrimination was accomplished by processes within the central auditory system.

Subsequent measures have established the presence of sharply tuned filters within the peripheral auditory system. Intracellular recordings from single auditory nerve fibers have shown that the majority are sharply tuned (Kiang et al., 1955). Intracellular measurements of d.c. responses of inner hair cells show tuning characteristics that match those of single nerve fibers (Russell and Sellick, 1978). Recent in vivo measurements of basilar membrane (BM) responses to pure tones have demonstrated iso-displacement tuning curves that are similar to neural tuning curves (Khanna and Leonard, 1982; Sellick, Patuzzi, and Johnstone, 1982). Even measurements of oto-acoustic emissions in the ear canal have provided evidence of sharply tuned mechanical responses of cochlear origin (Kemp, 1982).

It is now firmly established that the peripheral auditory system is capable of fine frequency analysis with frequency components distributed spatially along the length of the cochlear partition. Clearly most, if not all, of the sharp tuning required for that frequency analysis is accomplished mechanically within the cochlear partition. The details of the micromechanics of the cochlear partition are still not yet well understood. This makes cochlear micromechanics a subject of considerable interest to the hearing research community.

2. A SIMPLE MODEL FOR MECHANICS OF THE PARTITION

The input to cochlear micromechanics is a spatial distribution of forces on the cochlear partition due to fluid pressure gradients generated by cochlear macromechanics. The output of cochlear micromechanics is the radial displacement of the HB of the IHC which determines (upon transduction) the sensory input to the afferent auditory nerve fibers.

To understand the relationship between fluid pressure distribution in the cochlea and radial displacement of the HB, consider a radial cross-section through the cochlea, as illustrated schematically in Fig. 1. The appearance of such a cross-section is qualitatively similar from one end of the cochlea to the other. The morphological features, (such as width of the BM, areas of the scalae, etc.) vary gradually (in most mammals) from base to apex.

The predominate coupling in the spiral (longitudinal) dimension is through the fluid. Spiral coupling through the cochlear partition is relatively small since the radial stiffness is much greater than the spiral (longitudinal) stiffness (Voldřich, 1978). Thus, the cochlear partition can be modeled as an anisotropic plate (Allen and Sondhi, 1979) or even as a series of adjacent viscoelastic beams with gradually changing characteristics (Diependaal and Viergever, 1983).

The bending stiffness of the BM dominates the mechanics of the cochlear partition for most frequencies. The BM is narrow and stiff at the base to resonate at high frequencies; it is wide and compliant near the apex to resonate at low frequencies. When the instantaneous fluid pressure at a given position in the scala tympani exceeds the fluid pressure in the adjacent scala media the fluid pressure difference creates a force which pushes upward on the BM (see Fig. 1).

The resonate frequency of a given place on the BM represents a "cut off" frequency for signals traveling by that position (Lighthill, 1981). Traveling wave propagation is supported for frequencies below BM resonance, but not for frequencies above BM resonance. The pressure difference across the partition at a given place has a low-pass characteristic; it has almost constant amplitude below the BM resonant frequency and drops off abruptly above this frequency. Additional filtering is accomplished in the micromechanics of the cochlear partition.

Displacement of the BM is tightly coupled to the RL due to the stiff arches located between the IHC and OHC. An upward displacement of the BM pushes the RL radially inward (see Fig. 1) toward the inner edge of the TM. If the TM is rigid and tightly hinged, then the HB will be bent over such that the tips of the stereocilia move radially outward. Displacement of the HB in this direction causes depolarization of the hair cell (Hudspeth and Corey, 1977).

A model of cochlear mechanics with a point impedance characterization of the cochlear partition (Diependaal and Viergever, 1983) shows tonotopic organization and propagation of traveling waves along the cochlear partition (Peterson and Bogert, 1950; Zwislocki, 1950). Extensions of the basic model are required to explain the sharp tuning and the high sensitivity typical of cochlear responses.

3. THE NEED FOR A 'SECOND' FILTER

As techniques were developed over the past 20 years to measure *in vivo* BM motion, it became evident that normal *in vivo* responses are significantly different than the *post mortem* responses observed by von Békésy. At moderate sound levels, optimal *in vivo* responses are more sharply tuned and have larger amplitude than the corresponding _post mortem_ responses (Rhode, 1973). This sharp tuning and high sensitivity is difficult to maintain under the laboratory conditions required for its measurement; however, in a few cases, it has been possible to measure BM tuning curves that are similar to neural tuning curves (Khanna and Leonard, 1982; Sellick, Patuzzi, and Johnstone, 1982).

A 1 nm iso-displacement BM tuning curve based on measurements of Sellick, Patuzzi, and Johnstone (1982) is compared with a neural spike rate threshold tuning curve in Fig. 2; the neural tuning curve was obtained from the same animal species (guinea pig) in the same laboratory. The tip-to-tail ratio is less in the BM curve than it is in the neural curve, but there is good agreement in the tip segment. The evidence to establish an _exact_ correspondence between neural tuning and BM tuning does not yet exist (Sellick, Patuzzi, and Johnstone, 1983), but, clearly, the sharp tuning of the cochlea is observable in BM displacement.

Whether or not BM tuning is as sharp as neural tuning, we need a second filter to explain the dual nature of the "tip" and "tail" segments of a high frequency neural tuning curve (Davis, 1981). The tip segment of the tuning curve is vulnerable to hypoxia and ototoxins (Evans, 1974). The loss of sharp neural tuning observed as an elevation in threshold of the tip segment has been linked to damage in OHC (Evans and Harrison, 1975; Liberman and Dodds, 1984). We can define the "first" filter as the robust, low-pass characteristic associated with BM resonance. The "second" filter we will define as the physiologically vulnerable, sharply tuned characteristic associated with the tip segment of high frequency tuning curves. The question of whether the second filter is passive or active will be discussed in the next section.

One micromechanical model for the second filter assumes the tectorial membrane possesses an independent degree-of-freedom (Allen, 1980; Zwislocki and Kletsky, 1979a; Zwislocki and Kletsky, 1979b). In other words, the TM is not assumed to be rigid and tightly hinged, but capable of some bending and/or stretching. Because the length of the stereocilia (especially in the tallest strereocilia of the outer hair cells) increases from base to apex, (Lim, 1980), the stiffness of the HB decreases from base to apex (Strelioff and Flock, 1984). The mass of the TM, together with the stiffness of the HB creates a second resonant subsystem at each longitudinal position. This resonant TM can sharpen the tuning of the HB displacement if it is tuned to a frequency lower than the BM resonant frequency. In this case, the TM

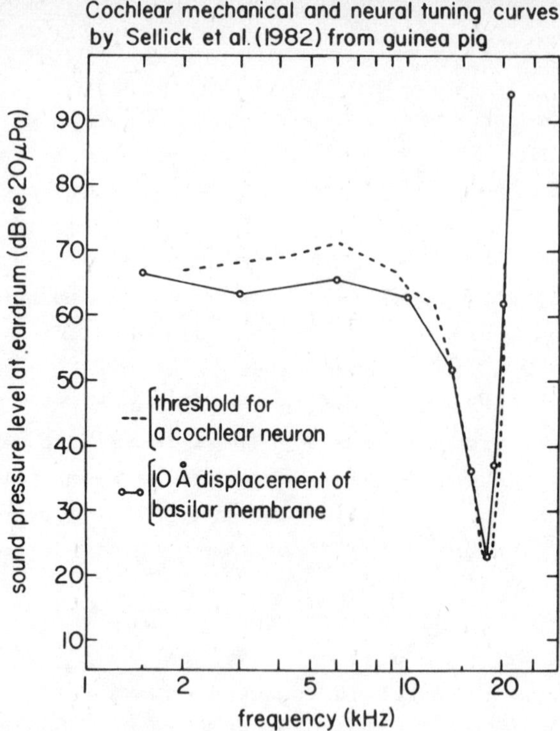

Figure 2. Comparison between between mechanical and neural tuning curves. The dashed line indicates the spike rate threshold for a spiral ganglion cell in a guinea pig. The connected open circles indicate a 1 nm iso-displacement condition determined by a Mössbauer source on the basilar membrane of a guinea pig. [Figure from Neely and Kim (1983). The data plotted are from Sellick, Patuzzi, and Johnstone (1982).]

moves in phase with the RL for frequencies below the TM resonant frequency and moves 180° out of phase with the RL motion between TM resonance and BM resonance. Thus, the resonant TM sharpens the tuning of the HB displacement by attenuating frequencies below the best or characteristic frequency (CF) of the HB.

The resonant TM hypothesis predicts a spectral zero in the transfer function between the frequency responses of BM displacement and HB displacement at a frequency about one half octave below CF (Allen, 1980). The characteristics of this spectral zero in the transfer function are: (1) an increase in the slope of the magnitude curve by 12 dB/octave and (2) an abrupt shift of 180° (toward leading phase) in the phase curve. This leading phase shift has been observed in the responses of a population of nerve fibers as a function of CF

(Kim, Siegel, and Molnar, 1979).

4. THE NEED FOR AN ACTIVE PROCESS

The resonant tectorial membrane hypothesis described above helps to explain the discrepancy (if any) between BM iso-displacement tuning curves and neural rate tuning curves, but something more is needed to explain the sensitivity of BM responses to low intensity sounds (Davis, 1983).

The displacements of the BM and HB are quite small at the threshold of hearing. Extrapolating downward from the curve in Fig. 2, we would expect to see BM displacement with an amplitude of about 0.1 nm at 0 dB SPL. We expect about the same amplitude of displacement fluctuations at the BM due to thermal noise (Bialek, 1983). The fact that the "signal" and the "noise" have about the same amplitude at 0 dB SPL is consistent with using this sound level to represent the threshold of hearing.

An active model that includes "negative damping" elements in the micromechanics of the cochlear partition can simulate BM tuning curves with the same sensitivity (0.1 nm at 0 dB SPL) as the one shown in Fig 2 (Neely and Kim, 1983). We have been unable to simulate BM tuning curves with this sensitivity in any passive model with acceptably small delay. Since the active model provides about 45 dB total power gain under the conditions required for this simulation, it seems unlikely that a passive model will ever achieve this level of sensitivity. Independent analysis of BM response curves (Boer, 1983) confirms the need for an active process in cochlear mechanics.

Another source of evidence for the existence of an active process in cochlear mechanics comes from measurements of spontaneous oto-acoustic emissions. The distribution of instantaneous ear canal pressure amplitudes in the presence of a strong spontaneous oto-acoustic emission has been shown to have "statistical properties in accord with a simple model of instabilities in an active filter" (Bialek and Wit, 1984). In other words, spontaneous oto-acoustic emissions look as if they are being generated by an active process. Furthermore, the presence of (stable) active filters has been hypothesized to play an important role in a reflection model of evoked oto-acoustic emissions, wherein the active filter would "reduce damping below its 'natural' value". The physiological vulnerability of the evoked oto-acoustic emission is "readily explainable with this model" (Kemp, 1982).

The need for an active process to explain measured

characteristics of the auditory system is matched by the need to assign a functional role to the outer hair cell subsystem (Kim, 1984). The OHC receive most of the efferent innervation from the auditory nerve and little of the afferent innervation. Certainly, the OHC are not designed to carry acoustic signal information to the brain. OHC are capable of a motile response to electrical stimulation (Brownell et al., 1985); IHC do not have this property. The OHC subsystem appears to have the characteristics of a motor system rather than a detection system.

The mechanical force generating properties of the OHC together with the resonant TM could provide the cochlea with an active, second filter. It appears that such an active, second filter is both possible and necessary to explain the high sensitivity and sharp tuning of the cochlea for low level sounds near the threshold of hearing.

ACKNOWLEDGEMENT

This work was supported by NIH grant no. NS20652.

REFERENCES

Allen, J. B. and Sondhi, M. M. (1979). "Cochlear macromechanics: Time domain solutions," J. Acoust. Soc. Am. 66, 123-132.

Allen, J. B. (1980). "Cochlear micromechanics - A physical model of transduction," J. Acoust. Soc. Am. 68, 1660-1679.

Békésy, G. von (1960). "Experiments in Hearing," , (McGraw-Hill, New York).

Bialek, W. (1983). "Thermal and quantum noise in the inner ear," in Mechanics of Hearing, edited by E. de Boer and M. A. Viergever, (Delft U. P., Delft, The Netherlands), 185-192.

Bialek, W. and Wit, H. P. (1984). "Quantum limits to oscillator stability: Theory and experiments on acoustic emissions from the human ear," Physics Letters 10A, 173-177.

Boer, E. de (1983). "On active and passive cochlear models - toward a generalized analysis," J. Acoust. Soc. Am. 73, 574-576.

Brownell, W. E., Bader, C. E., Bertrand, D., and Ribaupierre, Y. de (1985). "Evoked mechanical responses of isolated cochlear outer hair cells," Science 227, 194-196.

Davis, H. (1981). "The second filter is real, but how does it work?," Am. J. Otolaryngol. 2, 153-158.

Davis, H. (1983). "An active process in cochlear mechanics," Hearing Research 9, 1-49.

Diependaal, R. J. and Viergever, M. A. (1983). "Point impedance characterization of the basilar membrane in a three-dimensional cochlear model," Hearing Research 11, 33-40.

Evans, E. F. (1974). "Auditory frequency selectivity of the auditory nerve," in Facts and Models in Hearing, edited by E. Zwicker and E. Terhardt, (Springer-Verlag, New York), 118-129.

Evans, E. F. and Harrison, R. V. (1975). "Correlation between outer hair cell damage and deterioration of cochlear nerve tuning properties," J. Physiol. 256, 43-44.

Greenwood, D. P. (1960). "Critical bandwidth and the frequency coordinates of the basilar membrane," J. Acoust. Soc. Am. 33, 1344-1356.

Hudspeth, A. J. and Corey, D. P. (1977). "Sensitivity, polarity, and conductance change in the response of vertebrate hair cells to mechanical stimuli," Proceedings of the National Academy of Science USA 74, 2407-2411.

Kemp, D. T. (1982). "Evidence of mechanical nonlinearity and frequency selective wave amplification in the cochlea," Arch. Oto-Rhino-Laryngol. 224, 37-45.

Khanna, S. M. and Leonard, D. B. G. (1982). "Basilar membrane tuning in the cat cochlea," Science 215, 305-306.

Kiang, N. Y. -S., Wanatabe, T., Thomas, E. C., and Clark, L. F. (1955). "Discharge patterns of single auditory nerve fibers in the cat's auditory nerve," Research Monograph No. 35, M. I. T. , Cambridge, MA.

Kim, D. O., Siegel, J. H., and Molnar, C. E. (1979). "Cochlear nonlinearities in two-tone responses," Scandanavian Audiology Suppl. 9, 63-81.

Kim, D. O. (1984). "Functional roles of inner and outer hair cell subsystems in the cochlea and brainstem," in Hearing Science, edited by C. I. Berlin, (College-Hill, San Diego, CA), 241-262.

Liberman, M. C. (1982). "The cochlear frequency map for the cat: Labeling auditory nerve fibers of known characteristic frequency," J. Acoust. Soc. Am. 72, 1441-1449.

Liberman, M. C. and Dodds, L. W. (1984). "Single-neuron labeling and chronic cochlear pathology. III. Stereocilia damage and alterations to threshold tuning curves," Hearing Research 16, 55-74.

Lighthill, J. (1981). "Energy flow in the cochlea," J. Fluid Mechanics 106, 149-213.

Lim, D. J. (1980). "Cochlear anatomy related to cochlear micromechanics. A review," J. Acoust. Soc. Am. 67, 1686-1695.

Neely, S. T. and Kim, D. O. (1983). "An active cochlear model showing sharp tuning and high sensitivity," Hearing Research 9, 123-130.

Peterson, L. C. and Bogert, B. P. (1950). "A dynamical theory of the cochlea," J. Acoust. Soc. Am. 22, 369-381.

Rhode, W. S. (1973). "An investigation of post-mortem cochlear mechanics using the Mössbauer effect," in Basic Mechanisms in Hearing, edited by A. R. Møller, (Academic, New York), 49-67.

Russell, I. J. and Sellick, P. M. (1978). "Intracellular studies of the hair cells in the mammalian cochlea," J. Physiol. 284, 261-290.

Sellick, P. M., Patuzzi, R., and Johnstone, B. M. (1982). "Measurement of basilar membrane motion in the guinea pig using the Mössbauer technique," J. Acoust. Soc. Am. 72, 131-141.

Sellick, D. M., Patuzzi, R., and Johnstone, B. M. (1983). "Comparison between the tuning properties of inner hair cells and basilar membrane motion," Hearing Research 10, 93-100.

Spoendlin, H. (1979). "Sensori-neural organization of the cochlea," J. Laryngol. Otol. 93, 853-977.

Strelioff, D. and Flock, Å. (1984). "Stiffness of sensory-cell hair bundles in the isolated guinea pig cochlea," Hearing Research 15, 19-28.

Voldřich, L. (1978). "Mechanical properties of basilar membranes," Acta Otolaryngol. 86, 331-335.

Weiss, T. F. (1982). "Bidirectional transduction in vertebrate hair cells: A mechanism for coupling mechanical and electrical processes," Hearing Research 7, 353-360.

Zwislocki, J. J. (1950). "Theory of acoustical action in the cochlea," J. Acoust. Soc. Am. 22, 778-784.

Zwislocki, J. J. and Kletsky, E. J. (1979a). "Tectorial membrane: A positive effect on frequency analysis in the cochlea," Science 204, 639-641.

Zwislocki, J. J. and Kletsky, E. J. (1979b). "Micromechanics in the theory of cochlear mechanics," Hearing Research 2, 505-512.

ON THE ROLE OF FLUID INERTIA AND VISCOSITY IN STEREOCILIARY TUFT MOTION: ANALYSIS OF ISOLATED BODIES OF REGULAR GEOMETRY

Dennis M. Freeman and Thomas F. Weiss
Dept. of Elec. Eng. and Comp. Sci. and Research Laboratory of Electronics
Massachusetts Institute of Technology, Cambridge, MA. 02139;
Eaton-Peabody Laboratory of Auditory Physiology
Massachusetts Eye and Ear Infirmary, Boston, MA. 02114

ABSTRACT

We assume that cochlear fluids are Newtonian and show that for physiological stimuli: fluid compressibility is negligible; convective non-linear inertial forces are small; and both viscous and linear inertial forces are appreciable. We examine the frequency dependence of viscous and inertial fluid forces that act on oscillating, isolated bodies of regular geometry. The mechanical admittance of such a body submerged in a fluid and supported by springs shows a geometry-dependent resonance. This resonance has a quality (Q_{3dB}) that is less than 1 for an infinitesimally thin plate vibrating in its plane, but can be arbitrarily large both for a sphere and for a circular cylinder oscillating in a direction perpendicular to its long axis. From considerations of hair cells with free-standing stereocilia in the alligator lizard cochlea we conclude that stereociliary tufts in the cochlea could be resonant mechanical systems.

INTRODUCTION

Stereociliary tuft displacement is the mechanical input to hair cells in the acoustico-lateralis system. In a system where we have extensive observations, the alligator lizard cochlea, stereociliary tuft motion is frequency selective and clearly contributes to the frequency selectivity and tonotopic organization observed in cochlear hair-cell and nerve-fiber responses (Weiss & Leong, 1985). To determine which mechanisms might give rise to this frequency selectivity, we have examined the theory of fluid forces on isolated vibrating bodies of regular geometry in viscous, incompressible, Newtonian fluids (Stokes, 1851). We examine the motion of such bodies suspended by ideal springs to determine if a passive, mechanical resonance is possible for stereociliary tufts.

EQUATIONS OF MOTION FOR COCHLEAR FLUIDS

We assume that cochlear fluids have a uniform density and a viscosity that is both uniform and independent of direction, i.e. we assume they are Newtonian. The equation of motion of such fluids can be described in terms of the variables listed below:

$\bar{u}$	particle velocity	cm/sec	x	distance	cm
P	pressure	dynes/cm^2	t	time	sec
ρ	density	gm/cm^3	c	speed of sound	cm/sec
μ	viscosity	gm/cm-sec	f	frequency	Hz
$\nu = \mu/\rho$	kinematic viscosity	cm^2/sec	$\omega = 2\pi f$	angular frequency	rad/sec

Fluid compressibility is negligible.

Fluid compression is one of several factors that can contribute to spatial variations of fluid velocity. If the gradient in velocity resulting from compression is small compared to the total gradient in velocity, the flow is approximately *incompressible*, and the analysis is simplified. Let U_c represent a characteristic particle-velocity difference between two points separated by L_c. The magnitude of the ratio of the compression-generated velocity gradient to total velocity gradient is less than the larger of U_c^2/c^2 and $\omega^2 L_c^2/c^2$ (Batchelor, 1967; p. 167). The length of the entire sensory epithelium in the alligator lizard (4×10^{-2} cm) can be used as a conservative value for L_c. The velocity of the basilar membrane with high intensity acoustic stimulation (100 dB SPL) is roughly 1 cm/sec (Peake & Ling, 1980) which provides a value for U_c. The physiological range of excitation frequencies extends to $\omega \approx 10^5$ rad/sec. We assume the velocity of sound in endolymph equals that in water, $c \approx 1.5\times 10^5$ cm/sec. Thus $U_c^2/c^2 \approx 4\times 10^{-11}$ and $\omega^2 L_c^2/c^2 \approx 7\times 10^{-4}$. Therefore, the compression-generated velocity gradient is likely to be less than 1/1000 of the total and cochlear fluids can be assumed incompressible.

Viscous forces and linear inertial forces are important.

The equations of motion for an incompressible, Newtonian fluid are:

$$\frac{D\bar{u}}{Dt} \equiv \frac{\partial \bar{u}}{\partial t} + \bar{u}\cdot\nabla\bar{u} = -\frac{1}{\rho}\nabla P + \nu\nabla^2\bar{u} \quad ; \quad \nabla\cdot\bar{u} = 0 . \qquad (1)$$

The expansion of the material derivative, D/Dt, includes a nonlinear term that derives from the movement of fluid *particles* relative to a frame of reference that is fixed in space. Since such motion convects momentum, $|\bar{u}\cdot\nabla\bar{u}|$ is often called the *convective nonlinearity*.

The relative magnitudes of the terms of Equation (1) can be evaluated by expressing this equation in dimensionless variables. Let U represent the peak velocity magnitude, and L represent the smallest distance over which the velocity changes significantly (e.g. by a factor of e). Scaling all velocities by U, all distances by L, and time by $1/\omega$, we define

$$\hat{u} = \bar{u}/U \quad ; \quad \hat{x} = x/L \quad ; \quad \hat{t} = \omega t . \qquad (2)$$

Equation (1) becomes

$$\frac{L^2\omega}{\nu}\frac{\partial \hat{u}}{\partial \hat{t}} + \frac{UL}{\nu}(\hat{u}\cdot\hat{\nabla}\hat{u}) = -\frac{L}{\rho\nu U}\hat{\nabla}P + \hat{\nabla}^2\hat{u} , \qquad (3)$$

where $\hat{\nabla}$ denotes derivatives with respect to the new spatial scale. Equation (3) defines three dimensionless parameters,

$$R_l = L^2\omega/\nu \quad ; \quad R_{nl} = UL/\nu \quad ; \quad P_{ref} = \rho\nu U/L , \qquad (4)$$

where P_{ref}, a scale factor for pressure, is a consequence of the choice of scale factors in Equation (2), and the two Reynold's numbers, R_{nl} and R_l, characterize the relative importance of inertia to viscosity for the nonlinear and the linear inertial terms, respectively.

When the boundary conditions are oscillatory, a spatial scale is also defined by the Stokes' boundary layer thickness

$$\delta = \sqrt{2\nu/\omega} , \tag{5}$$

whose significance is discussed below. We express R_l in terms of δ as follows:

$$R_l = 2\,(L/\delta)^2 . \tag{6}$$

The velocity gradients that result from oscillation of an isolated body in an infinite fluid typically extend over distances (L) that are comparable to δ. For such isolated bodies, R_l is neither very large nor very small and neither viscous nor linear inertial terms can be ignored. To evaluate R_{nl}, let X represent the peak fluid-particle displacement, i.e. $X = U/\omega$. For $L = \delta$,

$$R_{nl} = 2\,(X/\delta) . \tag{7}$$

Thus, for motions of isolated bodies, the magnitude of the nonlinearity in Equation (1) is proportional to the ratio of peak body displacement to boundary layer thickness. In the alligator lizard cochlea, displacements of free-standing stereocilia are estimated to be about 0.2 μm at 1 kHz for levels of about 80 dB SPL (Frishkopf & DeRosier, 1983; Holton & Hudspeth, 1983). At this frequency and level, δ is roughly 20 μm and R_{nl} is approximately 0.02. Thus there is a physiologically important range of levels for which R_{nl} is less than 0.02.

In summary, fluid compression is negligible, the convective nonlinearity is negligible except perhaps at high sound-pressure levels, and both viscous and linear inertial forces of fluid origin are appreciable. Therefore, in subsequent analyses, we have used Equation (1) and omitted the convective nonlinear term.

FLUID FORCES ON ISOLATED BODIES OF REGULAR GEOMETRY

The mechanisms by which fluids generate both inertial and viscous forces on moving bodies can be understood by considering bodies of regular geometry. We assume that the bodies are supported by massless springs as shown in Figure 1 (a). First imagine the bodies in a perfect vacuum. If the bodies are perturbed, they will oscillate forever at a natural frequency determined by their mass and the compliance of the springs. The differences in geometry of these bodies play no role in determining their motion. Next imagine the bodies immersed in a fluid that is both incompressible and inviscid. To the extent that the plate is infinitesimally thin -- the plate's motion, which is parallel to its surface, displaces no fluid. Since the fluid is inviscid, there is no frictional force on the plate. Hence, the motion of the plate is the same as in a vacuum. The cylinder however, in order to move at all, must push some fluid out of its path (Figure 1 (b)). In contrast to the plate, the cylinder experiences an inertial force due to the mass of the fluid. That is, the springs must accelerate an effective mass that consists of the mass of the cylinder plus the mass of some fluid. Motion of a sphere similarly displaces fluid, however the resulting pattern of fluid motion is different, and therefore the mass of fluid that is effectively coupled to a sphere differs from that coupled to a cylinder. Because there is no frictional force,

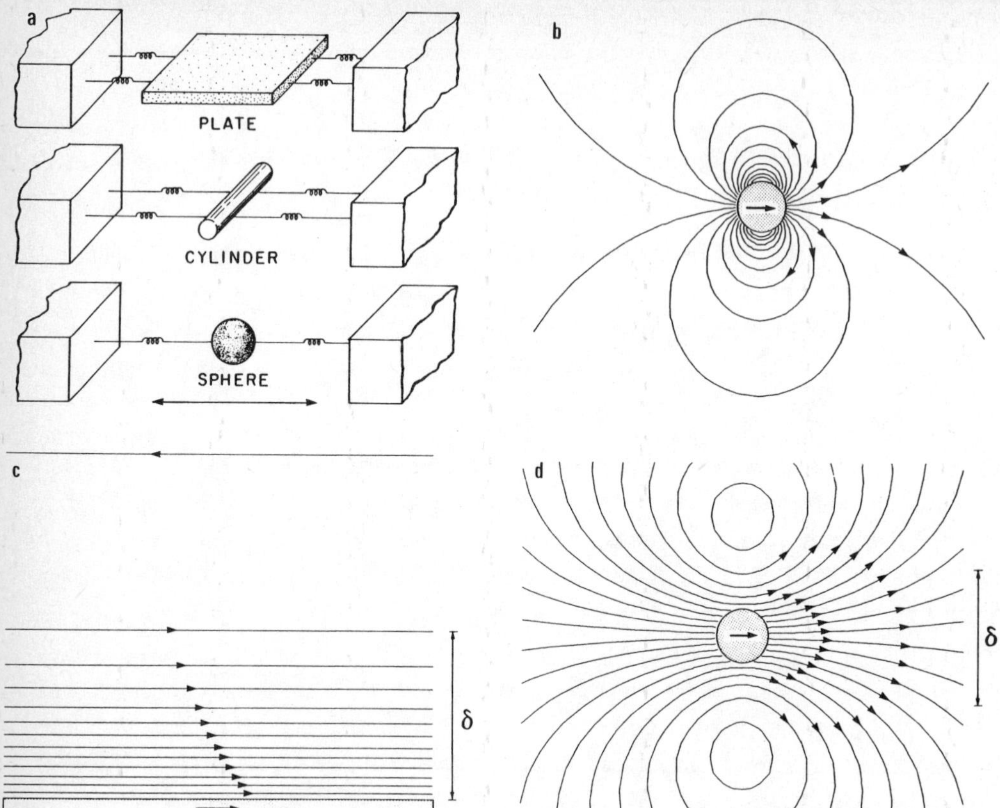

Figure 1: (a) Bodies of regular geometry -- a flat plate, a circular cylinder, and a sphere -- are connected by springs to rigid supports. Motion of these bodies in the direction of the arrow is considered when the bodies are immersed in various fluids. Streamlines (lines drawn parallel to the direction of fluid motion with a density proportional to the magnitude of the fluid velocity) depict the fluid motion that results from sinusoidal oscillation of: a cylinder in an inviscid fluid (b), a plate in a viscous fluid (c), and a cylinder in a viscous fluid (d). For clarity, the springs and supports have been omitted in (b), (c), and (d). The streamlines are obtained from exact solutions of the equations of motion for the fluid (Stokes, 1851). These streamlines change with time. The arrow on each streamline in (b), (c), and (d) indicates the direction of fluid particle motion at the time of maximum body velocity in the direction indicated by the arrow on the body. The vertical arrows in (c) and (d) indicate the thickness of the boundary layer δ.

the cylinder and sphere will oscillate forever if perturbed, but at frequencies that are lower than their resonant frequencies in a vacuum. In summary, immersing these bodies in an inviscid fluid generates a geometry dependent inertial load on the bodies.

Now consider how a viscous fluid affects the motion. When the plate is displaced (Figure 1 (c)), the fluid in its neighborhood is dragged along -- but this fluid also has inertia. Thus, while an inviscid fluid has no effect on the motion of the plate, a viscous fluid produces both viscous and inertial forces. These forces give rise to a layer of fluid, the boundary layer, that is sheared by the relative motion of the plate and the distant, motionless fluid. Thus, the boundary layer increases the effective mass of the plate, but as we shall see, the inertial and viscous forces on the plate are equal. Thus the system is highly damped so that a transient displacement of the plate is quickly damped out and a high quality resonance does not occur. With the cylinder (Figure 1 (d)) and the sphere, the boundary-layer effect is combined with the inertial effect that occurs in an inviscid fluid. Fluid near the cylinder is entrained by the viscous forces to move with the cylinder, and the motion of the resulting somewhat larger effective cylinder pushes fluid out of its path. We shall see that unlike the plate, the sharpness of the resonance for the cylinder and sphere, is not limited and is determined by their dimensions as well as by the boundary layer thickness.

Impedance of fluid load on isolated bodies of regular geometry

The forces of fluid origin that act on each of the bodies in Figure 1 can be characterized quantitatively by an impedance. For the plate, the impedance is defined as the ratio of the force exerted by the fluid on a unit area of the plate to the plate velocity, and is

$$Z_p \; = \; \mu\sqrt{\pi f/\nu}\;(1+j)\,, \tag{8}$$

where $j = \sqrt{-1}$. For a sphere of radius a, the impedance is defined as the ratio of the force exerted by the fluid on the sphere to the sphere velocity. Let $f_a = \nu/\pi a^2$, then (Stokes, 1851)

$$Z_s \; = \; 6\pi a\,\mu\;\left((1+\sqrt{f/f_a})+j\sqrt{f/f_a}\;(1+2/9\sqrt{f/f_a})\right)\,. \tag{9}$$

For a cylinder of radius a, the impedance is defined as the ratio of the force exerted by the fluid per unit length of cylinder to the cylinder velocity. No simple mathematical expression is available for Z_c, but it can be evaluated by series approximations (Stokes, 1851).

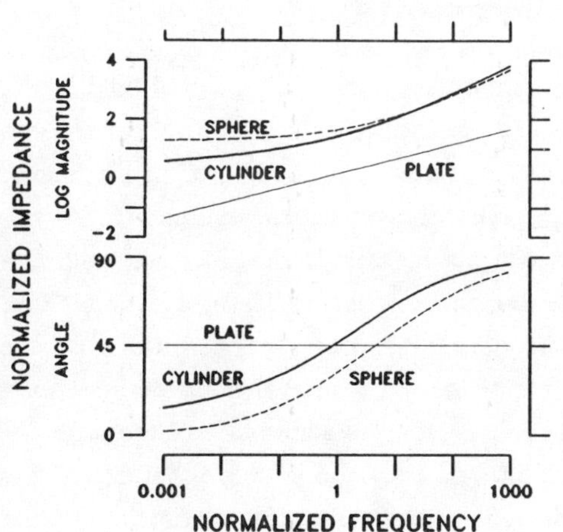

Figure 2: Normalized impedance of the fluid load on a plate, a cylinder and a sphere as a function of normalized frequency f/f_a. Since $f/f_a = (a/\delta)^2$, these plots also show the dependence of the impedance on the ratio of the radius (of the cylinder and sphere) to the boundary layer thickness. The upper panel shows the common logarithm of the impedance magnitude; the lower panel shows the angle of the impedance in degrees. In order to plot the impedance of the plate in the same coordinates as the other bodies we note that Equation (8) can be written as $(a/\mu)Z_p = \sqrt{f/f_a}\,(1+j)$. The curves labelled PLATE describe $(a/\mu)Z_p$. The curves labelled CYLINDER describe $(1/\mu)Z_c$. The curves labelled SPHERE describe $(1/(a\mu))Z_s$.

Z_p, Z_c, and Z_s have magnitudes that increase with frequency and angles that vary between 0° and 90° (Figure 2). Although these impedances have both viscous (resistance) and inertial (mass) components, none of the three is the impedance of a constant resistance in series with a constant mass. The impedance of the plate has a magnitude that increases at 10 dB/decade and an angle that is 45° for all frequencies, i.e the viscous and inertial components of the impedance have the same magnitude at each frequency and both have a magnitude that increases as $\sqrt{f}$. Therefore, as frequency increases the resistance increases while the mass *decreases*. The mass decreases because the boundary-layer thickness decreases with increasing frequency (Equation (5)) and hence the mass of fluid entrained by the plate decreases. In contrast to the plate, both the cylinder and sphere behave as constant resistances at low frequency and as constant masses at high frequencies. Consider the impedance of the sphere. At low frequencies Z_s approaches the constant resistance $6\pi a \mu$; a result known as *Stokes' law*. As frequency increases, Z_s approaches $j2\pi f\,(2/3\,\pi a^3 \rho)$ which is the impedance of a mass equivalent to half the mass of fluid that has the volume of the sphere. The impedance has an angle of 45° at a normalized frequency that equals 4.5, i.e. at a frequency for which the boundary-layer thickness equals about twice $(3/\sqrt{2})$ the radius of the sphere. The impedance of the cylinder has some similarities to that of the sphere. As frequency increases, Z_c approaches $j2\pi f\,(\pi a^2 \rho)$ (Stokes, 1851) the impedance of a mass that *equals* the mass of fluid whose volume equals that of the cylinder. For the low frequencies shown, the impedance approaches that of a resistance. The impedance has an angle of 45° at a normalized frequency of 1, i.e. at a frequency for which the boundary-layer thickness just equals the radius of the cylinder. To summarize, when the dimensions of the cylinder or sphere are much larger than the boundary layer thickness, the fluid exerts a predominantly inertial force on the body. When the boundary layer thickness exceeds these dimensions, the fluid exerts a predominantly viscous force.

Frequency-selective motion of isolated bodies of regular geometry

Motion of the bodies in Figure 1 is determined by the body masses, by forces of fluid origin, by forces from the spring attachments, and by externally applied forces. These systems can resonate. Figure 3 shows the ratio of body velocity to externally applied force (a mechanical admittance). Each of the admittances displayed in Figure 3 shows frequency selectivity. At sufficiently low frequencies in all cases, the admittance has a magnitude that increases with frequency at a rate of 20 dB/decade and an angle that approaches 90°, i.e., the admittance approaches the admittance of the spring. As frequency increases, the magnitude of the impedance of the spring decreases and the magnitude of the impedance of both the body mass and fluid load (Figure 2) increases. Hence, at high frequencies, the admittance of the plate system has a magnitude that decreases at 10 dB/decade of frequency and an angle that approaches −45°. This is the admittance of the fluid load. The situation is more complex for the cylinder and sphere. As frequency increases, there is a frequency at which the magnitude of the admittance of the spring equals that of the combined body mass and fluid load. If at that frequency, the combined (body mass and fluid load) impedance is predominantly resistive, then the system admittance is broadly tuned as occurs for the smaller bodies. Alternatively, if at that frequency the combined impedance is predominately inertial, then the system admittance shows sharp

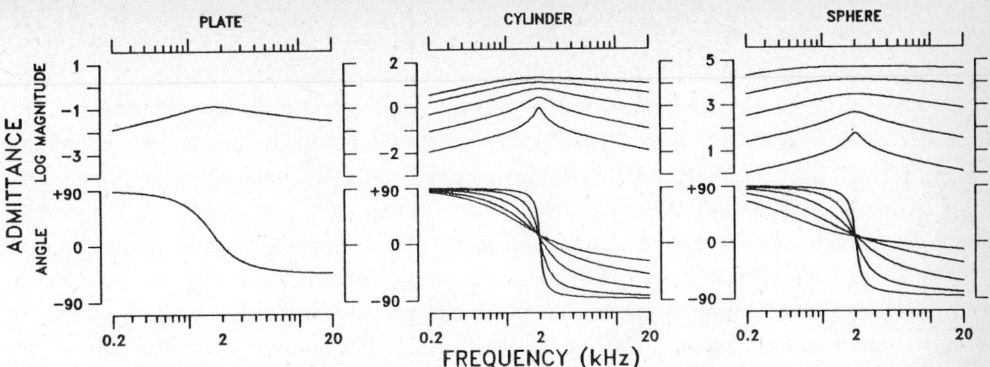

Figure 3: Frequency selectivity of spring-loaded bodies in fluid. Magnitude and angle of the: plate velocity that results from applying one unit of force per unit area of the plate; cylinder velocity that results from applying one unit of force per unit length of the cylinder; sphere velocity that results from applying one unit of force to the sphere. The different curves shown for the cylinder and sphere correspond to different radii -- 1, 3.16, 10, 31.6, and 100 μm. The smaller radii correspond to magnitudes and angles that have more broadly tuned frequency responses. The density of both the cylinder and sphere is equal to the density of the fluid. The spring constants where chosen so that the response magnitudes had peaks at 2 kHz. The spring constants were: for the plate -- 10^5 dynes/cm^3; for the cylinder (in order of increasing radius) -- 3.6×10^2, 7.9×10^2, 2.6×10^3, 1.4×10^4, and 1.1×10^5 dynes/cm^2; for the sphere (in order of increasing radius) -- 5.6×10^{-2}, 0.39, 3.6, 52, and 1.2×10^3 dynes/cm.

tuning as occurs for the larger bodies. The sharpness of tuning increases with increasing radius for both the cylinder and sphere (Figure 4). Q_{3dB} is less than 1 for the plate but exceeds 1 for cylinders with radii greater than 10 μm and for spheres with radii exceeding 18 μm.

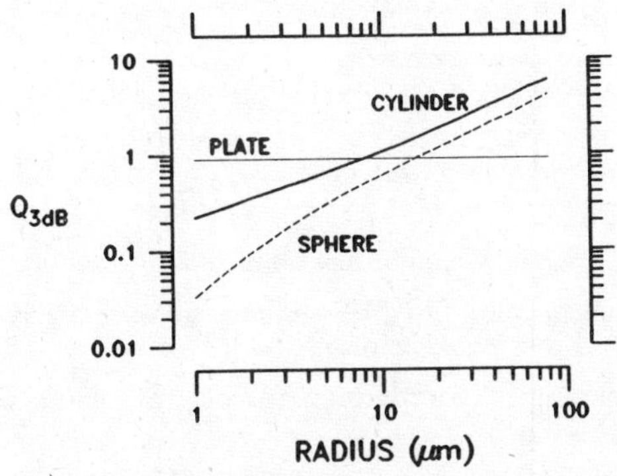

Figure 4: Quality of tuning as a function of radius. The dependence of Q_{3dB} on radius for responses shown in Figure 3 is plotted for the cylinder (thick line) and for the sphere (dashed line). Q_{3dB} for the plate is 0.9 (thin line). Q_{3dB} is defined as the ratio of the frequency of maximal response divided by the bandwidth 3 dB below the maximum admittance magnitude.

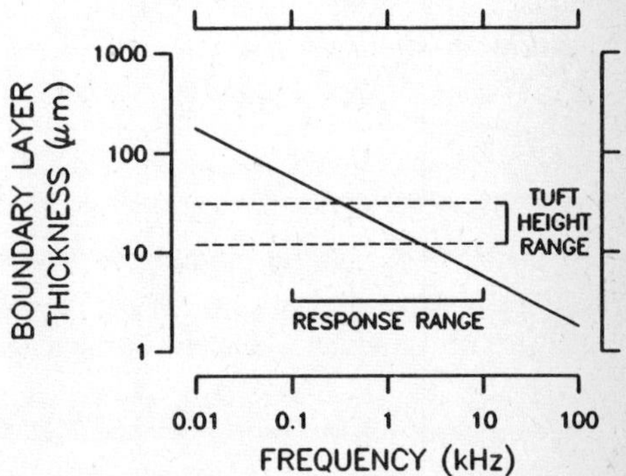

Figure 5: Boundary layer thickness δ in water as a function of frequency. Indicated are both the range of tuft heights (the length of the longest stereocilium in the tuft) for hair cells that have free-standing stereocilia in the alligator lizard cochlea (Mulroy, 1974) and the frequency range of physiological responses obtained from hair cells and cochlear neurons projecting to these hair cells (Holton & Weiss, 1983).

DISCUSSION

Newtonian fluids impose both viscous and inertial forces on isolated bodies. The magnitude of the inertial force exceeds that of the viscous force if the boundary layer thickness is smaller than the dimensions of the body (Figure 2). The potential for a high quality resonance of an isolated, spring-loaded body of arbitrary geometry can be assessed qualitatively by comparing its dimensions to the boundary layer thickness. Since tuft heights are comparable to boundary layer thickness (Figure 5), appreciable viscous and inertial fluid forces will act on stereociliary tufts. Furthermore, isolated bodies with dimensions comparable to these tufts can resonate (Figure 4) with frequency selectivities comparable to the frequency selectivity observed in responses from alligator lizard cochlear neurons (Holton & Weiss, 1983). Thus it appears possible that stereociliary tufts are mechanically resonant systems. However, the sharpness of frequency selectivity depends on geometry, and tufts differ appreciably from the isolated bodies of regular geometry that we have considered. First, tufts are not solid bodies but are composed of tens of stereocilia spaced about 1 μm apart. However, the inter-stereociliary distance is much smaller than the boundary layer thickness indicated in Figure 5. We therefore expect that viscous forces alone will tend to couple stereocilia so that they will move in unison, i.e. on fluid dynamic grounds we expect that the tuft *will* act as a rigid body. Second, tufts are not isolated bodies. They protrude from the receptor surface which has considerable spatial extent. The effect of this surface on the motion of the tufts has not been considered here. Also, the distance between tufts of neighboring hair cells is of the order of 10 μm and is thus comparable to the boundary layer thickness. Therefore, it is likely that appreciable mechanical coupling through the intervening fluid exists between neighboring tufts. If we are to further understand the mechanics of even the relatively simple free-standing stereociliary tufts, we will need to better understand both the effects of the receptor surface and of neighboring tufts on the motion of a stereociliary tuft.

ACKNOWLEDGMENTS

This work was supported by NIH grants. We thank W.T. Peake for comments on the manuscript.

REFERENCES

Batchelor, G.K. (1967): *An Introduction to Fluid Dynamics*. Cambridge Univ. Press, London.

Frishkopf, L.S. & DeRosier, D.J. (1983): Mechanical tuning of free-standing stereociliary bundles and frequency analysis in the alligator lizard cochlea. Hearing Res. *12*, 393-404.

Holton, T. & Hudspeth, A.J. (1983): A micromechanical contribution to cochlear tuning and tonotopic organization. Science *222*, 508-510.

Holton, T., & Weiss, T.F. (1983): Frequency selectivity of hair cells and nerve fibres in the alligator lizard cochlea. J. Physiol. *345*, 241-260.

Mulroy, M.J. (1974): Cochlear anatomy of the alligator lizard. Brain, Behavior, and Evolution *10*, 69-87.

Peake, W.T. & Ling, A. Jr. (1980): Basilar-membrane motion in the alligator lizard: Its relation to tonotopic organization and frequency selectivity. J. Acoust. Soc. Am. *67*, 1736-1745.

Stokes, G.G. (1851): On the effect of the internal friction of fluids on the motion of pendulums. Trans. Camb. Phil. Soc. *9*, 6-106.

Weiss, T.F. & Leong, R. (1985): A model for signal transmission in an ear having hair cells with free-standing stereocilia: III. Micromechanical stage. (Submitted for publication).

ROLE OF PASSIVE MECHANICAL PROPERTIES OF OUTER HAIR CELLS IN DETERMINATION OF COCHLEAR MECHANICS

David Strelioff
UCLA School of Medicine
Los Angeles, CA 90024

ABSTRACT

The feasibility of a resonant mechanical system in the cochlea, consisting of outer hair cells (OHCs) and the tectorial membrane (TM), playing a significant role in frequency selectivity has been investigated. As proposed initially by Zwislocki (J. Acoust. Soc. Am. 67, 1679 -1688, 1980), such a resonant OHC-TM system, with stiffness provided by the sensory hairs of the three rows of OHC and mass by the overlying TM, would be mechanically coupled to the basilar membrane (BM) via the sensory hairs and would vibrate in a plane parallel to the reticular lamina. Since the TM mass is estimated to be 1/5 - 1/4 of the mass of the organ of Corti, OHC-TM vibrations would have a significant effect upon BM frequency selectivity if OHC-TM resonant frequency along the length of the cochlea were close to the BM frequency at the same location.

Determination of the length and stiffness of OHC sensory hairs in the isolated guinea pig organ of Corti (Strelioff and Flock, Hearing Res., 15, 19-28, 1984) has made possible quantitative computations of OHC-TM resonant frequencies for the simple case where the effects of the TM attachement to the limbus are neglected. Computed OHC-TM resonant frequencies, assuming no variation in TM mass along the length of the cochlea, range from 1.2 kHz at the apex to 22 kHz at the base. These computed frequencies, based upon data from in vitro preparations, are sufficiently close to BM frequencies to demonstrate the feasibility of Zwislocki's proposal. As more data on the mass and elasticity of the TM and the characteristics of its attachment to the limbus become available, it will be possible to more accurately evaluate the role of OHC-TM micromechanics in BM frequency selectivity.

1. INTRODUCTION

The micromechanical properties of the structures that make up the organ of Corti have received a great deal of attention since the time that Mössbauer techniques made possible the measurement of basilar membrane (BM) motion at physiological stimulus levels (Johnstone et al, 1967; Rhode, 1971). These measurements demonstrated that the basilar membrane is much more sharply tuned than had been assumed on the basis of von Békésy's measurements (1960) and raised the possiblity that the sharp tuning observed in auditory-nerve fiber responses (Kiang et al, 1965) was due to the mechanical properties of the cochlea rather than to more central filtering mechanisms (e.g., Evans and Wilson, 1973). Some of the most recent theories regarding frequency implicate the tectorial membrane (TM) and the outer hair cells (OHCs) (Zwislocki, 1980; Allen, 1980; Turner and Nielsen, 1983; Strelioff et al, 1985).

Initially, when deflection of stereocilia on cochlear hair cells was first attributed to shearing forces between the cuticular plate and the TM, the mechanical properties of both the TM and the stereocilia were not considered in detail (Davis, 1958). The TM was regarded as a relatively stiff plate that pivoted about its attachment to the limbus and easily bent the stereocilia during displacement of the BM. No radial movement of the TM was considered. Some elastic properties of the TM have been investigated by von Békésy (1960) and Frommer (1982). Their findings indicate that the TM stiffness, for compression in the direction perpendicular to the TM surface, varies from about 1×10^2 to 1×10^4 dyne/cm. Although no data are available for TM stiffness for compression and elongation in the direction parallel to the TM surface, it is probable that it is of the same order of magnitude.

Recently, it has been shown that stereocilia act as stiff levers which pivot at their elastic attachements to the cuticular plates of hair cells (Flock, 1977; Flock and Strelioff, 1984). In addition, it has been shown that the combined stiffness of all the stereocilia in a third-row OHC sensory-hair bundle increases from about 0.8 dyne/cm at the apex to an estimated 1590 dyne/cm at the base (Strelioff and Flock, 1984). This stiffness gradation of OHC sensory hairs is consistent with the suggested role of OHCs in influencing BM tuning properties. The feasibility of such a role for OHCs has been investigated by determining, as a function of distance from the stapes, the natural frequency of a resonant mechanical system consisting of the mass of the TM and the stiffness of the OHC sensory hair bundles.

2. METHODS

The OHC-TM system was modeled as a mass resting on top of vertical rods which pivot about their elastic attachments to the apical surfaces of OHCs (Fig. 1). Due to its attachment to each row of OHCs, the TM is

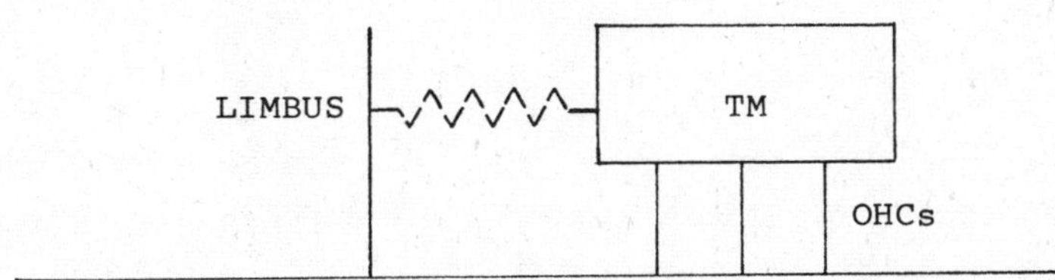

Figure 1. Schematic drawing of OHC-TM system illustrating the TM and OHC sensory hairs. The compliant attachment of the TM to the limbus is shown as a dashed spring.

confined to move parallel to the surface of the reticular lamina. It was assumed that the attachment to the limbus was sufficiently compliant so that its stiffness could be neglected in the calculations of resonant frequency. The inertial mass of the stereocilia was assumed to be negligible in comparison to that of the TM. Since some of the fluid within the subtectorial space would move with the TM and the stereocilia, one-half of the mass of this fluid is added to the TM mass. Finally, the damping that is undoubtedly present in the fluid movement and TM deformation is not included because active processes, most likely within the OHCs (e.g. Neely and Kim, 1983), probably provide sufficient energy input into the system to cancel such damping effects. Thus, the natural frequency, $f_o(d)$, of the OHC-TM system, as a function of distance, d, from the stapes, is given by the equation

$$f_o(d) = \frac{[S(d)/m]^{1/2}}{2\pi}$$

where S(d) is the combined stiffness of the three rows of OHC sensory hairs and m is the mass of the TM. Both S(d) and m are for a basilar membrane length equivalent to the diameter of one OHC (i.e. 10 μm).

The combined stiffness of the three rows of OHCs, S(d), was determined by adding together the stiffness of each row of OHCs as computed from the best-fit lines to the experimentally measured hair-bundle stiffness in the guinea-pig cochlea (Strelioff and Flock, 1984, Fig. 3, Table IV). The measured stiffness was the force required to linearly displace part of the hair bundle by a maximum distance of 1.0 μm. Since the data given in that table were for deflection of only a part of the hair bundle, the values were multiplied by a correction factor of 5.6 in order to determine the stiffness of the entire hair bundle. The stiffness, $S_n(d)$, at a distance, d, from the stapes, of entire hair bundles in the nth row was computed as

$$S_n(d) = S_o \exp(-k_s d)$$

where $S_o(d)$ is the intercept and k_s is the exponent of the best-fit line (Fig. 2, top). The intercepts and the exponents of the equations were 358, 230, and 1590 dyne/cm and 0.264, 0.230, and 0.406 mm^{-1} for OHC rows 1, 2, and 3, respectively.

Since reliable estimates of the width and thickness of the TM as a function of distance from the stapes are not available and the density of the TM is unknown, it was necessary to estimate these values. For the present calculations, the estimate of 1.1×10^{-7} gm per 10 μm length (e.g., Zwislocki, 1980a) and no variation with distance from the stapes was used for the computations.

3. RESULTS AND DISCUSSION

The computed natural frequency of the TM-OHC system increases by a factor of 18.3 from 1.2 kHz at the apex to 22 kHz at the base of the cochlea (Fig. 2, bottom). When compared with the measured resonant frequency of the guinea-pig basilar membrane in *in vitro* preparations (von Békésy, 1960), the BM resonant frequency increases by a factor of 171 from 0.13 kHz at the apex to 22.8 kHz at the base. Frequency estimates in *in vivo* preparations (Wilson and Johnstone, 1972; Dallos, 1973) are are about an octave higher and parallel to those of von Bekesy. It can be seen that the computed frequencies do not change as much with location as the measured ones and that they are higher than the measured values in the apical regions of the cochlea. Considering the numerous approximations that were required for the computations, it can be concluded that the agreement between the computed TM-OHC frequencies and the measured BM frequencies is close enough to make effective mechanical coupling between the TM-OHC system and the basilar membrane very feasibl (Zwislocki, 1980b).

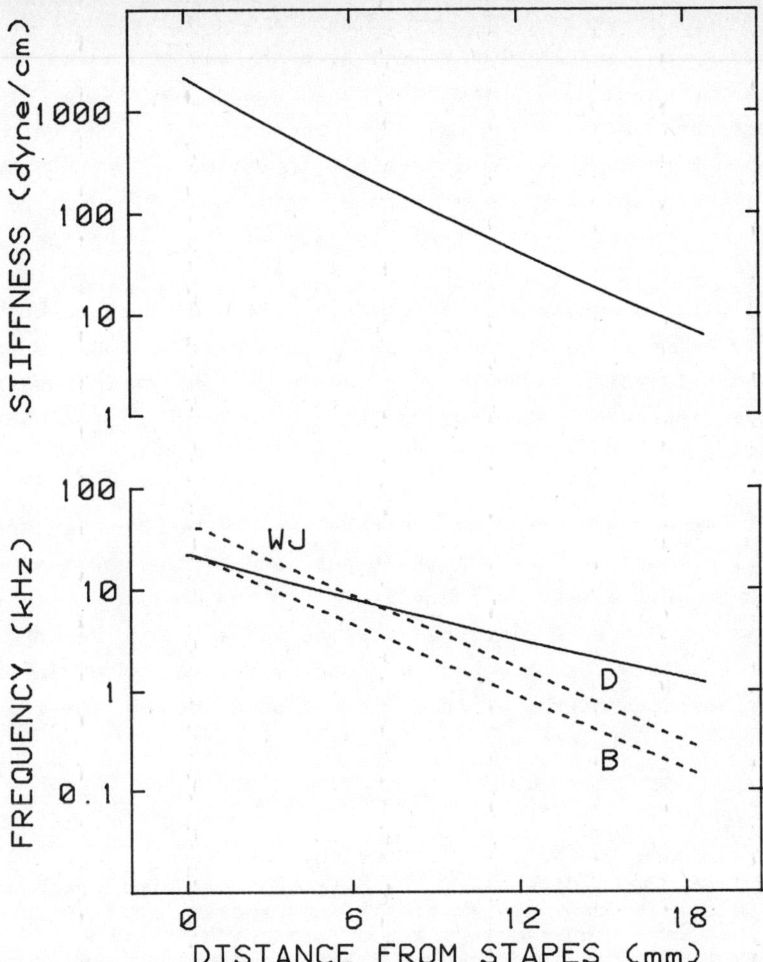

Figure 2. Top: Combined stiffness of all three rows of OHCs as a function of distance from the stapes. Bottom: Comparison of computed resonant frequency of the OHC-TM system (solid line) and measured best frequency for basilar membrane motion (dashed lines). Upper dashed line (B) is based upon von Békésy's (1960) _in vitro_ data. Lower dashed line is based upon capacitive probe measurements (WJ) by Wilson and Johnstone (1972) and cochlear microphonic measurements (D) by Dallos (1973) on _in vivo_ preparations.

The reasons for discrepancy between computed and measured values are (i) inaccuracy of the TM dimensions and density, (ii) inaccuracy of the sensory hair bundle stiffness data, and (iii) difference between _in vivo_ and _in vitro_ prepartions. Including an increase in TM mass with distance from the stapes would have increased the computed frequency range by a factor equal to the square root of the TM mass increase. An increase in TM mass by a factor of 90 would be required to produce the correct frequency variation. Due to the difficulties in making the measurements, the hair-bundle stiffness data have a considerable amount of scatter. Furthermore, since the measurements were made for hair-bundle deflections exceeding the physiological range, the values may not accurate. At present, it is impossible to estimate how these measurement errors may have affected the stiffness estimates. Since it has been demonstrated that efferent stimulation affects the mechanical properties of cochlear partition (Brown et al, 1983; Mountain, 1980; Siegel and Kim, 1982), it is possible that in vitro stiffness measurements may not accurately reflect conditions in the living animal. Thus, when some or all of these factors are taken into consideration, it is possible that much closer agreement between computed and measured frequencies may be found.

It is essential that more accurate data be obtained on the micro-mechanical properties of the TM. In particular, the mass and elasticity of the TM as well as the characteristics of the TM attachment to the limbus, as functions of distance from the stapes, are required before models of the TM-OHC system can be properly tested. Such information on passive properties of the organ of Corti could then be applied to the study of models with active properties.

ACKNOWLEDGEMENTS

I wish to thank Åke Flock for his contributions to all aspects of this work and, in particular, for initiating the development of the isolated organ of Corti preparation. The experimental data used in these computations were acquired in collaboration with Åke Flock at the Karolinska Institutet in Stockholm. Research support was provided by NIH grants NS 00255, NS 08193, and NS 09823 and by NSF grant BNS 84 08486.

REFERENCES

Allen, J.B., "A Cochlear Micromechanical Model of Transduction." In Psychological, Physiological and Behavioral Studies in Hearing, G. van den Brink and F.A. Bilsen, Eds., Delft University Press, The Netherlands, pp. 85-95, 1980.

Békésy, G. von, Experiments in Hearing, McGraw-Hill, New York, 1960.

Brown, M.C., Nuttall, A.F., and Masta, R.I., "Intracellular Recording from Cochlear Inner Hair Cells: Effects of Stimulation of the Crossed Olivocochlear Efferents." Science 222, pp. 69-72, 1983.

Dallos, P., "Cochlear Potentials and Cochlear Mechanics." In Basic Mechanisms in Hearing, A.R. Moller, Ed., Academic Press, New York, pp. 335-372, 1973.

Davis, H., "Mechanisms of the Inner Ear." Ann. Otol. Rhinol. Laryngol. 67, pp. 644-655, 1958.

Evans, E.F. and Wilson, J.P., "The Frequency Selectivity of the Cochlea." In Basic Mechanisms of Hearing, A.R. Møller, Ed., Academic Press, New York, pp. 519-554, 1973.

Flock, Å., "Physiological Properties of Sensory Hairs in the Ear." In Psychophysics and Physiology of Hearing, E.F. Evans and J.P. Wilson, Eds., Academic Press, London, pp. 15-25, 1977.

Flock, Å. and Strelioff, D., "Studies on Hair Cells in Isolated Coils from the Guinea Pig Cochlea." Hearing Res. 15, pp. 11-18, 1984.

Frommer, G.H., "Observations of the Organ of Corti under In Vivo-like Conditions." Acta Otolaryngol 94, pp. 451-460, 1982.

Johnstone, B.M. and Boyle, A.J.F., "Basilar Membrane Vibration Examined with the Mössbauer Technique." Science 158, pp. 389-390, 1967.

Kiang, N.Y-S., Watanabe,T.,Thomas, E.C., and Clark, L.F., Discharge Patterns of Single Fibers in the Cat's Auditory Nerve, M.I.T. Research Monograph No. 35, Technology Press, Cambridge, 1965.

Mountain, D.C. "Changes in Endolymphatic Potential and Crossed Olivocochlear Bundle Stimulation Alter Cochlear Mechanics." Science 210, pp. 71-72, 1980.

Neely, S.T. and Kim, D.O., "An Active Cochlear Model Showing Sharp Tuning and High Sensitivity. Hearing Res. 9, pp. 123-130, 1983.

Rhode, W.S., "Observations of the Vibration of the Basilar Membrane in Squirrel Monkeys Using the Mössbauer Technique." J. Acoust. Soc. Amer. 49, pp.1218-1231, 1971.

Siegel, J.H. and Kim, D.O., "Efferent Neural Control of Cochlear Mechanics? Olivocochlear Bundle Stimulation Affects Cochlear Biomechanical Nonlinearity." Hearing Res. 6, pp. 171-182, 1982.

Strelioff, D. and Flock, Å., "Stiffness of Sensory-cell Hair Bundles in the Isolated Guinea-Pig Cochlea." Hearing Res. 15, pp. 19-28, 1984.

Strelioff, D., Flock, Å., and Minser, K.E., "Role of Inner and Outer Hair Cells in Mechanical Frequency Selectivity of the Cochlea." Hearing. Res., Accepted for Publication.

Turner, R.G., Jr. and Nielsen, D.W., "A Simple Model of Cochlear Micromechanics in the Mammal and Lizard." Audiology 22, pp. 545-559, 1983. 22, 545-559.

Wilson, J.P. and Johnstone, J.R., "Capacitive Probe Measures of Basilar Membrane Vibration." Symposium on Hearing Theory 1972, IPO Eindhoven, Holland, pp. 172-181, 1972.

Zwislocki, J.J., "Five Decades of Research on Cochlear Mechanics." J. Acoust. Soc. Am. 67, pp. 1679-1688, 1980a.

Zwislocki, J.J., "Two Possible Mechanisms for the Second Cochlear Filter." In Psychological, Physiological and Behavioral Studies in Hearing, G. van den Brink and F.A. Bilsen, Eds., Delft University Press, The Netherlands, pp.16-33, 1980b.

THRESHOLDS OF AUDITORY SENSITIVITY AND AUDITORY FATIGUE: RELATION WITH COCHLEAR MECHANICS.

A. DANCER and R. FRANKE
French-German Research Institute of Saint-Louis
68301 SAINT-LOUIS, France

P. CAMPO
I.N.R.S. Avenue de Bourgogne B.P. 27
54500 VANDOEUVRE, France

ABSTRACT

The shape of the curve of the auditory sensitivity thresholds as a function of frequency is determined by the transfer function of the outer and middle ear. For a constant acoustic signal at the input to the cochlea the auditory sensitivity threshold is nearly constant. From theoretical results, it seems that the threshold of the neural responses appears always for the same velocity of the cochlear partition all along the cochlea.

Concerning the auditory fatigue, we applied to the guinea pig's cochlea a constant acoustic energy (pure tones of frequency from 2 kHz to 11,3 kHz) during 20 minutes and we measured the TTS by electrocochleography twenty minutes after the end of the exposure (from 2 to 32 kHz by half-octave steps). For each animal, the level of the stimulus at the input to the cochlea was adjusted by means of cochlear microphonic measurements.

a) In order to obtain a constant TTS for each frequency of stimulation, the acoustic level at the input to the cochlea must decrease by about 4 dB/octave.

b) A constant TTS at 16 kHz and at 22.6 kHz, whatever the stimulation frequency is, is obtained for an acoustic level at the input of the cochlea decreasing by about 10 dB/octave.

Given the facts that, for a constant acoustic input to the cochlea :

- the amplitude of the displacements of the cochlear partition near the CF doubles each time that the frequency is halved and consequently the velocity of the cochlear partition is constant,
- the amplitude of the displacements of the cochlear partition at the base of the cochlea is constant for frequencies lower than CF and consequently the velocity of the cochlear partition at the base doubles with the frequency,

the difference of slope between case a) and b), i.e. 6 dB/octave seems to indicate that the velocity of the cochlear partition is an important parameter which plays a significant role in the coming out of the auditory fatigue.

1. INTRODUCTION

The input signal to the cochlea is constituted by the pressure variations produced in the basalmost part of scala vestibuli by the stapes displacements in the oval window.

These pressure variations depend on the characteristics of the acoustic stimulus at the entrance to the ear and on the transfer function of the external and the middle ear (taking into account the cochlear input impedance).

a) Hearing sensitivity

It is known that in the cat (Dallos 1973) and in the guinea pig (Fig. 1) for example, the shape of the hearing sensitivity curve is determined by the shape of the transfer function which connects the pressure signal in the free field to the acoustic input signal to the cochlea. Thus, for a constant sound level versus frequency at the input to the cochlea, the hearing threshold would be constant too.

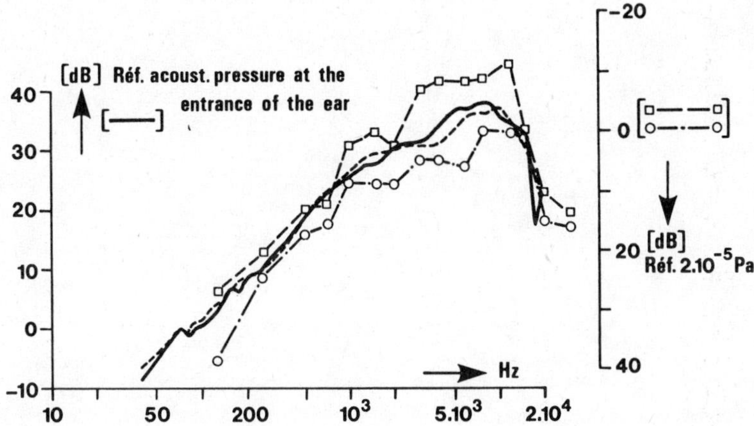

Fig. 1.- Comparison of : a) the auditory sensitivity in albino (o—·—o) and pigmented (□——□) guinea pigs (Prosen et al., 1978) and : b) the differential cochlear microphonic (-----, arbit. ref.) and the differential acoustic pressure in the first turn of the cochlea (———) (Dancer and Franke, 1980).

According to theoretical studies (Zwislocki, 1948; Peterson and Bogert, 1950), it seems that near the point of maximum amplitude of the traveling wave, the lower the frequency is, the larger are the displacements. Halving the frequency induces approximately a doubling of the amplitude of the displacement.

This suggests that the excitation threshold of the hair cells is attained for the same velocity of the cochlear partition all along the cochlea.

It can be seen from the work of Johnstone et al. (1982)(fig. 2) that the isovelocity curve of a given point of the cochlear partition and the tuning curve of a single unit connected to this point are superposable. This can be observed whether the active processes are acting or not.

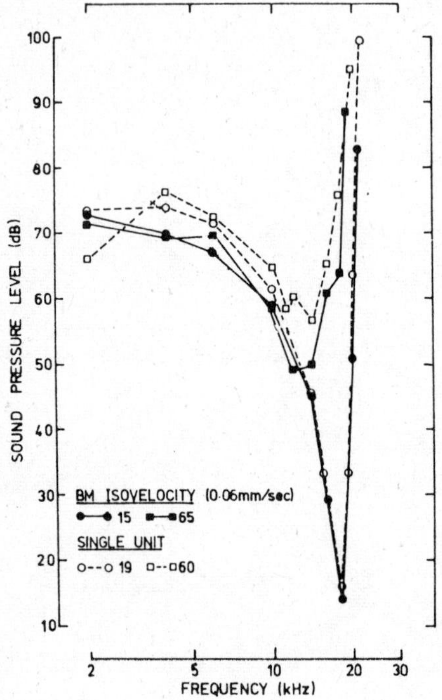

Fig. 2.- Comparison of neural and basilar membrane tuning curves (Johnstone et al., 1982).

b) TTS

According to Liberman (1982), the pure TTS would be associated to a vacuolization of the synaptic pole of the inner hair cells. Thus, this TTS would be mainly related to a metabolic exhausting.

In this respect : could the TTS be correlated to the same parameter than the one responsible for the initiation of the responses, i. e. the velocity of the cochlear partition ?

2. EXPERIMENTAL RESULTS

In order to test this hypothesis we realized the following experiment : pure tones from 2 to 11.3 kHz with half octave steps were applied to guinea pigs ears. The level of the signal entering the cochlea was kept constant versus frequency, by means of a compensation based of the recording of the cochlear microphonic response in each animal. The TTS was measured by electrocochleography using a round window electrode, 20 minutes after the end of the exposure.

Fig. 3 shows the TTS observed after a 20 minutes exposure to a 5.6 kHz pure tone, at different levels (reference : arbitrary value for the input signal to the cochlea). The maximum amplitude of the TTS corresponds to a frequency located half an octave higher than the stimulus frequency. If the amplitude of stimulation rises beyond a given level, relevant TTS appear in the high frequency range.

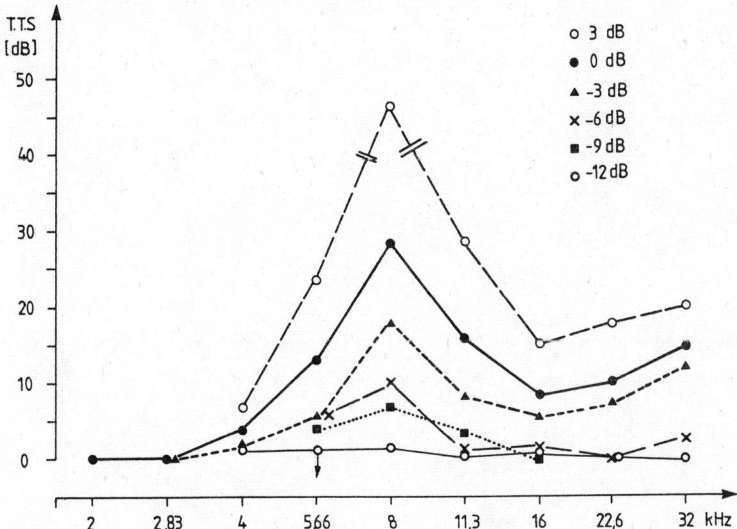

Fig. 3.- TTS measured 20 minutes after a 5.6 kHz exposure at different stimulation levels.

The stimulating levels needed to obtain a mean maximum TTS of 10 dB and 25 dB have been represented in fig. 4a. For a given value of mean TTS, the higher the frequency is, the lower is the level at the input of the cochlea. The slope of the least squares lines is about -4 dB/octave.

Fig. 4b shows, for each stimulating frequency, the levels needed to obtain a mean TTS of 5 to 11 dB at 16 kHz and of 9 to 15 dB at 22.6 kHz. As previously, the higher the stimulating frequency is, the lower is the required level. The slope of the least squares lines is about -10 dB/octave.

3. DISCUSSION

Let us consider at first the TTS observed at 16 and 22.6 kHz. The electrocochleographic responses corresponding to those frequencies originate from the base of the cochlea (fig. 4b). The traveling waves generated by fatiguing sounds of frequencies ranged between 2 and 11.3 kHz sweep in this part of the cochlea before reaching their maximum amplitude in the CF region. For a constant input signal to the cochlea, the basal part of the cochlear partition is submitted to a constant amplitude displacement for all frequencies lower than CF. This constant displacement induces a velocity which doubles for each doubling of the frequency (+6dB/octave).

On the other hand, for a constant input signal to the cochlea and at a cochlear location near CF, it is the velocity which remains constant all along the cochlear partition (in the same way as for the auditory sensitivity thresholds).

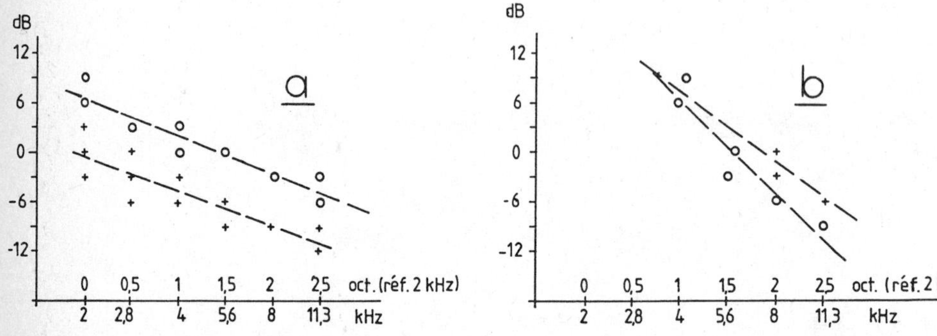

Fig. 4.- Iso-TTS as a function of level and stimulation frequency :
a) TTS max = 10 ± 6 dB (r^2 = 0.77; -4.2 dB/octave) + — +
TTS max = 25 ± 6 dB (r^2 = 0.90; -4.6 dB/octave) o — o o
b) TTS =12 ± 3 dB at 22.6 kHz (r^2 = 0.88; -9.2 dB/octave) + — + +
TTS = 8 ± 3 dB at 16 kHz (r^2 = 0.90; -11.3 dB/octave) o — o o

Since in the first case (basal zone) the velocity increases by 6 dB/octave (for frequencies lower than CF) whereas it remains constant in the second case (CF region), it seems logical that the stimulating levels needed to obtain iso-TTS curves show a slope difference of -6 dB/octave. This hypothesis seems to be confirmed by our results (fig. 4a and 4b).

With regard to the simple hypothesis issuing from the discussion about the auditory sensitivity levels, which implies that the TTS would be in relation only with the velocity of the cochlear partition, we observe a slope difference of 4 dB/octave. As a matter of fact, if the velocity were the only parameter coming into play, the observed iso-TTS curves would have :

- zero slope in the case of CF region
- -6 dB/octave slope in the case of the basal zone (for frequencies lower than CF)

instead of -4 and -10 dB/octave slopes.

The origine of the additional 4 dB/octave slope could be related to the number of stimulating periods, which for a constant exposure duration, is proportional to the frequency (6dB/octave).

4. CONCLUSION

As for the auditory sensitivity thresholds, it seems that the velocity of the cochlear partition plays a decisive role in the origin of the TTS. Nevertheless, this parameter cannot explain the whole phenomenon. Bioelectrical and metabolic processes are certainly also implicated.

REFERENCES

Dallos, P., "The auditory periphery." Academic press, New York and London, pp. 117 - 126, 1973.

Dancer, A. and Franke, R., "Intracochlear sound pressure measurements in guinea pigs." Hearing Research 2, pp.191 - 205, 1980.

Johnstone, B. M., Robertson, D. and Cody, A., "Basilar membrane motion and hearing loss." in : Hearing and hearing prophylaxis, Borchgrevink ed., Scand. Audiol. Suppl. 16, pp 89 - 93, 1982.

Liberman, M. J. and Mulroy, M. J., "Acute and chronic effects of acoustic trauma : Cochlear pathology and auditory nerve pathophysiology." in : New Perspectives on Noise-induced Hearing Loss, ed. by R. P. Hamernik, D. Henderson, R. Salvi, Raven Press, New-York, pp. 105 - 135, 1982.

Peterson, L. C. and Bogert, B. P., "A dynamical theory of the cochlea." J. Acoust. Soc. Am. 22, pp. 369 - 381, 1950.

Prosen, C. A., Petersen, M. R., Moody, D. B. and Stebbing, W. C., "Auditory thresholds and kanamycin-induced hearing loss in the guinea pig assessed by a positive reinforcement procedure." J. Acoust. Soc. Am. 63, pp. 559 - 566, 1978.

Zwislocki, J., "Theorie der Schneckenmechanik." (Thesis), Acta otolaryngol. Suppl. LXXII, pp. 50 - 51, 1948.

ANALYSIS OF STREAMING FLOW INDUCED IN THE TECTORIAL GAP

C. R. Steele and D. H. Jen

Division of Applied Mechanics, Stanford University, Stanford, CA 94305

ABSTRACT

Experiments with physical models and hair cell damage data indicate that a significant steady flow issues from the spiral sulcus under high intensity sound conditions. A mathematical model describing the sulcus streaming flow induced by waves travelling along the basilar membrane was formulated. Preliminary results obtained for the guinea pig cochlea reveal that the streaming flow through the tectorial gap is significantly sharper than the corresponding basilar membrane displacement envelope. The steady flow passes through the inner hair cell row; this suggests that streaming is a possible passive filtering mechanism.

INTRODUCTION

Steele (1973) pointed out the possibility that steady fluid motion in the tectorial gap induced by the travelling waves in the basilar membrane (BM) could act as a second filter at high sound intensity. Evidence that such flow exists comes from the results of Spoendlin (1971), who showed that damaged hair cells invariably turn away from the sulcus. This result is supplemented by data gathered in a physical model constructed by Cancelli *et al.* (1984), who observed substantial steady flow out of the sulcus.

The streaming flow in the sulcus is due to travelling waves and thus are similar to von Békésy eddies. Since the BM is linked to the organ of Corti (OC), and the OC is a sulcus boundary, the BM motion also induces streaming in the sulcus. Unlike scala streaming flow, the sulcus fluid can leak through the sensing inner hair cell (IHC) row and into the tectorial gap. This leakage flow is of primary interest.

The proposed model of tectorial gap leakage due to BM motion is shown in Fig. 1. The displacement of the OC tip is calculated from the BM envelope. It becomes a boundary condition in the sulcus model, which is based on the von Békésy eddy solution of Hallauer (1974). The boundary condition is modified for leakage using a model of the gap with IHC resistance replaced by equivalent depth.

A computer program was written to evaluate the Fourier series solution for streaming flow in the guinea pig cochlea. BM displacement envelopes were provided by Steele and Zais (1983) and Steele and Taber (1981).

A formulation of the model follows, including a rederivation of Hallauer's (1974) model for Cartesian coordinates. The results show streaming envelopes which are significantly sharper than the BM envelopes and compare this effect for various frequencies. The effect of tectorial gap height on sharpness and localization is also discussed.

MODEL DESCRIPTION

The first motion to be modelled is the transverse displacement of the OC tip due to BM motion. A kinematic analysis was done by Steele (1973), who used experimental data to establish an elastic deformation model of the OC. The present model assumes the OC to be approximately rigid and without longitudinal coupling so that $u_{oc} \propto w_{bm}$.

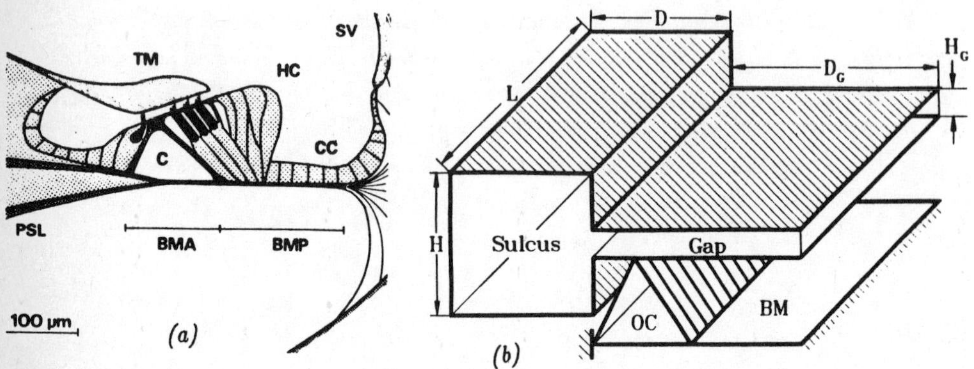

Fig. 1. (a) OC, sulcus apparatus of P. Duprasi *(from Bruns (1979))*
(b) Box model of sulcus and tectorial gap

The induction of streaming by a travelling wave is covered by Riley (1967). Hallauer's (1974) solution can be modified to model the sulcus streaming. The time-averaged Navier-Stokes equations in terms of dimensionless variables p=pressure and $\mathbf{v}$=velocity are

$$\nabla p - \frac{\pi^2 \nu}{H^2 \omega} \nabla^2 \mathbf{v} = \epsilon(-\frac{1}{2}\nabla \mathbf{v}^2 + \mathbf{v} \times \nabla \times \mathbf{v}), \tag{1}$$

where H=sulcus height, ω=frequency, and ν=kinematic viscosity form a dimensionless group. The variables $\mathbf{v}$ and p are expanded in a regular perturbation for small $\epsilon = \pi u_{oc}/H$:

$$(\mathbf{v}, p) = (\mathbf{v}_0, p_0) + \epsilon(\mathbf{v}_1, p_1) + O(\epsilon^2). \tag{2}$$

In this problem $H = 0.1mm$ and $u_{oc} \approx w_{bm} < 1\mu$ for even very high intensity sound. Thus, $\epsilon \approx .03$ and is indeed quite small. After substituting the expansions, the $O(\epsilon^0)$ terms have zero time average. We can split $\mathbf{v}_0 = \nabla\phi_0 + \nabla \times \underline{\psi}_0$, whereupon the $\underline{\psi}_0$ is negligible in the creeping flow region outside the boundary layer. The time average of the $O(\epsilon^1)$ terms is then

$$\frac{\pi^2 \nu}{H^2 \omega} \nabla^2 \mathbf{v}_1 = \nabla(p_1 + \frac{1}{2}(\nabla\phi_0 \cdot \nabla\phi_0)^s) \equiv \nabla P_1, \tag{3}$$

where s denotes the steady component. These equations are simplified by assuming parabolic profiles over the height H (for $Q = (\mathbf{v}_1, P_1)$),

$$Q(x, y, z) = \frac{3}{2}\bar{Q}(x, y)[1 - (\frac{2z}{H})^2]. \tag{4}$$

This leads to (with $\bar{\nabla} \equiv \mathbf{e}_x \frac{\partial}{\partial x} + \mathbf{e}_y \frac{\partial}{\partial y}$):

$$\bar{\nabla}^2 \bar{\mathbf{v}}_1 - \frac{12}{H^2} \bar{\mathbf{v}}_1 = \frac{H^2 \omega}{\pi^2 \nu} \bar{\nabla} \bar{P}_1. \tag{5}$$

From continuity ($\bar{\nabla} \cdot \bar{\mathbf{v}}_1 = 0$), we can take $\bar{\mathbf{v}}_1 = \bar{\nabla} \times \psi \mathbf{e}_z$. By taking the curl of Eqn. (5) and substituting, we obtain

$$\bar{\nabla}^4 \psi - \frac{12}{H^2} \bar{\nabla}^2 \psi = 0, \tag{6}$$

which is the well-known Hele-Shaw model equation.

The boundary conditions result from transferring the boundary layer solution of Eqn. (1) to the x-axis. From Hallauer (1974), the conditions are:

$$\frac{\partial \psi}{\partial y})_{y=0} \equiv U_H = \omega k(x) u_{oc}^2(x), \tag{7}$$

$$\psi)_{y=0} \equiv \psi_H = \omega u_{oc}^2(x) \left(\frac{1}{2\pi^2} + \frac{1}{16} \frac{q}{\sqrt{1+q^2}} + \frac{q}{\pi^2} \sum_{n=1}^{\infty} \frac{1}{(4n^2-1)^2 \sqrt{4n^2+q^2}} \right), \tag{8}$$

where $q = \frac{xk(x)}{T}$, x is the axial coordinate, ω is the frequency, $k(x)$ is the local wavenumber, and $u_{oc}(x)$ is the OC tip displacement envelope. As in Hallauer (1974), $T = 350$ is a BM taper parameter. Eqn. (6) with boundary conditions Eqns. (7) and (8) at the gap, and $\mathbf{v} = 0$ on the top and sides comprise the Hallauer model.

In order to apply the Hallauer model to the sulcus, leakage must be allowed on the lower boundary. The hair cell rows are replaced with equivalent additional gap depth to equate flow resistance, as in Frommer (1977). Poiseuille flow is assumed in the gap since the uniform gap height is much smaller than its other dimensions (Fig. 2). So assuming $D_G \ll L$, we can take

$$v_{lkg}(x) = -\frac{H_G^2}{12 \mu D_G} p(x) \tag{9}$$

where $p(x)$ is the local pressure, $v_{lkg}(x)$ is the gap leakage, and μ is the viscosity.

It is thus necessary to reconcile the boundary conditions of the Hallauer model and the gap leakage effect. Since the Hallauer model applies to creeping flow outside the boundary layer, the stream function identity $v = -\frac{\partial \psi}{\partial x}$ does not represent the leakage at the true boundary. However, since pressure changes very little across a boundary layer, we can take v_{lkg} as in Eqn. (9). This leakage must be added to the local v at the lower boundary (assuming a thin boundary layer), so that the boundary conditions become:

$$u)_{y=0} = U_H(x) \tag{10}$$

$$v)_{y=0} = -\frac{\partial \psi_H}{\partial x} - \frac{H_G^2}{12\mu D_G} p(x). \tag{11}$$

The final system model is shown in Fig. 3.

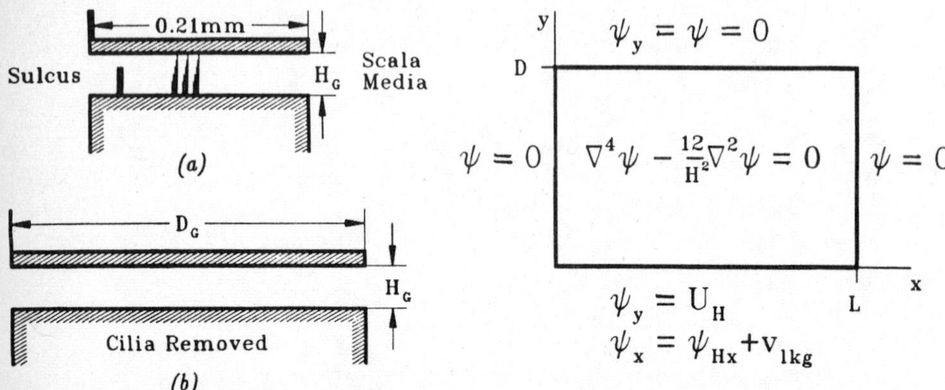

Fig. 2. (a) Flow model of tectorial gap
(b) Cilia replaced by equivalent gap depth (after Frommer (1977))

Fig. 3. Flow model of sulcus

Since the governing equation and boundary conditions are linear for the uniform gap streaming problem, a series solution is sought. An attempt at separation of variables revealed that a separable solution exists if the no-slip condition on the vertical walls is relaxed. This was allowed since the boundary conditions along the short boundaries ($D = 0.1mm$) do not have a significant effect over most of the domain ($L = 18mm$).

A solution of the Eqn. (6) that satisfies the homogeneous boundary conditions is

$$\psi(\hat{x}, \hat{y}) = \sum_{n=1}^{N} sin(n\pi\hat{x})[C_{1n}(sinh(n\pi\hat{y}) - \frac{n\pi}{\beta_n} sinh(\beta_n\hat{y})) + C_{2n}(cosh(n\pi\hat{y}) - cosh(\beta_n\hat{y}))] \tag{12}$$

where $\beta_n = \sqrt{(n\pi)^2 + \frac{12L^2}{H^2}}$, $\hat{x} = x/L$, and $\hat{y} = (y - D)/L$. The arrays of constants C_{1n} and C_{2n} are determined by the streaming and leakage conditions at the lower boundary (Eqns. (10) and (11)). By letting $\alpha_n = n\pi$, $c_1 = cosh(n\pi\hat{D})$, $c_2 = cosh(\beta_n\hat{D})$, $s_1 = sinh(n\pi\hat{D})$, and $s_2 = sinh(\beta_n\hat{D})$, we have

$$\alpha_n C_{1n}(c_1 - c_2) + C_{2n}(-\alpha_n s_1 + \beta_n s_2) = L\tilde{U}_{Hn} \tag{13}$$

$$C_{1n}(-\alpha_n(-s_1 + \frac{\alpha_n}{\beta_n} s_2) + k_p c_1) + C_{2n}(-\alpha_n(c_1 - c_2) - k_p s_1) = -\alpha_n \tilde{\psi}_{Hn} \tag{14}$$

where $k_p = \frac{H_G^2}{H^2 D_G}$, and $\tilde{U}_{Hn}$ and $\tilde{\psi}_{Hn}$ are sine components of U_H and ψ_H, which are known from Eqns. (7) and (8).

This gives a system of two linear algebraic equations in two unknowns for each n. Once C_{1n} and C_{2n} are determined, they are substituted to find $\psi(\hat{x},\hat{y})$ from Eqn. (12) and v_{lkg} from

$$v_{lkg}(\hat{x}) = -\frac{H_G^2}{H^2 D_G}\sum_{n=1}^{N} cos(n\pi\hat{x})[C_{1n} cosh(n\pi\hat{D}) - C_{2n} sinh(n\pi\hat{D})] \tag{15}$$

The results in the next section were obtained from a computer program used to find the terms and sum the series. Since we are primarily interested in localization and sharpening, the BM and leakage envelopes were normalized for comparison. The OC tip displacement was taken as equal to the BM displacement. Finally, the streaming flow was assumed not to affect the nominal BM motion.

RESULTS

The BM displacement envelopes for the guinea pig cochlea were provided by by Steele and Zais (1983). Wavenumber envelopes are from Zais (1985). Nominal dimensions of the system were taken from Fernandez (1952).

The comparison of streaming flow through the tectorial gap with BM displacement with a 4 kHz input is shown in Fig. 4. The plot demonstrates increasing leakage and localization with increasing tectorial gap height. The streamlines inside the sulcus in Fig. 6 show the physical basis for localization. An upper eddy increases in size as the gap height decreases. This pushes the leakage streamlines toward the gap and spreads them axially.

The effect of input frequency on streaming response is shown in Fig. 5. The steady gap leakage becomes larger as frequency increases despite the much smaller change in the corresponding BM displacement envelope size. The 2 kHz curve shows the inaccuracy due to relaxing the no-slip condition on the right side. The curve should reach zero by $x/L = 1$, so the separation solution becomes invalid near the apex below 2 kHz. The localization does not appear to be appreciably affected by frequency.

Although a spring lip at the edge of the tectorial gap is appropriate, it renders the problem nonlinear since

$$H_G = H_{G0} + \frac{p(x)}{S} \Longrightarrow v_{lkg}(x) = -\frac{1}{12\mu D_G}(H_{G0} + \frac{p(x)}{S})^2 p(x) \tag{16}$$

gives a nonlinear boundary condition. The present analysis assumes a uniform gap which corresponds to $s \to \infty$. Work is in progress on the solution to the spring lip problem.

Further work is necessary before quantitative conclusions can be drawn. More detailed measurements of the sulcus apparatus are necessary to determine the range of significance of the streaming effects. A more refined model must also consider interaction between streaming flow and BM motion.

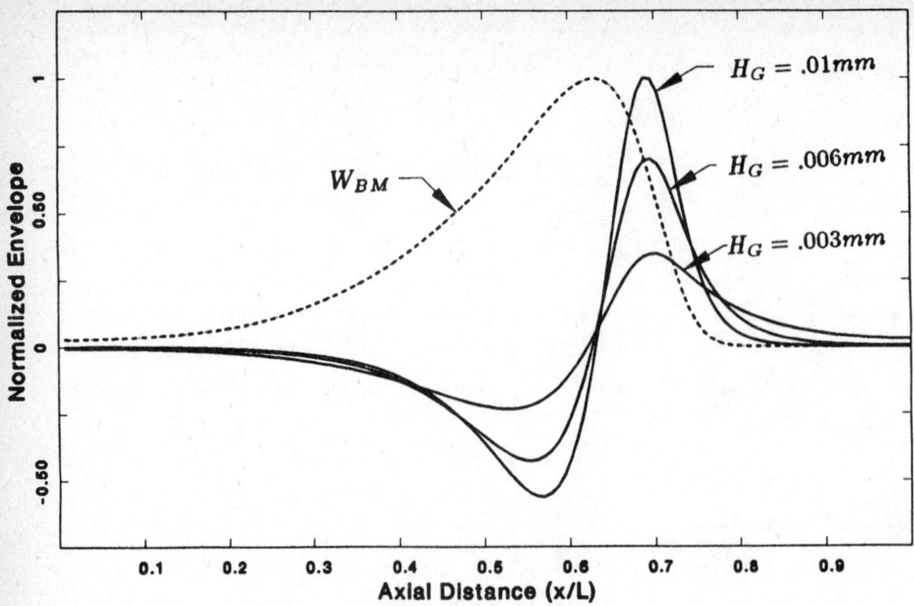

Fig. 4. Effect of gap height on streaming leakage at 4 kHz
- - - - BM displacement, ——— Gap leakage velocity

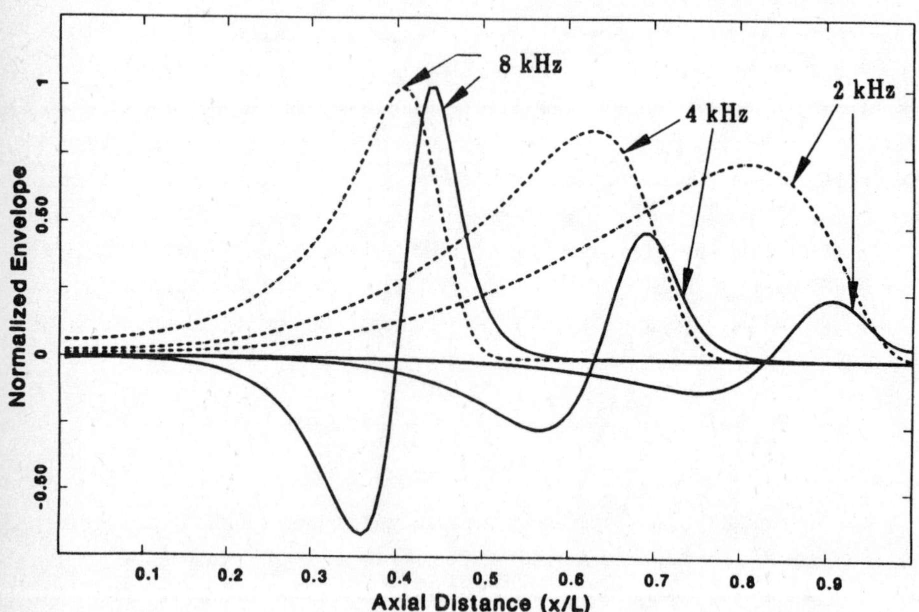

Fig. 5. Effect of frequency on streaming leakage with $H_G = .01mm$
- - - - BM displacement, ——— Gap leakage velocity

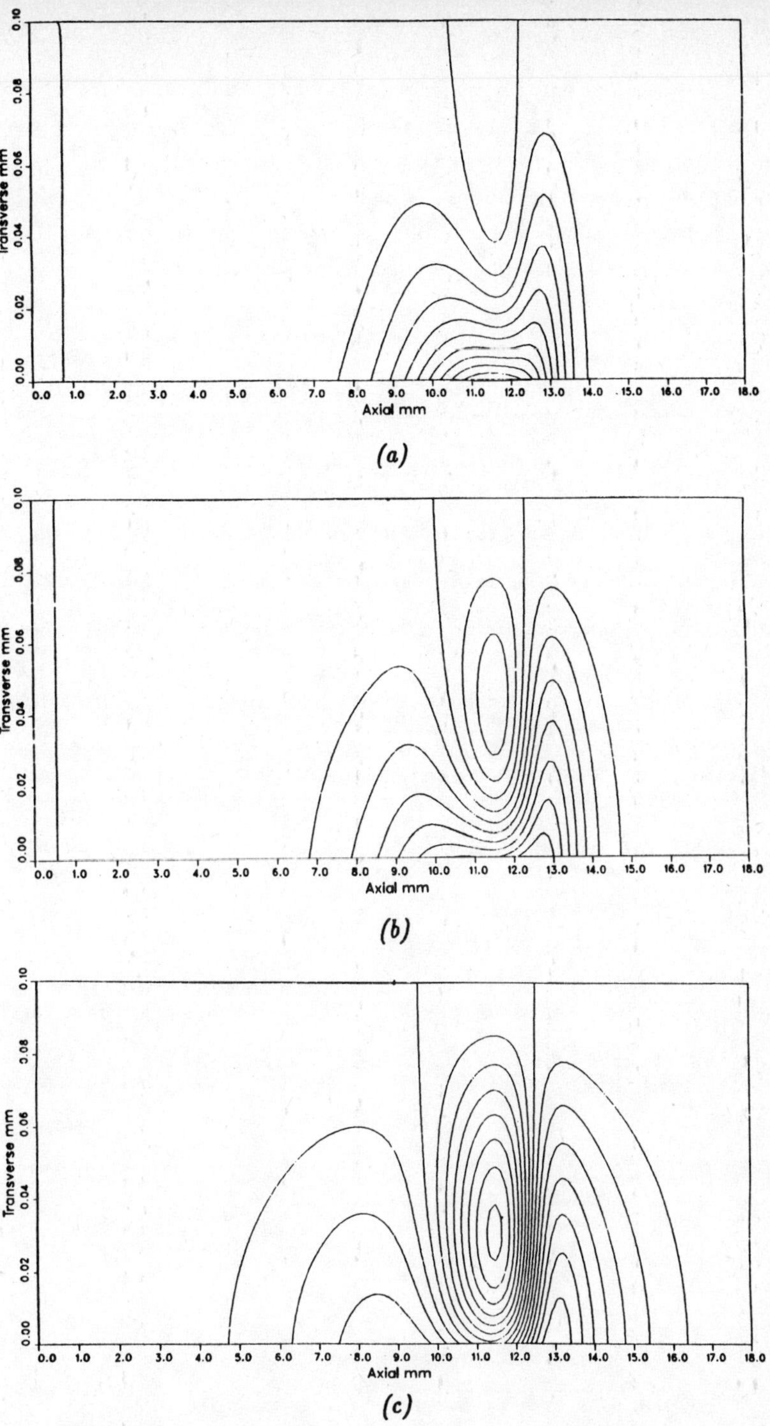

Fig. 6. Sulcus streamlines for various gap heights
(a) $H_G = .01mm$, *(b)* $H_G = .006mm$, *(c)* $H_G = .003mm$

CONCLUSION

A model of steady streaming in the spiral sulcus with a uniform tectorial gap shows that the envelope of fluid flow leaking out of the sulcus across the inner hair cell row is significantly sharper than the corresponding basilar membrane displacement envelope. The effect appears to be more pronounced for large tectorial gap and high frequency. The results suggest that for high enough sound intensity, the steady streaming may provide a passive, mechanically realizable filter between the basilar membrane vibration and inner hair cell stimulation.

REFERENCES

Bruns, V. (1979). "Functional anatomy as an approach to frequency analysis in the mammalian cochlea", *Verh. Dtsch. Zool. Ges.*, 1979, 141-54

Cancelli, C., D'Angelo, S., Malvano, R., and Masili, M. (1984). "Experimental results in a physical model of the cochlea", Politecnio di Torino, manuscript.

Fernandez, C. (1952). "Dimensions of the cochlea (guinea pig)", *J. Acoust. Soc. Am.*, **24**, 519-23.

Frommer, G. H. (1977). *Fluid Motion in the Reticular Lamina*, Ph.D. Thesis, Stanford University.

Hallauer, W. L. (1974). *Nonlinear Mechanical Behavior of the Cochlea*, Ph.D. Thesis, Stanford University.

Riley, N. (1967). "Oscillatory viscous flows. Review and extension", *J. Inst. Math. App.*, **3**, 419-34.

Spoendlin, H. (1971). "Primary structural changes in the organ of Corti after acoustic overstimulation", *Acta Otolaryngol.*, **71**, 166-76.

Steele, C. R. (1973). "A possibility for sub-tectorial membrane fluid motion", *Basic Mechanisms in Hearing* (A. R. Møller, ed.), Academic, New York.

Steele, C. R. and Taber, L. A. (1981). "Three-dimensional model calculations for the guinea pig cochlea", *J. Acoust. Soc. Am.*, **69**, 1107-11.

Steele, C. R. and Zais, J. (1983). "Basilar membrane properties and cochlear response", *Mechanics of Hearing* (E. de Boer and M. A. Viergever, eds.), Delft University Press, 29-37.

Zais, J. G. (1985). *Analytic Models of the Cochlea: Openings, Coiling, and the Processing of Speech*, Ph.D. Thesis, Stanford University.

ACTIVE FILTERING IN THE COCHLEA

ACTIVE FILTERING BY HAIR CELLS

David C. Mountain
Depts. of Biomedical Engineering and Otolaryngology
Boston University
Boston, MA 02115

ABSTRACT

Receptor cells in many organs of the acoustico-lateralis system use active filtering mechanisms to detect sensory stimuli. The mechanisms appear to differ, but in each case feedback theory can be used to describe the filtering process. Changes in membrane potential appear to regulate electrical and/or electrical-mechanical changes in the receptor cell which lead to band-pass filtering of the sensory stimulus. An important feature of all these filtering processes appears to be a delay in the feedback loop which leads to a highly underdamped system.

1. INTRODUCTION

Extensive information has recently become available on hair cell receptor potentials in three types of organs. They can be classified according to the mechanisms which determine their frequency response characteristics. The first class are those in which the cell's frequency response appears to be determined solely by the passive electrical and mechanical properties of the system. The hair cells with free standing stereocilia in the alligator lizard appear to belong to this class (Frishkopf, et al., 1982; Holton and Hudspeth, 1983). A second class are hair cells that have voltage and ion-dependent conductances in their membranes which lead to resonances in the cell membrane impedance. Examples of this class include the frog sacculus (Lewis and Hudspeth, 1983) and the turtle cochlea (Fettiplace and Crawford, 1978). The third class consists of hair cells that actively produce forces which interact with the passive mechanical elements of the sensory organ to produce active filtering.

The mammalian cochlea appears to belong to this last class of active filtering. The analysis of this system has been complicated by the fact that large numbers of hair cells appear to work together to produce the necessary amplification for the filtering process. It has been argued that negative damping elements must be present in the cochlear partition that can replace the stimulus energy which is lost through viscous forces (deBoer, 1983; Neely and Kim, 1983). Negative damping implies an active mechanical energy source which exerts a force in phase with basilar membrane velocity. The active process must be able to react quickly if it is to increase the sensitivity of the hearing organ at high frequencies. Some species can hear frequencies in excess of 100 kHz, which means that the force generating mechanism must act on a time scale measured in microseconds.

A group of recent theoretical and experimental studies have focused attention on the possiblity that a reverse transduction process is present in hair cells which alters the mechanical properties of the cell in response to electrical stimulation (Brownell _et al._, 1985; Mountain, 1980; Hubbard and Mountain, 1983; Mountain _et al._, 1983; Weiss, 1982). A feedback system in which the mechanical to electrical (forward) transduction process in turn activates a reverse (electrical to mechanical) transduction process could act as a negative damping element if the loop gain exhibits the proper delay or phase shift.

2. EXPERIMENTAL RESULTS

Electrical tuning has been shown to be present in hair cells from the frog sacculus, as well as from the turtle cochlea. In both systems, a delayed K^+ conductance appears to be involved in the electrical tuning. Lewis and Hudspeth (1983) have demonstrated that tuning in the frog hair cells is due to a combination of a voltage-dependent Ca^{++} conductance and a Ca^{++}-activated K^+ conductance. A depolarization of the cell membrane causes an increased flow of Ca^{++} into the cell which in turn activates an outward flow of K^+ which tends to repolarize the cell. The change in the K^+ current is delayed by three factors: the activation time of the Ca^{++} channels, the accumulation of Ca^{++} in the cell, and the activation time of the Ca^{++}-activated K^+ channels. In the turtle, Fettiplace and Crawford (1978) have suggested that a voltage-dependent K^+ conductance plays an important role in the hair cell tuning. They have presented a simple model in which the delayed K^+ conductance acts as an inductor in parallel with the cell membrane capacitance to produce a tuned circuit.

These electrically tuned hair cells are innervated by efferent fibers. Art and Fettiplace (1984) have found that stimulation of the efferent fibers produces a decrease in the sharpness of the tuning, but without a change in the best frequency of the filter. They have shown, in the simple (RLC) equivalent circuit model of the cell membrane, that a change in the resting membrane resistance will produce a similar effect.

In mammalian hair cells, electrical stimulation produces a number of mechanical effects. Hubbard and Mountain (1983, also Mountain et al., 1983) have injected sinusoidal electrical current into scala media and measured a variety of otoacoustic emissions at the tympanic membrane. These emissions disappear if the hair cells are damaged. Emissions can be observed at the frequency of the electrical current and often of the second harmonic as well. If an external acoustic stimulus is used, then modulation products or sidebands are observed. The most prominent are at the sum (upper sideband) and difference (lower sideband) of the acoustic and electrical frequencies. Other sidebands are often observed flanking the harmonics of the acoustic stimulus. If the sidebands are considered to be the result of a modulation process, then the ratio of their amplitude to that of the acoustic component they relate to can be used to determine a modulation ratio. For the fundamental of the acoustic stimulus, the modulation is generally less than 1%. However, for the harmonics of the acoustic stimulus, the modulation can reach 100%. This result suggests that, at least at low acoustic levels, the harmonic distortion is generated in the cochlea.

The magnitudes of the sidebands grow as the intensity of the acoustic stimulus is increased. Figure 1 shows the level dependence of the sidebands flanking the acoustic fundamental. For high acoustic frequencies, the sidebands show a growth and saturation which mimics the growth of the cochlear microphonic (CM) measured at the current injection site (Hubbard and Mountain, 1983). For acoustic frequencies well below the best frequency of the current injection site, non-monotonic behavior is often observed at high sound pressure levels.

An external acoustic stimulus can also alter the amplitude of the otoacoustic emission at the electrical frequency. For most electrical frequencies, the acoustic stimulus increases the amplitude

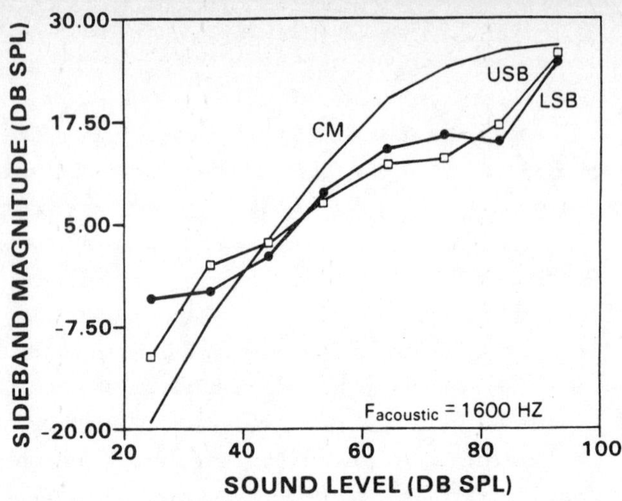

Fig. 1. Upper and Lower sidebands produced by a current of 10 microamperes peak, at 350 Hz. CM plotted on an arbitrary scale for comparison.

of the electrical response. This enhancement effect saturates at sound pressure levels similar to those which cause saturation of the CM. This interaction of electrical and acoustic stimuli again implies a role for the mechanical-to-electrical transducer of the hair cells in the generation of otoacoustic emissions.

The emission due to electrical stimulation alone is a bandpass function of electrical frequency. The high frequency cutoff varies with cochlear location and is approximately two octaves below the expected best frequency. In the basal turn of the gerbil this cutoff frequency is 4.5 kHz which corresponds to a rate limiting time constant of not more than 35 microseconds.

In vitro experiments with hair cells have also demonstrated electromechanical effects. Brownell *et al.* (1985) have observed that isolated outer hair cells will contract when electrically stimulated. The threshold for visual detection of the contractions was between 100 and 200 pA. If this threshold corresponds to movements of 0.1 to 0.2 micrometers, then the sensitivity of the electromechanical process is about 1 micrometer/nA. It is interesting to note that the basilar membrane motion at neural threshold is thought to be about 1 nm (Sellick *et al.*, 1982). A current of 1 pA would elicit an outer hair cell contraction of this magnitude. Holton and Hudspeth (1984) have calculated the conductance of a single mechanoreceptor channel to be 30 pS. Because the potential across the apical membrane of an outer hair cell is about 150 mV, the current due to a single channel opening would be 4.5 pA. If the outer hair cells in the intact organ behave the same way as they do *in vitro*, then the current through a single receptor channel should result in a significant mechanical response.

A direct comparison between the *in vivo* and *in vitro* experiments is not yet possible. The current levels used in the *in vivo* experiments were 100 times those used in the *in vitro* experiments. However, only a fraction of this current would pass through a single outer hair cell, and if reasonable assumptions are made concerning the current paths in the cochlea, then it is possible to show that the currents through a single cell are similar in both types of experiments. Unfortunately, it is not possible without the aid of a detailed cochlear model to compare the magnitude of the otoacoustic emissions to the size of the contractions observed in isolated hair cells.

3. MODEL RESULTS

The simulations described in this paper concentrate on studying the effects of voltage-dependent force production by hair cells. The first set of results are from a linear (small signal), one-dimensional cochlear model in which the hair cell receptor potential is assumed to be a linear function of cilia displacement and the force produced by the hair cells is assumed to be a linear function of membrane potential. The effects of cell membrane capacitance are included so that a phase shift will be present between the displacement of the cilia and the generation of the active force. The second set of results are from a low frequency model in which the cochlear mechanics are greatly simplified, but a much more realistic model of the hair cell electrical properties is used.

If these models are to increase cochlear sensitivity, then a depolarization of the outer hair cells must result in a force which displaces the hair cell cilia in the inhibitory direction (towards the inner hair cells). Such a force will result in a movement of the basilar membrane towards scala tympani. The phase shift due to the hair cell electrical properties results in the hair cell acting as a parallel combination of a frequency-dependent negative damping element and a frequency-dependent spring.

For the linear cochlear model, the cell membrane time sonstant was assumed to be inversely proportional to the passive natural frequency of the basilar membrane at a given cochlear location. The effect of increasing the gain of either the forward or reverse transduction process is to move the peak of the traveling wave apically and greatly increase its amplitude. The cell membrane time constant also affects the mechanical response. As the time constant is shortened, the peak becomes narrower and moves more apically.

Fig 2 shows the effect of changing the cell membrane conductance by an equal fraction in all cochlear regions. As the conductance increases, the magnitude of the peak drops rapidly, but does not shift in position until the magnitude of the response is greatly diminished. Stimulation of the cochlear efferents causes a reduction in the sensitivity of auditory nerve fibers without significantly altering the best frequency (Weiderhold, 1970). The results in fig. 2 suggest that the efferent effect may be mediated primarily by an increase in the hair cell membrane conductance. The change in membrane conductance reduces the forward transduction gain and simultaneously reduces the membrane time constant. The transduction gain and the membrane time constant have opposite effects on the peak location which cancel. The size of the traveling wave peak, however, is still reduced.

A common finding in models of this type is that the response of the active system is very sensitive to small changes in model parameters. For example, in fig 2, decreasing the membrane resistance by 50% resulted in a decrease of the peak response by about 55 dB. These models are also very close to instability, with only very small increases in gain needed to cause spontaneous oscillation.

In a second series of simulations, a much more detailed model of hair cell electrical properties was used. The forward transduction process was modeled using a transfer fuction similar to that derived by Corey and Hudspeth (1983) for a three state channel model. An equivalent circuit model for the stria vascularis was also included so that experiments in which the endocochlear potential (EP) has been

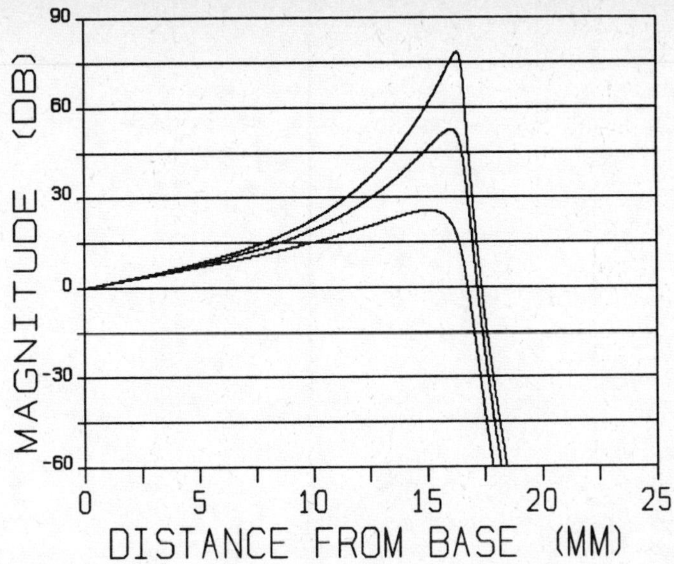

Fig. 2 Basilar membrane velocity as a function of cochlear location for three values of hair cell membrane resistance. The curves (in order of decreasing amplitude) are for 100%, 75%, and 50% of the normal value.

manipulated could be simulated. The simulations were limited to low frequencies so that only the stiffness of the basilar membrane needed to be considered. In models of this type the nonlinearity of the forward transduction process results in a nonlinear mechanical response. For example, a static shift in basilar membrane position can result from acoustic stimulation.

The injection of sinusoidal electrical current into scala media produces a variety of acoustic emissions which can be measured at the tympanic membrane. The model produces a basilar membrane response which includes all of the components observed in the otoacoustic emissions. Fig. 3 shows the results of a simulation of basilar membrane response to a combination of electrical and acoustic stimulation. The sideband components and the CM are plotted as functions of the level of the acoustic signal. The sidebands increase with acoustic level until the forward transduction reaches saturation as indicated by saturation of the CM. Non-monotonic

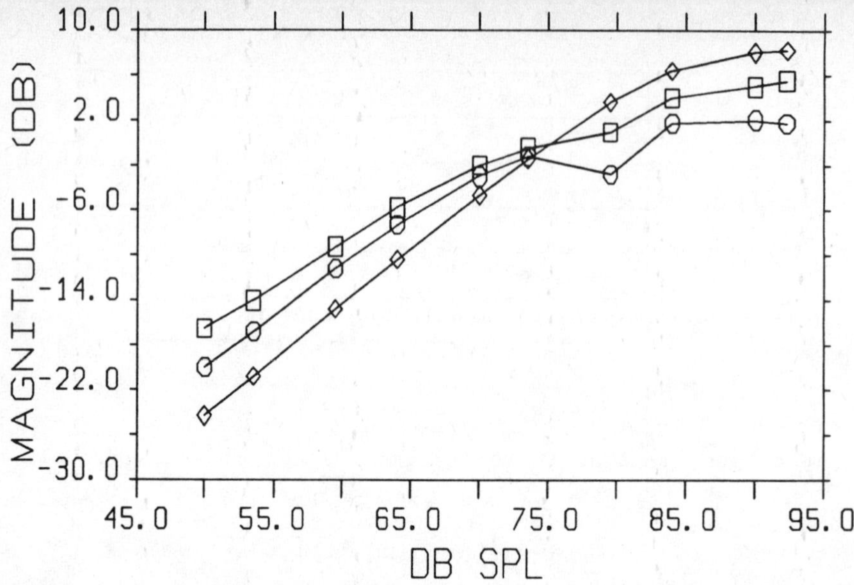

Fig. 3 Magnitude of the lower sideband (squares), upper sideband (circles), and CM (diamonds) as a function of acoustic stimulus level for the non-linear model.

behavior at high SPL is observed, as is also the case in the experimental data (fig. 1).

As mentioned above, the component which is at the fundamental electrical frequency can be enhanced by the presence of an external tone. This enhancement effect is also seen in the model results and saturates at levels of the external tone for which the CM measured at the electrical stimulation site also saturates.

4. DISCUSSION

The electrical-mechanical hypothesis for active filtering by cochlear hair cells has the potential to unify a diverse set of experimental results. One of the key features of the hypothesis is that the observed nonlinearity in cochlear mechanics may be due largely to the nonlinearity of the forward transduction process which has been well characterized. Two-tone suppression, cochlear distortion products, changes in basilar membrane tuning with acoustic level, and other nonlinear phenomena may all be due to driving the hair cell transducer into the saturation region. The picture is a complicated one because in active mechanical models many outer hair cells, perhaps distributed over several millimeters, contribute to the sensitivity and tuning of a single inner hair cell.

Efferent stimulation alters the tuning of hair cells which belong to both the electrical filtering class and the electromechanical filtering class. In both cases, modeling results suggest that the efferent effect is largely due to changes in the resting membrane conductance.

The electromechanical model for the cochlea may be an over-simplification. Voltage-dependent conductances also appear to be present in outer hair cells (Cody and Russell, personal communication). The contribution of these membrane properties to the tuning process remains to be determined. It is also possible that a voltage-dependent membrane characteristic could act as an automatic gain control which would help stabilize the feedback system.

ACKNOWLEDGEMENTS

I wish to thank A. Hubbard, E. LePage, T. McMullen, and H. Voigt for many helpful discussions. I thank M. Zagaeski for his help with some of the programing. This research was funded by NIH grant NS16589.

REFERENCES

Art, J.J., Fettiplace, R., and Fuchs, P.A., "Synaptic hyperpolarization and inhibition of turtle cochlear hair cells." J. Physiol., 356, pp. 525-550, 1984.

de Boer, E., "No sharpening? A challenge for cochlear mechanics." J. Acoust. Soc. Am., 73, pp. 567-573, 1983.

Brownell, W.E., Bader, C.R., Bertrand, D., and de Ribaupierre, Y., "Evoked mechanical responses of isolated cochlear hair cells." Science, 227, pp. 194-196, 1985.

Corey, D.P. and Hudspeth, A.J., "Kinetics of the receptor current in bullfrog saccular hair cells." J. Neurosci., 3, pp. 962-976, 1983.

Fettiplace, R. and Crawford, A.C., "The coding of sound pressure and frequency in cochlear hair cells of the terrapin." Proc. R. Soc. Lond. B., 203, pp. 209-218, 1978.

Frishkopf, L. S., DeRosier, D. J., and Egelman, E. H., "Motion of basilar papilla and hair cell stereocilia in the excised cochlea of the alligator lizard: relation to frequency analysis." Soc. Neurosci. Abstr., 8, pp. 40, 1982.

Holton, T. and Hudspeth, A.J., "A micromechanical contribution to cochlear tuning and tonotopic organization." Science, 222, pp. 508-510, 1983.

Holton, T. and Hudspeth, A.J., "Transduction current in saccular hair cells examined with the whole-cell voltage-clamp technique." Neurosci. Abstr., 10, pp. 10, 1984.

Hubbard, A.E. and Mountain, D.C., "Alternating current delivered into the scala media alters sound pressure at the eardrum." Science, 222, 510-512, 1983.

Lewis, R.S. and Hudspeth, A.J., "Frequency tuning and ionic conductances in hair cells of the bullfrog's sacculus." In Hearing- Physiological Bases and Psychophysics. Ed. by R. Klinke and R. Hartmann, Springer-Verlag, New York, 1983.

Mountain, D.C., "Changes in endolymphatic potential and crossed-olivocochlear-bundle stimulation alter cochlear mechanics." Science, 210, pp. 71-72, 1980.

Mountain, D.C., Hubbard, A.E., and Geisler, C.D., " Voltage-dependent elements are involved in the generation of the cochlear microphonic and the sound-induced resistance changes in scala media of the guinea pig." Hearing Res., 3, pp. 215-229, 1980.

Mountain, D.C., Hubbard, A.E., and McMullen, T.A., "Electromechanical processes in the cochlea." In Mechanics of Hearing. Ed. by E. de Boer and M.A. Viergever, pp.119-126. Delft University Press, Delft, 1983.

Neely, S.T. and Kim, D.O., "An active cochlear model shows sharp tuning and high sensitivity." Hearing Res., 9, pp. 123-130, 1983.

Sellick, P.M., Patuzzi, R., and Johnstone, B.M., "Measurement of basilar membrane motion in the guinea pig using the Mossbauer technique." J. Acoust. Soc. Am., 72, pp. 131-141, 1982.

Weiderhold, M.L., "Variations in the effects of electrical stimulation of the crossed olivocochlear bundle on cat single auditory-nerve-fiber responses to tone bursts." J. Acoust. Soc. Am., 48, pp. 966-977, 1970.

Weiss, T.F., "Bidirectional transduction in vertebrate hair cells: A mechanism for coupling mechanical and electrical processes." Hearing Res., 7, pp. 353-360, 1982.

DETERMINATION OF THE COCHLEAR POWER FLUX FROM BASILAR MEMBRANE VIBRATION DATA

Rob J. Diependaal[1][2], Egbert de Boer[2], Max A. Viergever[1]

(1) Delft University of Technology, Delft - The Netherlands
(2) Academic Medical Centre, Amsterdam - The Netherlands

ABSTRACT

We address the question of whether we can conclude just from basilar membrane (BM) vibration data that the cochlea is an active mechanical system. To this end we study an "inverse" problem for the cochlea in the framework of a short-wave model. Using the "inverse" problem formulation we compute the power flux through a channel cross-section from the BM velocity pattern. A rise in the power flux function indicates that the cochlea itself adds energy to the BM vibration. In order to avoid numerical errors as a result of too few data points, we have interpolated the BM velocity curves. The choice of interpolation method appears to influence the power flux function very much. Nonetheless, we conclude that the power flux method is able to determine from measured BM vibration patterns whether the underlying behaviour of the cochlea has been active or not.

I. INTRODUCTION

In recent years many investigators have been modelling the cochlea as a nonlinear and active system. Evidence to include active features in descriptions of cochlear functioning has been supplied by experimental results concerning acoustic emissions (pioneered by Kemp, 1978). Also, a comparison of the threshold auditory signal to the expected levels of quantum noise and thermal noise of the stereocilia provides arguments that the cochlea is an active system (Bialek, 1983). Measurements of Khanna and Leonard (1982), Sellick, Patuzzi and Johnstone (1982) and Robles, Ruggero and Rich (1984) seem to indicate that the active behaviour manifests itself at the level of basilar membrane (BM) vibration. These experimental data show a very sharp tuning of the BM as compared with all earlier data. In this paper we address the question of whether we can conclude just from BM vibration data that the cochlea is an active mechanical system. To this end we formulate the equations of a short-wave cochlear model in such a way that the power flux through a channel cross-section can be computed from a given BM vibration pattern. The power flux function is an important indicator of active behaviour of the cochlea, since an increase (decrease) in this function corresponds to a source (sink) of vibrational energy for the BM. We chose a short-wave model description of cochlear mechanics because this provides for a simple relationship between the power flux and the BM velocity. The long-wave case will be discussed in a future paper.

LIST OF MAIN SYMBOLS

x	coordinate along the basilar membrane (BM)
z	coordinate normal to the BM
j	imaginary unity
b, h, L	width, height, length of a channel of the cochlear model
$\beta(x)$	$= \beta_0 \exp(\beta_1 x)$, BM width
$\varepsilon(x)$	$= \beta(x)/b = \varepsilon_0 \exp(\varepsilon_1 x)$
ω	radian frequency of the stapes motion
ω_0	reference frequency
x_0	point of observation
ρ	fluid density
$p_{BM}(x,j\omega)$	trans-BM pressure
$v_{BM}(x,j\omega)$	BM velocity
$A(x,j\omega)$	amplitude of BM velocity
$\phi(x,j\omega)$	phase of BM velocity
$p(x,z,j\omega)$	fluid pressure in a channel of the cochlear model
$v_x(x,z,j\omega)$	fluid velocity in the x-direction
$E_x(x,z,j\omega)$	average acoustic intensity in the x-direction
$F_x(x,j\omega)$	power flux through a channel cross-section
$W(x,j\omega)$	amplitude of BM displacement
$\kappa(x,j\omega)$	wave number of BM velocity wave
α	cochlear map parameter

II. MODEL AND METHOD

For a fluid wave in a two-dimensional cochlear model the average acoustic intensity (energy per unit time per unit cross-sectional area) in the x-direction is given by

$$E_x(x,z,j\omega_0) = \frac{1}{2} \mathrm{Re}\,[p(x,z,j\omega_0)v_x^*(x,z,j\omega_0)]\ , \tag{1}$$

where * means complex conjugate.

We assume that the mechanical properties of the cochlea do not vary much within a wavelength of the wave propagated along the BM. This is the basis of the Liouville-Green (LG) approximation. It gives the following expressions for the pressure and the fluid velocity in the x-direction (see, e.g. Steele and Taber, 1979)

$$p(x,z,j\omega_0) = -\frac{2\omega_0^2\rho\varepsilon W}{\pi\kappa\sinh[\kappa h]} \cosh\,[\kappa(z-h)]\, e^{-j\kappa x}, \tag{2}$$

$$v_x(x,z,j\omega_0) = -\frac{2\omega_0\varepsilon W}{\pi\sinh[\kappa h]} \cosh\,[\kappa(z-h)]\, e^{-j\kappa x}. \tag{3}$$

To arrive at the total power flux, we substitute eqs. (2) and (3) into eq. (1) and integrate the result over the width and the height of the channel:

$$F_x(x,j\omega_0) = b\int_0^h E_x(x,z,j\omega_0)dz =$$

$$= \frac{b\omega_0^3\rho\varepsilon^2 W^2\{\sinh[2\,\mathrm{Re}(\kappa)h] + \frac{\mathrm{Re}(\kappa)}{\mathrm{Im}(\kappa)}\sin[2\,\mathrm{Im}(\kappa)h]\}}{\pi^2|\kappa|^2\{\cosh[2\,\mathrm{Re}(\kappa)h] - \cos[2\,\mathrm{Im}(\kappa)h]\}} e^{2\,\mathrm{Im}(\kappa)x} \quad . \tag{4}$$

Defining movements in the direction of scala vestibuli as positive (as opposed to the convention used in most of the experiments), we have

$$p_{BM}(x,j\omega_0) = -\,2p(x,0,j\omega_0). \tag{5}$$

Equations (2), (4) and (5) jointly yield

$$F_x(x,j\omega_0) = \frac{b}{16\omega_0\rho}\,\frac{\tanh[2\mathrm{Re}(\kappa)h] + \frac{\mathrm{Re}(\kappa)\sin[2\,\mathrm{Im}(\kappa)h]}{\mathrm{Im}(\kappa)\cosh[2\mathrm{Re}(\kappa)h]}}{1 + \frac{\cos[2\,\mathrm{Im}(\kappa)h]}{\cosh[2\mathrm{Re}(\kappa)h]}}\,|p_{BM}(x,j\omega_0)|^2 . \tag{6}$$

To get rid of the wavenumber κ, we take the limit for $\mathrm{Re}[\kappa h] \rightarrow +\infty$. Physically, this means we consider waves to be short as compared to the channel height. The differential equation describing the short-wave model reads

$$\frac{dp_{BM}}{dx} + \frac{16\omega_0\rho\varepsilon}{\pi^2}v_{BM} = 0, \text{ with } p_{BM} = 0 \text{ at } x = L. \tag{7}$$

We have identified the point $x = L$, the helicotrema, with the resonance location, because apical to this point the short-wave response is completely unrealistic (Viergever, 1980). The boundary condition in (7) implies that the travelling wave along the BM is extinct when reaching the resonance location.
When integrating (7), substituting the result into eq. (6) and taking the limit for $\mathrm{Re}\,[\kappa h] \rightarrow +\infty$, the following form for the power flux is found

$$F_x(x,j\omega_0) = \frac{16\omega_0\rho}{\pi^4 b}\left|\int_x^L \beta(\xi)v_{BM}(\xi,j\omega_0)d\xi\right|^2 . \tag{8}$$

From eq. (8) we can determine the power flux at any place on the BM once we know the BM velocity pattern for (all) positions apicalwards.
Since in all recent experiments the velocity is measured as a frequency response curve, we need a transformation from the frequency domain to the spatial domain in order to be able to apply the present method.

III. COCHLEAR MAP

For several species a frequency-position map is described in the literature. Such a map is a relation between characteristic frequency and characteristic place along the BM. Often that relation is formulated as place being a logarithmic (or nearly logarithmic) function of characteristic frequency (Robertson and Johnstone, 1979 (guinea pig); Eldredge et al., 1981 (chincilla);Liberman, 1982 (cat)). Viergever (1980) shows that, under certain restrictions on the mechanical parameters of the BM, the LG-solution of a two-dimensional cochlear model is frequency-shift invariant for that place to log frequency transformation only when the BM velocity is normalized to its value at the resonance place. His conclusion reads that for quantitative comparisons of model results and experimental data the place to log frequency transformation is not sufficiently accurate. Hence we try and improve the transformation by including the mentioned normalization factor. The customary place to log frequency transformation reads

$$t : x = x_0 + \frac{2}{\alpha} \ln \frac{\omega}{\omega_0} . \qquad (9)$$

Its inverse, given by

$$t^{-1} : \omega = \left[\omega_0^2 e^{\alpha(x - x_0)}\right]^{\frac{1}{2}} , \qquad (10)$$

transforms the BM vibration data from the frequency domain into the place domain. The error we make when applying (9) or (10) can be corrected approximately in the following way. In the region where the stiffness part dominates the BM impedance, the short-wave LG solution of the BM velocity is given by

$$v_{BM}(x, j\omega_0) = C_1\omega^2 e^{\alpha x} \exp\left[-j\{ \frac{\pi}{2} + C_2\omega^2 (e^{(\alpha + \varepsilon_1)x} - 1)\}\right] \qquad (11)$$

In the same region, the following quantities are invariant under transformation t:

$$A(x, j\omega_0) \overset{t}{\longleftrightarrow} A(x_0, j\omega) \qquad (12)$$

and

$$\frac{\phi(x, j\omega_0) + \frac{\pi}{2}}{\omega_0^2 (e^{(\alpha + \varepsilon_1)x} - 1)} \overset{t}{\longleftrightarrow} \frac{\phi(x_0, j\omega) + \frac{\pi}{2}}{\omega^2 (e^{(\alpha + \varepsilon_1)x_0} - 1)} . \qquad (13)$$

Expressions (12) and (13) define the corrected cochlear map to transform A and ϕ from the frequency into the place domain. Simulation studies have shown that this map gives improved results as compared with the place to log frequency transformation.

IV. NUMERICAL RESULTS

Before determining the power flux from actual measurement results we tested the method on computer-generated BM velocity patterns. In order to simulate measurement errors we introduced several disturbances to these patterns and studied their effects on the power flux calculations. For all disturbances considered, the resulting deviations in the power flux function are small. In consequence, the method is not very sensitive to small errors in the BM velocity.

We selected two different sets of experimental BM vibration data to which the method will be applied. The results of Johnstone and Yates (1974) were chosen to represent the older vibration data. The results of Robles et al. (1984) are the natural choice for the recent, mechanically sharply tuned data, since they are the only ones which present phase information in addition to amplitude response curves. Application of the power flux method requires a discussion of the following features.

a) Middle-ear transfer filter.

Robles et al. (1984) publish their amplitude data as sound pressure at the eardrum necessary to produce a certain BM velocity as a function of frequency. Their phase data are given as BM displacement with respect to rarefaction at the eardrum. Our power flux method expects as input the ratio of BM velocity and stapes velocity. Therefore, the data have to be modified by a middle-ear transfer filter. The filter represents stapes displacement over pressure at the eardrum. The amplitude of the filter we use, has a value of C_A mm/mPa at frequencies < f_A kHz, and decays at S_AdB/octave above f_A kHz. The phase of the filter has a value 0 at frequencies < f_p kHz and decays at a rate of S_p radians/octave above f_p kHz. The filter parameters C_A, f_A, f_p, S_A and S_p can be estimated from Guinan and Peake (1967) for the cat, but corresponding middle-ear characteristics for the chinchilla are not known to us. Therefore, we used several middle-ear parameter sets in the calculations. We found the power flux function to be quite sensitive to the choice of parameters in a quantitative sense, but not qualitatively.

We did not need a middle-ear transfer filter for the measurements of Johnstone and Yates (1974), because they already published their data as ratios of BM velocity and stapes velocity.

b) Interpolation of data points.

In order to avoid numerical errors in the power flux function as a result of too few data points, some interpolation of the BM velocity curves is mandatory. We considered two interpolation methods.

i) Cubic splines.

A cubic spline interpolation function is a twice continuously differentiable function through the data points. In each subinterval between two successive data points, the cubic spline function is a polynomial of the third degree. The number of parameters to be determined for a unique cubic spline function

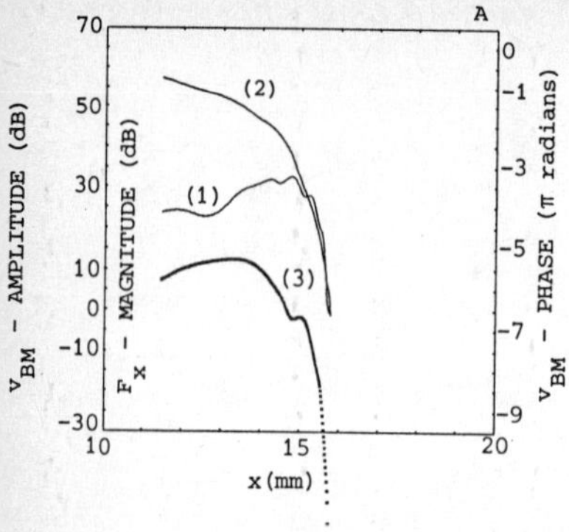

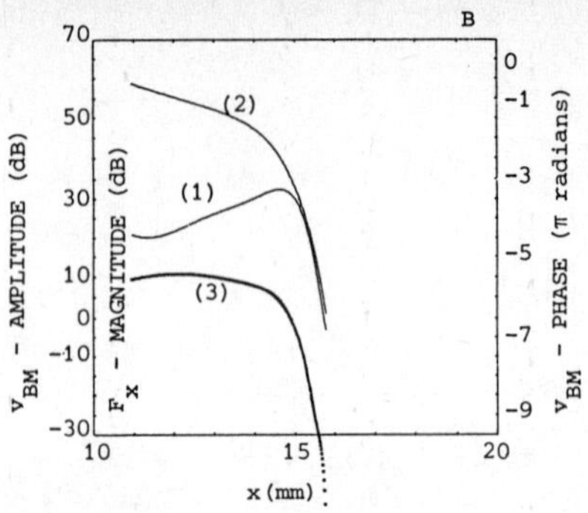

Fig. 1: Power flux calculated from the data of Johnstone and Yates (1974, fig. 3). Panel A: data interpolated by cubic splines. Panel B: data interpolated by one polynomial of 6th degree (amplitude) and of 3rd degree (phase), using the least-squares method. Parameters: $\rho = 1$ mg/mm^3, b = 0.5 mm, $\beta = 0.08 \exp(0.04x)$mm, $\varepsilon_1 = 0.04$ mm^{-1}, $\omega_0 = 2\pi.0.777$ kHz, $x_0 = 4.12$ mm, $\alpha = 0.55$ mm^{-1}. Cochlear map parameters are derived from Robertson and Johnstone (1979). Curves: (1): amplitude of BM velocity, (2): phase of BM velocity, (3): power flux.

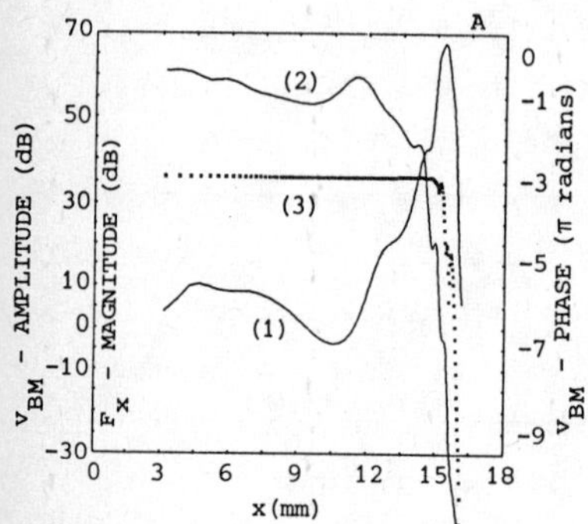

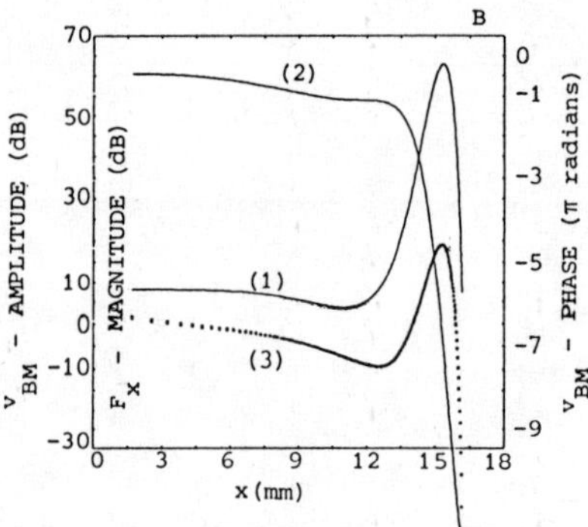

Fig. 2: Power flux function calculated from the data of Robles, Ruggero and Rich (1984, figs. 3 and 9). Panel A: data interpolated by cubic splines. Panel B: data interpolated by one polynomial of 4th degree (amplitude) and of 5th degree (phase), using the least-squares method. Parameters: $\omega_0 = 2\pi.0.33$ kHz, $x_0 = 3.5$ mm, $\alpha = 0.554$ mm^{-1}, $C_A = 4.10^{-8}$ mm/mP$_a$, $f_A = f_p = 3$ kHz, $S_A = 9.5$ dB/octave, $S_p = 0.3\pi$ radians/octave. Other parameters as in fig. 1. Cochlear map parameters are derived from Eldredge, Miller and Bohne (1981) and from private communication with Mario Ruggero. Curves: (1): amplitude of BM velocity, (2): phase of BM velocity, (3): power flux.

equals the number of data points minus two.

ii) Least-squares.

The interpolating function is one polynomial of fixed degree. The number of parameters which uniquely determines this function equals the degree of the polynomial plus one.

The power flux function appears to be much dependent on the choice of the interpolation method, as will be discussed below.

c) Cochlear map parameters.

The cochlear map has been defined in section III. It contains four parameters, viz. ω_0, x_0, α and ε_1. Variation of these parameters does not produce significant changes in the global shape of the power flux function.

Figure 1 presents the BM velocity pattern (after interpolation and cochlear map) and the calculated power flux for the data of Johnstone and Yates (1974). Figure 2 shows the BM velocity (modified by middle-ear transfer filter, interpolation and cochlear map) and the calculated power flux for data of Robles et al. (1984). In figs. 1A and 2A the data have been interpolated by cubic splines. The method of least-squares has been used to obtain the results as shown in figs. 1B and 2B. See for further details the legends to the figures.

The power flux function in fig. 1A is slightly different from that in fig. 1B, but their common main feature is the moderate rise of at most a few dB over the length of the BM. For the data of Fig. 2 the difference in the calculated power fluxes is very pronounced. In fig. 2A the power flux function hardly rises, whereas in fig. 2B it shows a strong increase in the region just before the BM velocity amplitude peak.

V. CONCLUSIONS AND DISCUSSION

We have developed a method to estimate the cochlear power flux function from basilar membrane vibration data. Under the restrictions and assumptions outlined in the preceding sections the results obtained suggest the following conclusions:

i) In the experiment of Johnstone and Yates (1974) the cochlea behaved as a passive filter (see figs. 1A and 1B).

ii) The measurements of Robles et al. (1984) do not allow a definitive conclusion about the cochlea being either an active or a passive system. The calculated power flux function is greatly dependent on how the experimental data are interpolated.

When the data points are interpolated with cubic splines, many parameters have to be determined. This corresponds to considering the cochlea as a system having many degrees of freedom. Using this procedure, we find no evidence of mechanical activity. On the contrary, the result of fig. 2A implies that the measured response can be mimicked using a passive model. This conclusion supports the statement made by Viergever and Diependaal (1985) that sharply tuned BM data can be matched using a passive distributed parameter model of the cochlea. Alternatively, we can interpolate the data by one polynomial of low degree using the least-squares method. This corresponds to considering the cochlea as a system having only a few degrees of freedom. The measurements of Robles et al. (1984) now imply that the cochlea is active at the level of BM vibration (fig. 2B), the region of the cochlea activity being situated just basalwards of the maximum of the BM velocity amplitude.

We suggest that any high-order interpolation method produces a response containing too much detail. In fact, many of the variations are within the range of inaccuracy of the data. The resulting variations in the power flux function are, consequently, not contingent on the data, but rather on the noise in the data. This makes a low-order interpolation method physically more realistic. Continuing this line of reasoning, we arrive at the conclusion that sharply tuned BM responses do imply mechanical activity at the BM level.

ACKNOWLEDGEMENT

This work has been supported by the Netherlands Organisation for the Advancement of Pure Research (ZWO).

REFERENCES

1. Bialek, W. (1983). Quantum effects in the dynamics of biological systems. Ph.D. diss. Lawrence Berkeley Laboratory and University of California, Berkeley.
2. Elredge, D.H., Miller, J.D. and Bohne, B.A. (1981). A frequency-position map for the chinchilla cochlea, J. Acoust. Soc. Am. 69, 1091-1095.
3. Guinan, J.J. Jr. and Peake, W.T. (1967). Middle ear characteristics of anesthetized cats, J. Acoust. Soc. Am. 41, 1237-1261.
4. Johnstone, B.M. and Yates, G.K. (1974). Basilar membrane tuning curves in the guinea pig. J. Acoust. Soc. Am. 55, 584-587.
5. Kemp, D.T. (1978). Stimulated acoustic emissions from within the human auditory system, J. Acoust. Soc. Am. 64, 1386-1391.
6. Khanna, S.M. and Leonard, D.G.B. (1982). Basilar membrane tuning in the cat cochlea, Science 215, 305-306.
7. Liberman, M.C. (1982). The cochlear frequency map for the cat: Labeling auditory-nerve fibers of known characteristic frequency, J. Acoust. Soc. Am. 72, 1441-1449.
8. Robertson, D. and Johnstone, B.M. (1979). Aberrant tonotopic organization in the inner ear damaged by kanamycin, J. Acoust. Soc. Am. 66, 466-469.
9. Robles, L., Ruggero, M. and Rich, N.C. (1984). Mössbauer measurements of basilar membrane tuning curves in the chinchilla. J. Acoust. Soc. Am. 76, S35.
10. Sellick, P.M., Patuzzi, R. and Johnstone, B.M. (1982). Measurement of basilar membrane motion in the guinea pig using the Mössbauer technique. J. Acoust. Soc. Am. 72, 131-141.
11. Steele, C.R. and Taber, L.A. (1979). Comparison of WKB and finite difference calculations for a two-dimensional cochlear model. J. Acoust. Soc. Am. 65, 1001-1006.
12. Viergever, M.A. (1980). Mechanics of the inner ear - a mathematical approach. Ph.D. diss., Delft University of Technology, Delft University Press.
13. Viergever, M.A. and Diependaal, R.J. (1985). Quantitative validation of cochlear models using the Liouville-Green approximation. Submitted to Hearing Research.

AN ISOLATED SOUND EMITTER IN THE COCHLEA: NOTES ON MODELLING

E. de Boer[1], Chr. Kaernbach[2], P. König[2] and Th. Schillen[2]

1. Acad. Medical Centre, Meibergdreef 9, 1105 AZ Amsterdam (Neth.)
2. Physikalisches Institut, Nussallee 12, Universität Bonn (Germany)

ABSTRACT

A most important question concerns the stability of active cochlea models. We have addressed this question by studying effects occurring when a small section of the cochlea is made active and the remainder is left passive. In our study we determine the limiting value of the length Δl of the active section for which the system becomes unstable. The cochlea is modelled as a long-wave structure and described as a nonuniform transmission line. The principal parameter involved is the characteristic impedance of this line. At the resonance frequency of the emitter the characteristic impedance has a phase angle of $\pi/4$ radians (45°). The real part damps the emitter, and mainly determines the limiting length Δl. The imaginary part is inductive and causes the emitter, if it becomes unstable, to oscillate at a frequency somewhat below its natural resonance frequency. It is found that for a given degree of activity, the length Δl must remain larger than a certain value for the system to remain stable. Several remarkable properties result from the analysis, these give really new insight into the requirements that an active cochlea model must meet. We have generalized the concept of characteristic impedance and have subjected the predictions of the analytical theory to test by carrying out numerical calculations in the time domain.

LONG-WAVE MODEL AND TRANSMISSION-LINE ANALOGY

There are several reasons why modern models of the cochlea contain active elements. Two of the most important ones are 1) some ears emit continuous sounds and 2) it is impossible to simulate experimentally determined basilar-membrane (BM) response functions by purely passive models. These two requirements do not necessarily lead to postulating active elements with the same properties. For instance, to replicate an acoustic emission we require an oscillator, i.e. a resonating device that becomes unstable at its resonance frequency. To simulate BM response functions it is necessary to boost the model response in the frequency region below its resonance peak. The present paper contains the first few steps of a study directed at reconciling these two trends in considering the cochlea as an active device.

Let us start with modelling the cochlea as a passive structure in which a short section acts as an active resonator. We will next try to find the limit of instability of the system. We use here the long-wave (or one-dimensional) model ([1,4]) of the cochlea, mainly because this model can be studied by analytical methods and because the formulation of a numerical solution in the time domain is fairly simple. The basilar membrane is to be described by its impedance $\zeta(x)$ and the coupling between adjacent points in the x-direction occurs only by way of the fluid. The long-wave model is analogous to an electrical transmission line and tools of research can therefore be borrowed from the field of transmission-line theory ([2]). Fig. 1-a shows the transmission line analogous to the cochlear model we want to study. The segments L_1 and L_3 of the line represent passive parts of the model. The section L_2, of length Δl, is to be active and the main goal of this work is to find the limiting value of Δl beyond which the system becomes unstable, i.e. where L_2 starts to act as an emitter of sound.

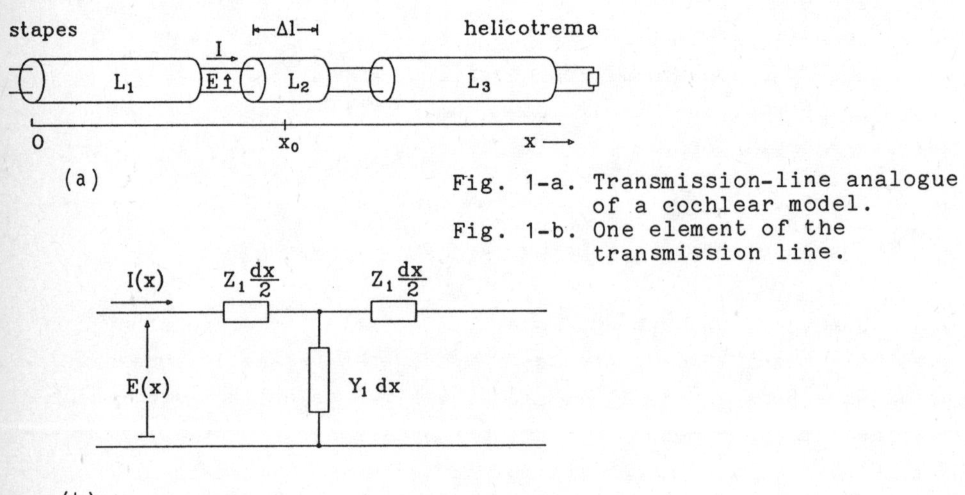

Fig. 1-a. Transmission-line analogue of a cochlear model.

Fig. 1-b. One element of the transmission line.

Transmission lines have distributed series impedance and shunt admittance. For the problem at hand we define the parameters of the line in such a way that they directly relate to the input impedance of the cochlea. To this aim we let the current I(x) represent the (point) velocity $v_x(x)$ of the cochlear fluid (which in the long-wave model is directed solely in the x-direction). The voltage E(x) represents the pressure p(x) in scala vestibuli, i.e., half the trans-membrane pressure. Infinitesimally small elements of the transmission line are shown by Fig.1-b, Z_1dx is the series impedance and Y_1dx the shunt admittance. We will take L_2 as so short that it is described by a lumped circuit of the same form but with dx replaced by the finite length Δl. For the present type of analogy Z_1 must equal $i\omega\rho$ where ω is the radian frequency and ρ the fluid density. The shunt admittance coefficient Y_1 must be equal to $2\,h_e^{-1}\zeta^{-1}(x)$ where h_e is the effective height (i.e., scalar area divided by BM width) and $\zeta(x)$ is, as stated above, the BM impedance.

The long-wave model has the inherent property that very little internal reflection occurs. Hence we expect that the analogous transmission line, when it is cut at a certain location, provides a nearly perfect impedance match on both sides of the cut. That is, each part should be matched to its ´characteristic impedance´. This is the impedance with which a finite line can be loaded without altering the distribution of voltages and currents from those in an infinitely long line. For an electrical line with constant parameters the characteristic impedance Z_o is given by

$$Z_o = (Z_1 / Y_1)^{\frac{1}{2}} \qquad (1)$$

For a homogeneous line with no external reflections, the quotient E(x)/I(x) is equal to the CCImp everywhere. For the line representing a cochlear model the parameters are not constant. We will use the term ´classical characteristic impedance´ (abbreviated CCImp) for Z_o computed from eq. (1) evaluated locally.

If impedance matching in the cochlear model is really so perfect, the active section L_2 is loaded on both sides by the CCImp evaluated at its end points. To find out whether this is true, the local impedance $p(x)/v_x(x)$ was calculated as a function of x. The model equation reads:

$$\frac{d^2 p(x)}{dx^2} - 2\, i\, \omega\, \rho\, p(x) \,/\, [\, h_e\, \zeta(x)\,] = 0 \qquad (2)$$

We used an exponential function for the BM stiffness:

$$\zeta(x) = i\omega M_1 + R(x) + S_0 \exp(-\alpha x)/i\omega \qquad (3)$$

Here M_1 is the BM´s mass constant, S_0 the stiffness constant at the stapes location $x = 0$, and α the space constant of stiffness. The resistance function R(x) is also an exponential function:

$$R(x) = \delta\, (S_0\, M_1)^{\frac{1}{2}} \exp(-\alpha x/2) \qquad (4)$$

where δ is the damping constant. Numerical values are the same as in [1]. The number of mesh points was 350. On a VAX 11/750 computer, the required cpu time for the solution is 0.11 seconds.

Fig. 2 presents typical impedance functions, LImp and CCImp. The excitation frequency is 1000 Hz, the figure depicts the x-region around the location of resonance $x = x_0$ for this frequency (indicated by an arrow). Excitation occurs at the left side of the cochlea (x = 0). Curves 1-R and 1-I show the real and imaginary parts, respectively, of the CCImp. Note that at the resonance location these two parts are equal because at resonance $\zeta(x)$ is real. Curves 2-R and 2-I show, in the same way, the LImp, i.e., the local impedance $p(x)/v_x(x)$.

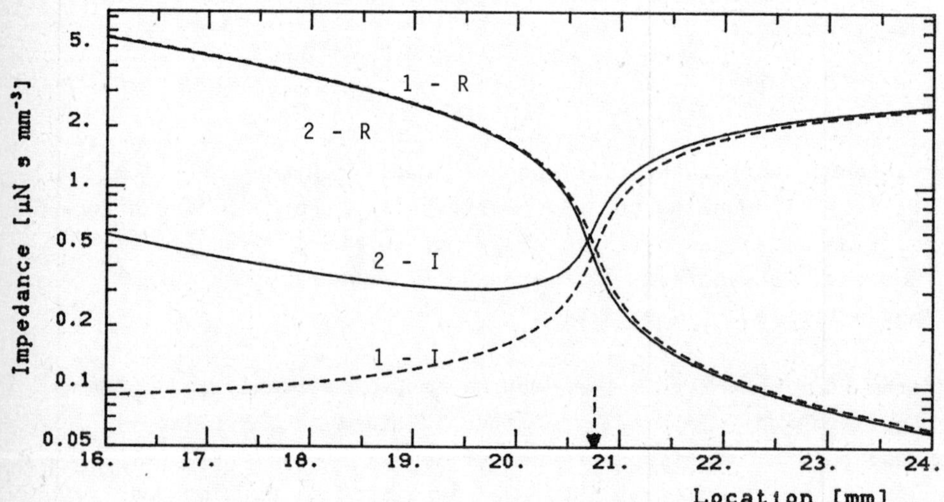

Fig. 2. Impedance levels along the line.

It is seen that curve pairs 1 and 2 differ appreciably from one another. However, in the region of resonance CCImp and LImp are not too far apart, their magnitudes and slopes differ by not more than 20 percent. Further computations show that LImp can be taken as the impedance of L_3 as it is loading L_2 at the right side. For the section L_1 the situation is more complicated. There will be reflections inside L_1 when we attempt to drive it from the right. This is because the cochlear model has a 'preference' for waves to the right (i.e., in the direction of increasing x). Hence the impedance loading L_2 at the left side cannot be equal to the LImp of Fig. 1, and indeed it is found to show clear evidence of standing waves. Over the resonance region, however, this load impedance remains fairly close the LImp, the deviations being of the same order of magnitude as those between LImp and CCImp. The loading at the stapes has, remarkably enough, little influence.

CONDITION FOR STABILITY

In view of the foregoing we can assume that the active section, L_2 of Fig.1-a, is loaded on both sides by the appropriate CCImp. Now we make the length Δl of L_2 very small so that L_2 can be represented as in Fig. 1-b with its impedance parameters evaluated at $x = x_o$. It should be well noted at this point that the shunt impedance $(Y_1 \Delta l)^{-1}$ of L_2 increases in magnitude when Δl decreases, this turns out to have important consequences. For the active section L_2 the imaginary part of $\zeta(x)$ is the same as for L_1 and L_3 but the real part is negative. At and around the location $x = x_o$ we choose the latter as:

$$\text{Re}\{\zeta(x_o)\} = -\varepsilon\,\omega\,M_1 \qquad (5)$$

The positive constant ε describes the degree of activity just as δ describes the degree of damping. Note that in the circuit of Fig. 1-a the pressure remains continuous at the boundaries of the active section whereas the BM velocity is discontinuous. This is because we visualize the local activity of L_2 as a jump in the impedance of the BM.

Since L_2 is loaded by impedances with a substantial inductive component oscillation will occur somewhat below the resonance frequency of L_2. For very small Δl the shunt reactance of L_2 is large compared to the other reactances so that the deviation remains small. To formulate an approximate condition for oscillation (the boundary between stability and instability) it is thus sufficient to consider only the real parts of the pertinent impedances at the resonance point $x = x_o$. To make the system stable the following relation must be obeyed:

$$\mathrm{Re}\,[-\varepsilon\,\omega\,M_1 + (Y_1\Delta l)^{-1}\,] > 0 \tag{6}$$

This can be reformulated as the final condition for stability:

$$\Delta l > 2\,\varepsilon\,(h_e\,M_1/\rho\,\delta)^{\frac{1}{2}} \tag{7}$$

For the typical parameter values we have been using the limiting value of Δl is approximately equal to 6 ε [mm].

We next discuss this result in relation to the problem of modelling a sound emitter in the cochlea. Interpreting eq. (7) as one producing ε for given Δl we see that an active unit of length $\Delta l = 10$ μm (about one hair cell wide) will be stable as long as ε remains below approximately 1.6 10^{-3}. For a larger value of ε the unit will go into spontaneous oscillation. Physically, this is easily understood: creating a larger negative resistance in a circuit increases the danger of oscillation.

The converse property is less easily understood: eq. (7) effectively tells us that, when an active section is made sufficiently long, the system becomes stable. This runs counter to physical intuition. The reason for this paradox lies in the fact that the magnitude of the negative resistance is inversely proportional to Δl. When Δl increases from a very small value, there comes a point where the negative resistance becomes too small to compensate the positive resistance component of $Z_o/2$. The system then remains stable. We should also remember that the representation of the active section L_2 as a T-network of the form of Fig. 1-b is correct only for values of the length Δl that are small compared to one wavelength.

The validity of the derivation was checked by developing a numerical solution of the long-wave cochlear model in the time domain (for details of the method employed see [3]). Given basilar membrane displacement u and velocity w, the fluid pressure p(x,t) is calculated from

$$\frac{d^2 p(x,t)}{dx^2} - 2\rho\, p(x,t)/(h_e * M_1) = -2\rho[R(x)w+S(x)u]/(h_e * M_1) \quad (8)$$

where S(x) stands for $S_0\exp(-\alpha x)$. From p(x,t) the BM acceleration is derived and the result is employed in a fourth-order Runge-Kutta method to update u and w. The program requires 0.2 ms per mesh point and time step. Convergence or divergence of w as a function of time is used as the stability criterion for given values for ε and Δl. Fig. 3 shows a number of representative results (see legend). In a general sense the numerical results agree well with the prediction provided by eq. (7) (the dashed line). What is most important: the figure confirms that larger values of Δl generally lead to more stable systems (the largest values of Δl obviously fall outside the scope of the derivation).

DISCUSSION, EXTENSION OF THE THEORY

The main findings of this study are

1) The minimal length Δl for a stable system is proportional to the degree of activity ε (at least for small Δl).

2) The impedances affecting the stability of a potential emitter depend surprisingly little on the loading conditions at the stapes.

To study the second point somewhat more, we generalized the concept of characteristic impedance. The generalized characteristic impedance function, let us call it $Z_{oo}(x)$, should have the following property. When the network of Fig. 1-b, of length dx, is loaded with $Z_{oo}(x+dx)$, the system must have $Z_{oo}(x)$ as its input impedance. This requirement immediately leads to the following Riccati equation for $Z_{oo}(x)$:

$$\frac{d}{dx} Z_{oo}(x) = Y_1 Z_{oo}^2(x) - Z_1 \quad (9)$$

This equation has been solved numerically and the $Z_{oo}(x)$ function is found to agree very well with the LImp. Via a circuitous route (to be published elsewhere) the equation can be solved analytically by the LG method. The resulting impedance function only involves the local wavenumber, defined as $k(x) = (-Z_1 Y_1)^{\frac{1}{2}}$, and its derivative with respect to x. This approximate solution also gives a surprisingly good account of the course of the LImp as a function of x. This result once more confirms that internal reflections occur due to inhomogeneity and that these cause the observed differences between the CCImp and LImp. They do not affect acoustic emissions very much, however. Why this is so, remains difficult to explain. Directions of further study are inspired by the numerous simplifications and restrictions made in the present analysis such as: long waves, imbedding in a passive structure, second-order dynamics of the emitter, etc. Improvements on these points will be taken up later.

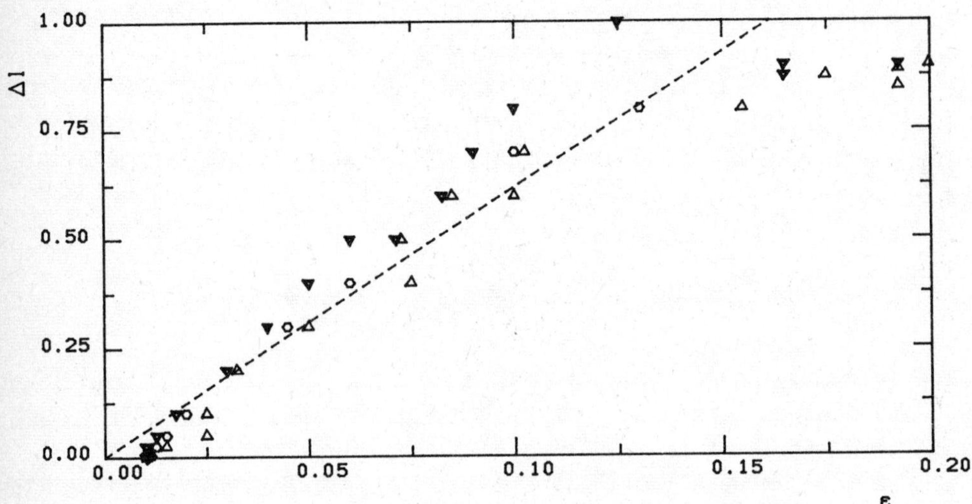

Fig. 3. Classification of time-domain solutions.
Stable (▽) -- constant amplitude (○) -- unstable (△).

REFERENCES

[1] de Boer, E. (1980) Auditory physics. Physical principles in auditory theory. Part I. Physics Reports 62, 87-174.

[2] de Boer, E. (1984) Auditory physics. Physical principles in auditory theory. Part II. Physics Reports 105, 141-226.

[3] Duifhuis, H., Hoogstraten, H.W., van Netten, S.M., Diependaal, R.J. and Bialek, W. (1985). Modelling the cochlear partition with coupled Van der Pol oscillators. This volume, pp. 290-297.

[4] Zweig, G., Lipes, R. and Pierce, J.R. (1976). The cochlear compromise. J. Acoust. Soc. Amer. 59, 975-982.

STABILITY OF ACTIVE COCHLEAR MODELS: NEED FOR A SECOND TUNED STRUCTURE?

Bernd Lütkenhöner and Dieter Jäger
ENT Clinic and Institute for Applied Physics, University of Münster
D-4400 Münster, Federal Republic of Germany

ABSTRACT

The question has been investigated whether or not it is possible to design a stable active model of the cochlea without using a second tuned structure. First a "learning" feedback model with negative damping in the cochlear partition is considered. This model, however, turned out to be unstable. The second model is characterized by a constant negative damping of the longitudinal fluid motion. An analysis of the energy flow shows that this simple model is stable as long as reflections at the helicotrema can be neglected. Consequently a second tuned structure seems not to be essential for the stability of a model.

1. INTRODUCTION

Considering recent experimental findings (see e.g. the survey given by Davis, 1983) there seems to be no alternative to the assumption that the cochlea is an active mechanical system. This view is strongly supported by a theoretical analysis of de Boer (1983). Since the pioneering work of Kim et al. (1980) who assumed a negative damping in the cochlear partition several active mechanical systems have been proposed (see e.g. the proceedings of the 1983 Mechanics of Hearing Symposium, ed. by de Boer and Viergever). All these theories, however, are more or less speculative. A decisive criterion of a model is its stability. The easiest way to test the stability is comparing the numerical time-domain and frequency-domain solution. Doing this Kim et al. found their model to be unstable (see also Diependaal and Viergever, 1983). An alternative method is a frequency-domain test described by Koshigoe and Tubis (1983). In contrast to Kim et al. whose model has only one degree of freedom in the partition mechanics, Neely (1981) introduced a second degree of freedom associated with the bending motion of the outer hair cell cilia. He showed that the parameters of his model can be chosen such that the model is stable. Since his model requires a second tuned structure, it is related to the concept of a 'second cochlear filter' suggested by Evans and Wilson (1973). A fundamental difference, however, is the active, bidirectional coupling in Neely's model. The present paper is devoted to the question whether or not a second tuned structure is essential to guarantee stability of an active cochlear model.

2. BASIC THEORY

For the sake of simplicity the following investigation is based upon the usual one-dimensional cochlear model extended by a longitudinal damping element. Fig. 1(A) shows an equivalent electrical circuit corresponding to a short section of the model. If it is assumed that the cross-sectional area of the scalae is constant, then the displacement ψ of the basilar membrane (BM) is given by the partial differential equation

$$\frac{\partial^2}{\partial x^2}\left(m\frac{\partial^2\psi}{\partial t^2} + r\frac{\partial\psi}{\partial t} + s\psi\right) - \frac{2\rho_0}{h}\frac{\partial^2\psi}{\partial t^2} - 2r_L\frac{\partial\psi}{\partial t} = 0. \quad (1)$$

The quantities m, r, and s correspond to mass, resistance and stiffness of the BM; ρ_0 is the density of the perilymph, while h is the effective height defined as the quotient of cross-sectional area of either scala and width of the BM. The quantity r_L corresponds to the damping of the longitudinal fluid motion. A frequency-domain formulation is obtained by means of the substitution $\psi(x,t)=\Psi(x)\cdot\exp(i\omega t)$. The pressure difference between the scalae is $2P=-i\omega Z\Psi$, where

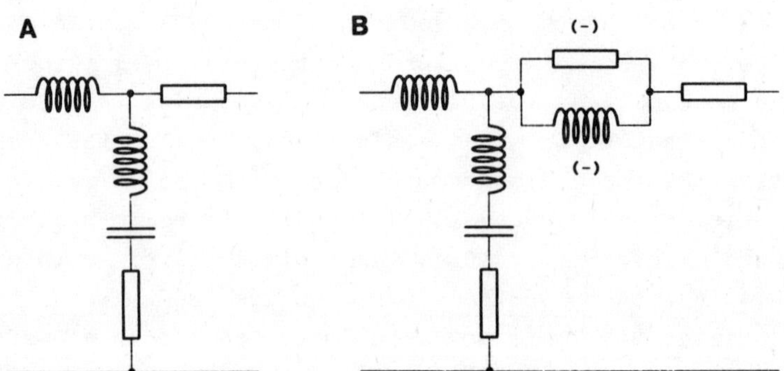

Figure 1. (A) Equivalent electrical circuit of an one-dimensional cochlear model. (B) Modified active model. Negative components are marked by the symbol (-).

$Z=i\omega m+r+s/i\omega$ is the impedance of the BM. Thus in the frequency-domain the model may be written as

$$\left(\frac{d^2}{dx^2}+K^2\right) P = 0, \tag{2}$$

where

$$K^2(x) = -2i\omega\rho_0(1-ig) / hZ(x), \qquad g=r_L h/\omega\rho_0. \tag{3}$$

3. FEEDBACK MODEL WITH NEGATIVE DAMPING IN THE COCHLEAR PARTITION

The problem inherent in the model of Kim et al. (1980) is that the active component of the impedance has to be defined explicitly for each frequency. But how does the cochlea "know" the actual stimulus frequency, and what will happen, if the stimulus consists of more than one frequency? To avoid the assumption of a second tuned structure (Neely, 1981) in the following a "learning" feedback model with only one degree of freedom and without a longitudinal resistance ($r_L=0$) is investigated: It is assumed that the resistance r is made up of a positive resistance r_0 and a negative resistance r_a given as

$$r_a(x,t) = -\int_0^L c(x,u)\cdot\Psi_\tau(u,t)\,du. \tag{4}$$

In this equation c is a weighting factor while the quantity Ψ_τ is a measure of the amplitude of the BM displacement ψ. Rather arbitrary the relation between Ψ_τ and ψ is

$$\Psi_\tau(x,t) = \int_{-\infty}^{t} \max[0,\psi(x,\theta)]\cdot\exp(-(t-\theta)/\tau)\,d\theta. \tag{5}$$

Roughly speaking these mathematical assumptions mean: There is an element with some kind of memory at each location x which continuously estimates the amplitude of the BM displacement ψ. The estimated amplitude is transmitted to other locations where it is used to update the value of the negative resistance r_a. The weighting factor c in equation (4) has to be chosen such that a damping profile results similar to that given by Kim et al. (1980). Instead of (4) only the special case $r_a(x,t)=-c_0\Psi_\tau(x+\Delta x,t)$ is considered in the following.

Fig. 2 shows the result of a computer simulation. The calculation has been started using the frequency domain solution obtained for the case $r_a=0$. This implies that the stimulus is switched on at time $t=-\infty$, while the negative resistance is switched on at time $t=0$. Fig. 2(B) shows that at time t=12 ms a damping profile is obtained

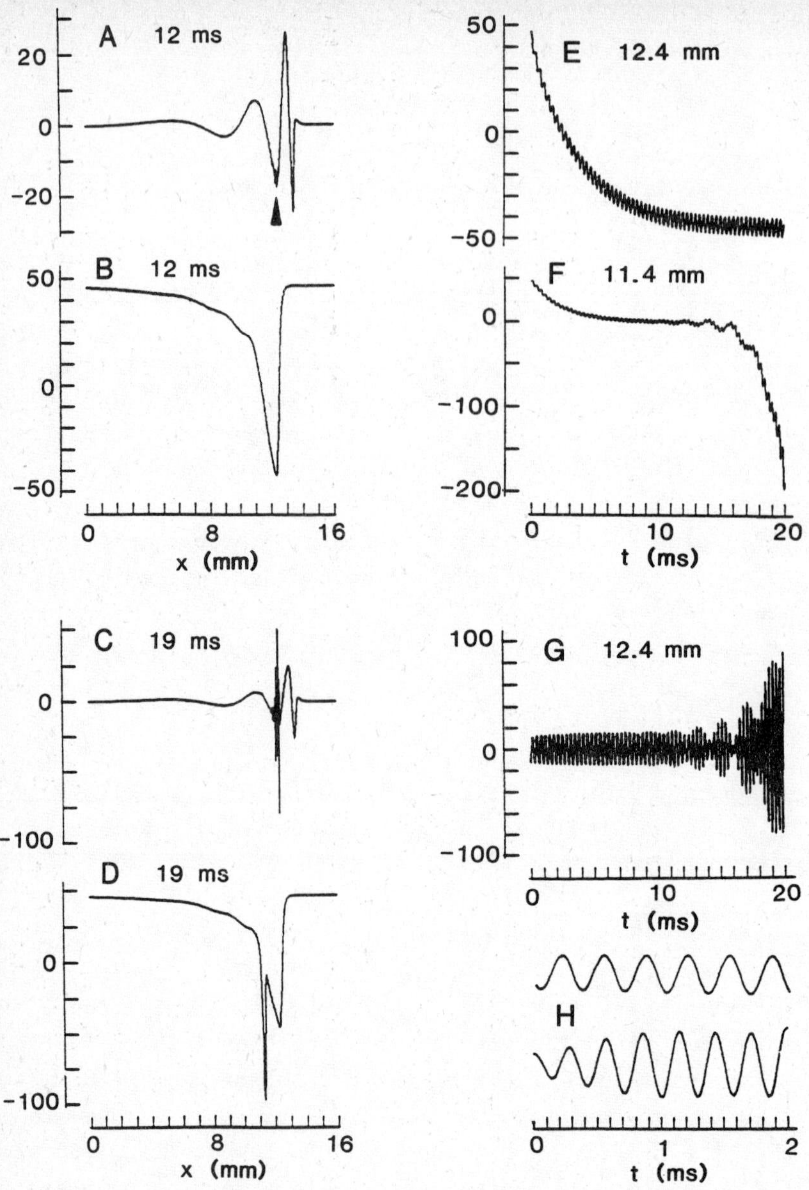

Figure 2. Computer simulation for the "learning" feedback model (arbitrary units). (A) BM displacement at time t=12 ms. (B) Resistance r at time t=12 ms. (C) BM displacement at time t=19 ms. (D) Resistance r at time t=19 ms. (E) Resistance at location $x_r-\Delta x$. (F) Resistance at location $x_r-2\Delta x$. (G) BM displacement at location $x_r-\Delta x$. (H) First (upper trace) and last (lower trace) 2 ms of (G) on an enlarged scale. For the numerical calculations a modification of the algorithm described by Diependaal and Viergever (1983) has been used. Parameter values: m_0=0.05 g/cm²; $r_0=15\pi$ g/cm²s; s=10⁹exp(-3x) dyn/cm³; h=0.1 cm; r_L=0; Δx=0.1 cm; stimulus frequency 3 kHz.

which is qualitatively identical to that of Kim et al. (1980). This means that the negative resistance of the model has automatically been tuned to the stimulus frequency. Now the question arises whether the model is stable or not. At time t=12 ms the displacement of the BM (fig. 2(A)) is inconspicuous, except for a small disturbance marked by an arrow. This disturbance coincides with the minimum of the resistance located at $x_r-\Delta x$ where x_r denotes the resonance location corresponding to the zero crossing of the imaginary part of the impedance Z. Within the next few milliseconds the disturbance rapidly increases; it has become a dominating feature at t=19 ms (fig. 2(C)). The resistance r now has a second minimum at $x_r-2\Delta x$ (fig. 2(D)). The time course of the resistance at $x-\Delta x$ (fig. 2(E)) and the resistance at $x-2\Delta x$ (fig. 2(F)) is completely different. In the first case the resistance is approximately constant after an exponential decay (the small oscillations reflect the periodicity of the stimulus), while in the second case a rapid decrease is found starting at about t=16 ms. This rapid decrease of the resistance at $x-2\Delta x$ is due to an infinite increase of the BM displacement at $x-\Delta x$ (fig. 2(G)). In summary: The location $x-\Delta x$ is a primary focus of unstability which establishes a secondary focus at $x-2\Delta x$.

To understand these results consider the energy flow. The energy of the sinusoidal stimulus first passes a region with small positive damping where the energy flow is only little affected, then enters a region with negative damping where a considerable amplification occurs, and finally reaches the region near the resonance location where the damping is positive again and all energy is dissipated, since the velocity of the travelling wave goes to zero. The model hence should be stable. The crucial point, however, is that the negative damping is not specific to the stimulus frequency, but affects any other frequency as well. Of special interest are frequencies with a resonance location lying within the region of negative damping. Such frequencies always exist, irrespective of the stimulus, due to ubiquitous noise. Obviously a frequency component with negative damping at its resonance location is infinitely amplified, i.e. the model is unstable. It takes about 12 ms in the present case until the frequency with resonance location at $x_r-\Delta x$ (about 3.5 kHz) becomes noticeable as a small disturbance in fig. 2(A) or an interference in fig. 2(G). This component is dominating at t=20 ms as demonstrated in fig. 2(H) which shows the first (upper trace) and the last (lower trace) 2 ms of fig. 2(G) on an enlarged scale. While in the upper trace 6 cycles are found corresponding to 3 kHz (the stimulus frequency), in the lower trace 7 cycles are found corresponding to 3.5 kHz. - Finally it shall be stressed that the results described above are typical for an active model with only one degree of freedom in the partition mechanics. Similar results have been obtained e.g. for the active model of Kim et al. (1980).

4. MODEL WITH NEGATIVE DAMPING OF THE LONGITUDINAL FLUID MOTION

The next model which has been investigated is extremely simple. The basic idea should be regarded, however, more as a mathematical formalism than as a physiological concept: It is assumed that the longitudinal fluid motion undergoes a constant and frequency-independent negative damping, i.e. $r_L<0$. To demonstrate the main properties of the model a frequency-domain calculation has been carried out for a stimulus of 3 kHz. Fig. 3(A) shows the pressure P(x) obtained for

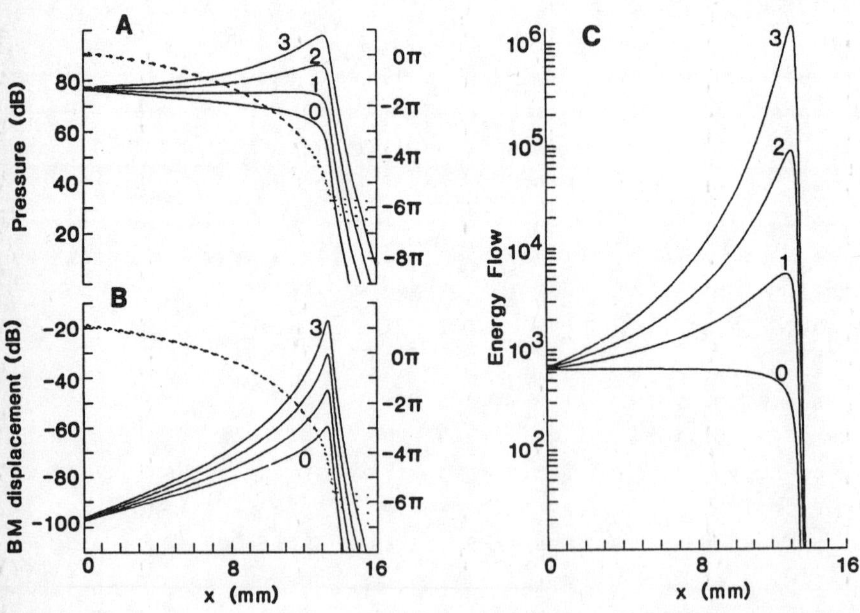

Figure 3. Model with negative longitudinal damping (frequency-domain solution by means of Hamming´s modified predictor-corrector method). (A) Pressure P. (B) BM displacement Ψ. (C) Energy flow. Parameter values: $m_0=0.05$ g/cm²; $r_0=15\pi$ g/cm²s; $s=10^9\exp(-3x)$ dyn/cm³; h=0.1 cm; g=0, -0.2, -0.4, -0.6; stimulus frequency 3 kHz.

4 different values of the parameter g (the symbols 0-3 correspond to g=0, -.2, -.4, -.6); fig. 3(B) shows the BM displacement ψ. The magnitude has been drawn as a solid line, the phase as a dashed line. Fig. 3(C) shows the energy flow. The figure demonstrates that, depending on the parameter g, a considerable gain can be obtained (about 40 dB for g=-.6) which gives rise to a significant peak of $|P(x)|$. Furthermore, a steep drop of the energy flow can be observed beyond the resonance location, irrespective of the parameter g. Provided that the length of the cochlea is infinite, similar results can be obtained for any other frequency. Since the model is linear, stability is guaranteed for any kind of stimulus.

A time-domain simulation with a model of finite length turned out to be unstable. This kind of unstability is due to the resonant amplification of low-frequency waves reflected at the helicotrema. To avoid this problem the active mechanism must be deactivated at low frequencies. A simple solution is shown in fig. 1(B) where a negative inductance is connected in parallel to the negative resistance. In the asymptotic case $\omega \to 0$ this inductance acts as a short circuit. By means of an additional positive resistance a positive longitudinal damping is achieved for all frequencies below a certain upper limit. It should be noticed that some care is necessary to avoid unstability due to numerical inaccuracies. For our investigations a nonlinear function has been used to devide the X-axis into segments (high density of segments at the base, lower density near the helicotrema). The stimulus was fed to the BM via a "primitive" middle ear model. The resulting system of differential equations was solved by means of a fourth-order Runge-Kutta integration procedure.

5. DISCUSSION

Two kinds of unstability have been observed. As a rule of thumb a model shows an unstability of the first kind, if a frequency exists for which the impedance of the cochlear partition has a negative real part. An unstability of the second kind is to be expected if the active mechanism is not specific to high frequencies, but also acts on low-frequency waves reflected at the helicotrema (compare e.g. the problem of helicotrema resonance as dicussed by Lighthill, 1980). These two kinds of unstability constitute a severe restriction for the development of active cochlear models. Nevertheless stability can be achieved by means of a very simple model (fig. 1(B)). This model clearly demonstrates that with respect to stability an active model of the cochlea does not necessarily require a second tuned structure as proposed by Neely (1981). On the other hand the concept of negative damping of the longitudinal fluid motion is not as different from the concept of negative damping in the cochlear partition as it may seem at first glance: As long as only the pressure P is considered the negative longitudinal resistance r_L can be absorbed as well in the impedance of the cochlear partition, since equation (3) may be rewritten as $K^2=-2i\omega\rho_0/hZ_e$ where Z_e denotes the "equivalent impedance" defined as $Z_e=Z/(1-ig)$. If $g<0$, Z_e is qualitatively very similar to the impedance obtained as a solution of the inverse problem considered by de Boer (1983), at least in the vicinity of the resonance location. Thus the model of fig. 1(B) can be considered as the (presumably) most simple stable time-domain realization of de Boer´s frequency-domain model.

REFERENCES

de Boer, E., "No sharpening? A challenge for cochlear mechanics." J. Acoust. Soc. Am. 73, pp. 567-573, 1983

de Boer, E., Viergever, M.A. (eds.), "Mechanics of Hearing." Martinus Nijhoff Publishers, Delft University Press, 1983

Davis, H., "An active process in cochlear mechanics." Hearing Research 9, pp. 79-90, 1983

Diependaal, R.J., Viergever, M.A., "Nonlinear and active modelling of cochlear mechanics: a precarious affair." In: de Boer and Viergever (eds.), loc. cit., pp. 153-160, 1983

Evans, E.F., Wilson, J.P., "The frequency selectivity of the cochlear." In Basic Mechanisms in Hearing, ed. by A. R. Moller, Academic Press, New York, pp. 519-554, 1973

Kim, D.O., Neely, S.T., Molnar, C.E., Matthews, J.W., "An active cochlear model with negative damping in the partition." In Psychophysical, Physiological and Behavioural Studies in Hearing, ed. by G. van den Brink, F. Bilsen. Delft University Press, pp. 7-14, 1980

Koshigoe, S., Tubis, A., "Frequency-domain investigations of cochlear stability in the presence of active elements." J. Acoust. Soc. Am. 73, pp. 1244-1248, 1983

Lighthill, J., "Energy flow in the cochlea." J. Fluid Mech. 106, pp. 149-213, 1981

Neely, S. T., "Fourth-order partition dynamics for a two-dimensional model of the cochlea." Doctoral dissertation, Washington University, St. Louis, MO, 1981

CHANGES IN SPONTANEOUS AND EVOKED OTOACOUSTIC EMISSIONS AND CORRESPONDING PSYCHOACOUSTIC THRESHOLD MICROSTRUCTURES INDUCED BY ASPIRIN CONSUMPTION.

Glenis R. Long, Arnold Tubis, and Kenneth Jones.
Purdue University
West Lafayette, IN 47907.

ABSTRACT

Spontaneous otoacoustic emissions, delayed evoked emissions, synchronous evoked emissions and psychoacoustical threshold microstructure were monitored (in two subjects) before, during and after the consumption of 3.9g of aspirin per day for three consecutive days (12 doses of three 325mg tablets every 6 hours). Spontaneous emissions followed a pattern similar to that found by McFadden and Plattsmeir (1984). Evoked emissions were also reduced by aspirin consumption but persisted longer and recovered sooner. Reduction of psychoacoustic threshold microstructure associated with the emissions followed much the same time course as the evoked emissions. In most instances the reduction of threshold microstructure began with a lowering of threshold maxima (with threshold minima remaining relatively constant) and ended with all thresholds elevated.

1. INTRODUCTION

Since Kemp (1978) first demonstrated that sounds of cochlear orgin (otoacoustic emissions) can be detected in the human ear canal, research in many different laboratories (reviewed in Zwicker and Schloth, 1984) has lead to the description of three different types of otoacoustic emissions: 1) spontaneous emissions (SOAE), which occur in the absence of external stimulation; 2) delayed evoked emissions, stimulated by clicks or brief sinusoidal signals; and 3) synchronous evoked emissions, stimulated by tonal signals and having the same frequency as the external signal. The three different types of emission appear to be closely related to one another and to the psychoacoustic threshold microstructure (unique consistent patterns of threshold maxima and minima), but the exact relationship is difficult to specify, particularly since the measures have mostly been obtained in different individuals at different times. The best way to examine the relationship between these phenomena would be to find some way of modifying one or more of them in the same individual across time.

Recently aspirin consumption has been found to reduce delayed evoked otoacoustic emissions (Johnson and Elberling, 1982) and reversibly reduce and subsequently eliminate spontaneous evoked emissions (McFadden and Plattsmeir 1984). Johnson and Elberling's subject consumed 10g of aspirin in 24 hrs, while McFadden and Plattsmeier's subjects consumed 3.9g per day for 3.75 consecutive days; consequently, it is not possible to compare the effects on the different emissions. Neither investigator measured synchronous evoked emissions or threshold microstructure. McFadden and Plattsmeir did monitor audiometric thresholds in two of their subjects but did not relate them to evoked emissions or detailed measures of threshold microstructure.

The present study was undertaken in the hope that simultaneous measurement of all three types of otoacoustic emissions and the corresponding psychoacoustic threshold microstructure would help to increase our understanding of the interrelations of these phenomena and suggest constraints in the forms of models used to describe cochlear activity.

2. METHODS

Otoacoustic emissions were measured in, and external sounds presented to, the ear canal using Knowles transducers (models EA-1843 and BT-1752 respectively) connected to a Grason-Stadler otoadmittance meter earpiece. Otoacoustic emissions were amplified by a locally designed amplifier and analyzed using a Wavetex 5820A Spectrum analyzer. Spontaneous emissions were analyzed using spectral averaging with 2.5 Hz line spacing leading to a noise floor of approximately -10 dB SPL. Delayed evoked emissions were stimulated by 20 dB SL, 30 μs pulses and (after removal of the first 6 ms and filtering from 500 Hz to 5.5 kHz) time averaged (256 averages, 80 ms window) before analyzing spectrally.

Synchronous evoked emissions cannot be measured easily but an indication of their relative strengths was obtained by measuring the fluctuations of the rms level in the ear canal at 10 dB SL (measured by a Bruel and Kjaer heterodyne slave filter 2010, 3.16 Hz bandwidth) relative to the level at 60 dB SL (see Zwicker and Schloth, 1984).

Psychoacoustic thresholds near spontaneous and evoked emissions were obtained using a discrete frequency tracking technique. A continuous sequence of tone pulses (250 ms tones, 20 ms rise/fall, separated by 250 ms silence) decreased in intensity (1 dB/s) until the subject pushed one button and then increased in intensity at the same rate until another button was pushed. Four reversals of direction were obtained and averaged at one frequency before the frequency was incremented by 5 Hz. McFadden and Platsmeir measured thresholds at 8 test frequencies during aspirin administration. For comparison purposes we measured similar frequencies in this study. The same tracking procedure was used except that the frequency increments were 500 Hz and 8 reversals were measured. The average of the last 6 reversals was taken as the estimate of threshold.

The subjects were the first author and one other female subject who was payed for participating in the study. Measures were obtained from both subjects before during and after consumption of three 325 mg tablets at 6 hr intervals for a total of 12 doses (3 days). More frequent measurements were obtained from the first author (2 or 3 times a day during aspirin intake and recovery). At least two measurements were obtained of each type of otoacoustic emission per session. Each session took about 2 hrs to complete. To reduce the impact of time as a factor, measurements were always obtained in the following order: spontanous emissions, delayed evoked emissions, synchronous evoked emissions, and finally, thresholds.

3. RESULTS AND DISCUSSION

The effect of the aspirin on the spontaneous emissions (three in the right ear and one in the left ear of GL and four in the right ear and three in the left ear of CH) is very similar to that reported by McFadden and Plattsmeier, 1984 (see Figs. 1a-2a, note different scale on Fig 2). The smaller emissions tended to disappear into the noise floor within 24 hours and the larger ones within two days. Recovery was more variable and the relative size of emissions during recovery differed from that seen before aspirin consumption. The pre-aspirin state had not returned when the study was completed 4 days after termination of aspirin intake. All emissions returned at a somewhat lower frequency.

The delayed evoked emissions in GL's right ear (Fig. 1b) remained at pre-aspirin levels for 24 hours after the beginning of aspirin consumption but then gradually disappeared into the noise floor, recovering very rapidly after termination of aspirin intake. The delayed evoked emission in GL's left ear also stayed at pre-aspirin levels for 24 hours but then rapidly disappeared into the noise floor. Recovery was also extremely rapid. An additional measurement of delayed evoked emissions using 40 dB SL clicks was obtained at 1.67 days after the beginning of aspirin consumption to determine if the reduced evoked emissions indicated a reduced threshold or an overall reduction of the evoked emission. The delayed emissions evoked by the 40 dB SL clicks were 1-2 dB greater than the those evoked by the 20 dB SL clicks, still well below pre-aspirin levels. Furthermore, behavioral thresholds at this time (see below) were not elevated. These data are not in conflict with Johnson and Elberling's (1982) study as both aspirin dose and measuring technique were different.

Unfortunately, measurements were only obtained from CH every 24 hrs during the study (except twice on the first day of recovery), consequently different rates of change in the level of spontaneous vs evoked emissions were not always detected. Some of the evoked emissions persisted after the spontaneous emissions were no longer detectable and these emissions recovered sooner.

Synchronous evoked emissions (Figs. 1c and 2c) followed a time course similar to that of the delayed evoked emissions: the major contrast being the greater difference between the sizes of the individual emissions. Not all the synchronous evoked emissions were completely abolished. Such emissions were estimated by summing the maximum upward and downward fluctuation of ear canal sound pressure immediately above and below the frequency of the emission, a measure which is less susceptible to noise floor problems than the procedures used to measure spontaneous or delayed evoked emissions.

Detailed plots of psychoacoustical threshold microstructure from GL's right ear and CH's left ear at different stages of aspirin consumption are displayed in Fig. 3a and b. The effect of aspirin was to first reduce the threshold maxima while leaving threshold minima relatively unchanged or somewhat reduced. Finally, all thresholds were elevated, severely reducing, but not eliminating, the microstructure. The frequency of the threshold minima in Fig. 3b are somewhat lower after aspirin consumption, consistent with a downward shift in the frequency of these emissions during the study. In order to provide summary plots of the change over time we averaged 3 points at each threshold maxima and 3 points at each threshold minima. Fig. 3c tracks the changes of the threshold minima near 1895 Hz and the neighboring threshold maxima during the experiment. The initial

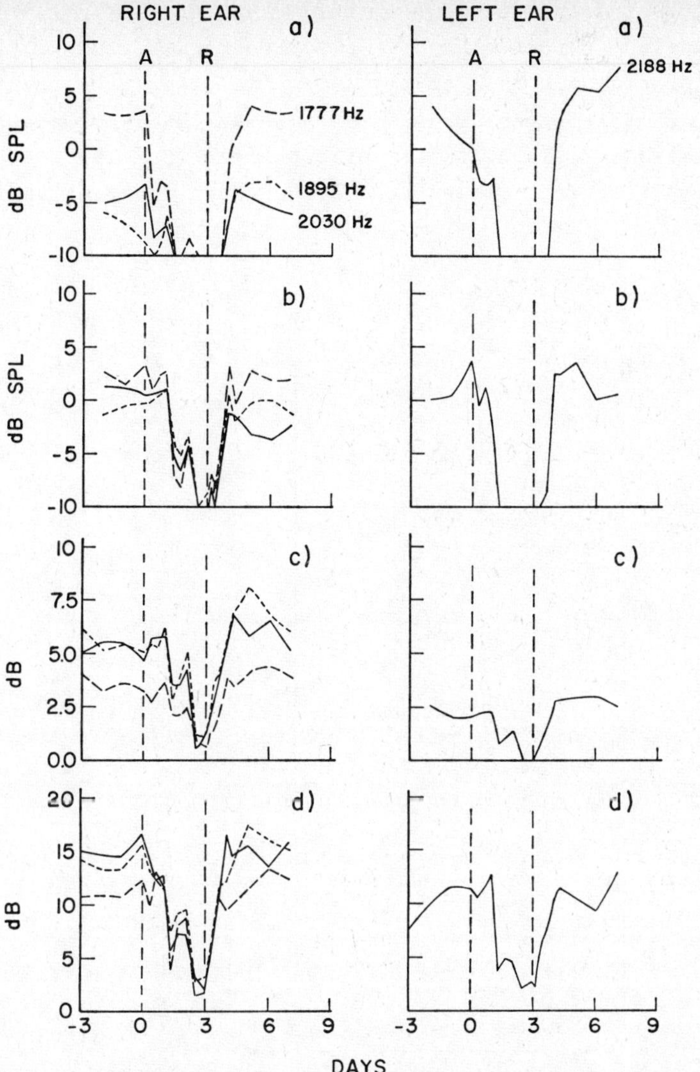

Fig. 1 (a) Spontaneous emissions, (b) delayed evoked emissions, (c) synchronous evoked emissions, and (d) the depth of threshold microstructure, as a function of time relative to the beginning of aspirin consumption in the right and left ears of GL. Each emission and its associated threshold microstructure is indicated by a different line. The frequency of each emission is indicated in a. A indicates beginning of aspirin intake and R the beginning of recovery.

impact of the aspirin is to reduce thresholds, particularly at threshold maxima, before elevating all thresholds. A similar trend is seen during recovery. This effect is not so clear from threshold tracks from GL's left ear, (Fig. 3d) in which the spontaneous and evoked isolated emissions rapidly disappeared. Similar trends were seen at the 8 test frequencies used by McFadden and Plattsmeier (1984)

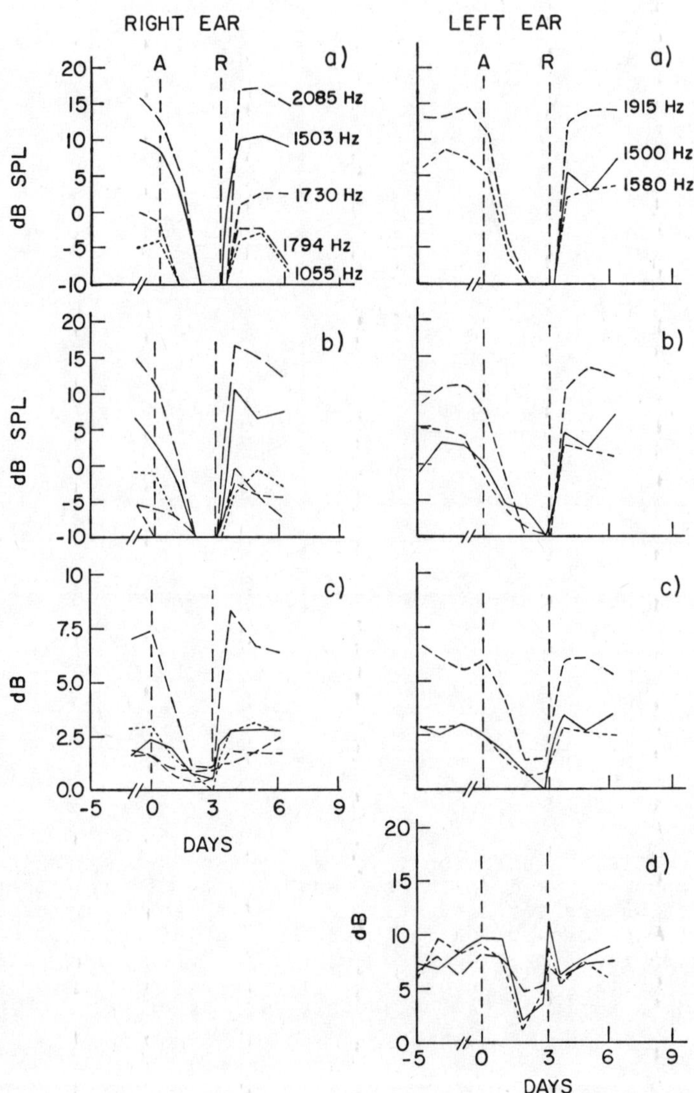

Fig. 2 Same as Fig. 1 for CH. Detailed thresholds were not obtained from the right ear.

when the test frequency fell in a region of threshold microstructure. An apparent reduction of threshold at some frequencies with aspirin intake was also found by McFadden and Plattsmier. We interpret the lowering of threshold maxima as stemming from a reduction of masking or other interference, of the externally presented tone by the otoacoustic emissions (Long et al., 1984).

A measure of the depth of the threshold microstructure (Figs. 1d and 2d) was obtained by subtracting the mean threshold of 3 adjacent frequencies at a threshold minima from the mean of three thresholds at each of the two neighboring maxima. The change in depth of threshold

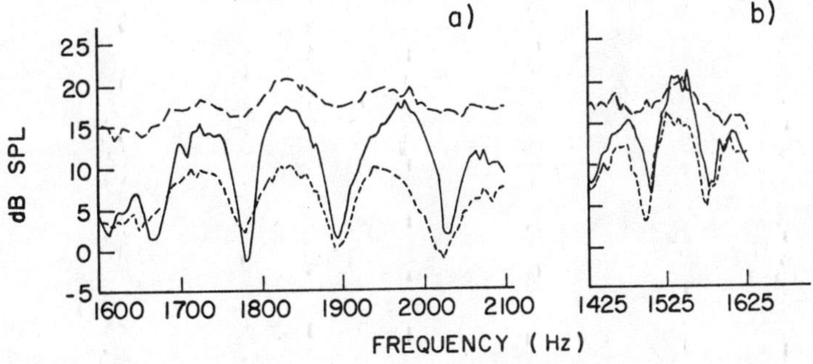

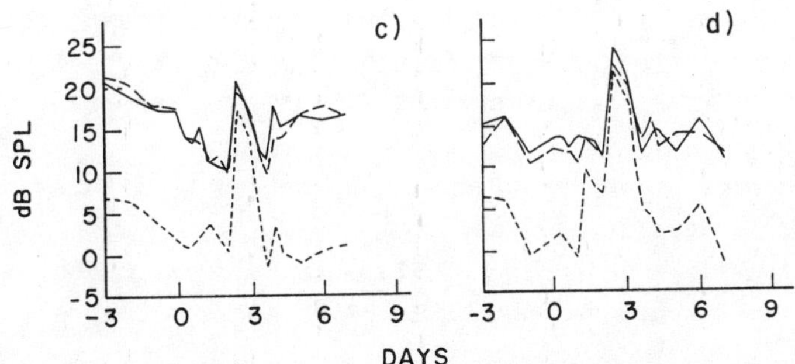

Fig. 3 (a) Threshold tracks from GL's right ear measured at the start of aspirin intake (solid line), and after 51 (short dashed line) and 61 hrs (long dashed line) of aspirin consumption. (b) threshold tracks from CH's left ear at start of aspirin consumption (solid line), 25 (short dashed line) and 49 hrs of aspirin. (c) Thresholds from a threshold minima (near 1895 Hz) and 2 adjacent threshold maxima (near 1835 and 1960 Hz) in GL's right ear as a function of time relative to the beginning of aspirin intake. (d) same as C for a threshold minima near 2190 Hz and threshold maxima near 2125 and 2225 Hz in GL's left ear.

microstructure follows much the same function of time as the evoked emissions. It is interesting to note that a small recovery after 1.67 days of aspirin consumption was reflected on all panels of Fig. 1, further indication of the interdependence of these measures.

The apparent abolition of spontaneous and delayed evoked emissions (within the constraints of the noise floor of the measuring system) and the substantial reduction of synchronous evoked emissions and behavioral threshold microstructure all support the assumption that aspirin intake diminishes the strength of active cochlear mechanisms. We are presently exploring the implications for models of active cochlear feedback forces (Koshigoe and Tubis, 1983) in which the effects of aspirin are associated with parametic changes in the model.

This work was supported by NINCDS Grant NS 22095-1 and Purdue University.

REFERENCES

Johnson, N.J., and Elberling, C. "Evoked acoustic emissions from the human ear. I. Equipment and response parameters," Scand. Audio. 11, 3-12, 1982.

Kemp, D.T. "Stimulated acoustic emissions from within the human auditory system," J. Acoust. Soc. Am. 64, 1386-1391, 1978.

Koshigoe, S., and Tubis, A. "A non-linear feedback model for outer-hair-cell stereocilia and its implications for the response of the auditory periphery," in: Mechanics of Hearing, edited by E. deBoer and M.A. Viergever, (Martinus Nijhoff Publishers, The Hague), pp. 127-134, 1983.

Long, G.R., "The microstructure of quiet and masked thresholds." Hearing Res. 15, pp.73-87, 1984.

Long, G.R., Tubis, A., and Jones, K., "Is threshold microstructure a manifestation of tone-on-tone masking?" J. Acoust. Soc. Am. Suppl. 1 76, S 95, 1984.

McFadden, D. and Plattsmier, H.S., "Aspirin abolishes spontaneous otoacoustic emissions." J. Acoust. Soc. Am. 76, pp. 443-448, 1984.

Zwicker, E., and Schloth, E. "Interrelation of different otoacoustic emissions," J. Acoust. Soc. Am. 75, 1148-1154, 1984.

STATISTICAL PROPERTIES OF A STRONG SPONTANEOUS OTO-ACOUSTIC EMISSION

Hero P.Wit
Institute of Audiology
Postbox 30.001
9700 RB Groningen
The Netherlands

ABSTRACT

Several properties of a spontaneous oto-acoustic emission are summarized (frequency spectrum, amplitude distribution, suppression, phase-lock). The emission is generated by an active filtering process in the inner ear. It can be described by a simple electronic model. Characteristics of the emission signal are compared with those of spontaneous voltage fluctuations in a hair cell.

1. SPONTANEOUS OTO-ACOUSTIC EMISSION

We investigated the relatively strong spontaneous oto-acoustic emission (20 dB SPL in the closed ear canal) from the right ear of a normal hearing male subject (age 23 years). This emission was measured with a sensitive microphone (Wit et al.,1981), connected to the ear canal with a short tube. The average amplitude spectrum of this emission at 1415 Hz is given in fig.1. It was measured with a Princeton Applied Research 4512 spectrum analyzer. From this amplitude spectrum alone it

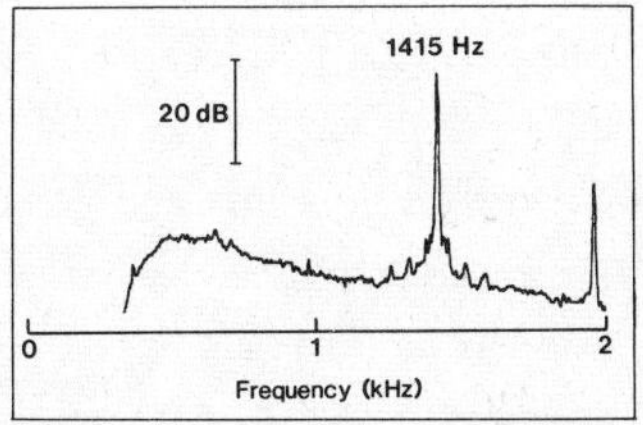

Figure 1. Amplitude spectrum of spontaneous oto-acoustic emission.

is impossible to decide whether the narrow-band emission is generated by some extraneous source of noise which is filtered by the ear, or that it is the result of an (unstable) active filtering process. The probability distributions for the emission amplitude, however, will be very different in these two cases (Bialek,1983): in a system where zero amplitude is a stable point, the probability distribution for the amplitude must have a local maximum at this point, and will generally be roughly Gaussian. But a system for which zero amplitude is an unstable point will have a local minimum in the probability distribution at zero. Figure 2 gives the amplitude distribution for the investigated emission; it is clearly in favour of an active filtering process in the ear. The amplitude distribution was measured with a Datalab DL 4000 system, after filtering of the amplified microphone signal with a Bruel and Kjaer 2020 slave filter with a bandwidth of 31.6 Hz.

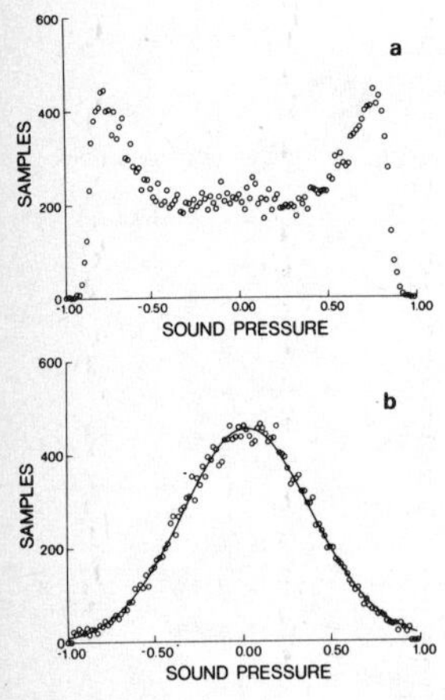

Figure 2.
Amplitude distribution for a frequency-band around 1415 Hz. (6 minutes time sample), containing the oto-acoustic emission signal (a), and for a frequency band adjacent to it with Gaussian least-squares fit (b).

A simple model for an active filter can be obtained by applying a feedback force with the proper phase to a mass-on-spring oscillator. Parameters in the (two-dimensional diffusion) model that describes such a system (Bialek & Wit, 1984) are rms amplitude fluctuations, the time constant for relaxation of the amplitude fluctuations and fluctuation in the time between zero crossings of the oscillator. Figure 3 gives an example of the determination of one of these parameters. After rectification and low-pass filtering of the emission signal its spectrum was measured with the PAR spectrum analyser, thus providing the power spectru of the envelope of the emission signal. This spectrum consists of two components, a white noise which arises from the background fluctuations and a Lorentzian whose width measures the time constant for relaxation of the amplitude fluctuations of the emission signal. Measurement of the other parameters yielded numbers that were in excellent agreement with the diffusion model (Bialek & Wit, 1984).

The above mentioned parameters were measured several times. On some occasions the envelope spectrum did not have the simple shape as given in fig.3, but an extra maximum appeared at a frequency different

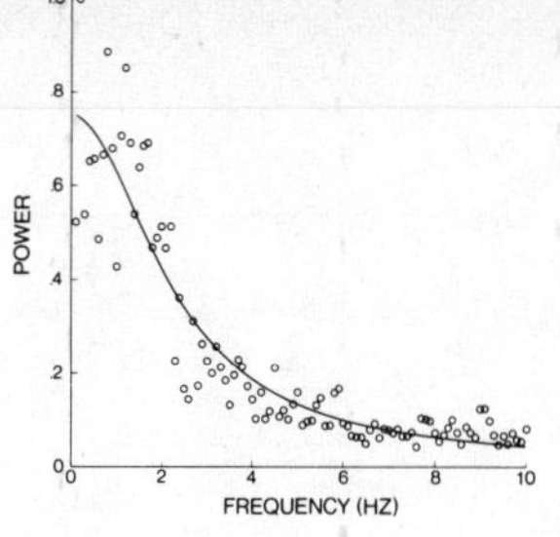

Figure 3.
Power spectrum of the envelope of bandpass filtered (100 Hz.bandwidth) oto-acoustic emission signal, with Lorentzian least-squares fit (solid line).

from zero. Although the appearance of this extra peak (fig.4) puzzled us for some time, we later found out that its frequency corresponded exactly with the value $(2f_1-f_2)-f_e$, in which f_1 and f_2 are frequencies of other emissions than the one that we have investigated in detail (with frequency f_e). One of these extra emissions (with frequency 1.94 kHz) can be seen in fig.1.

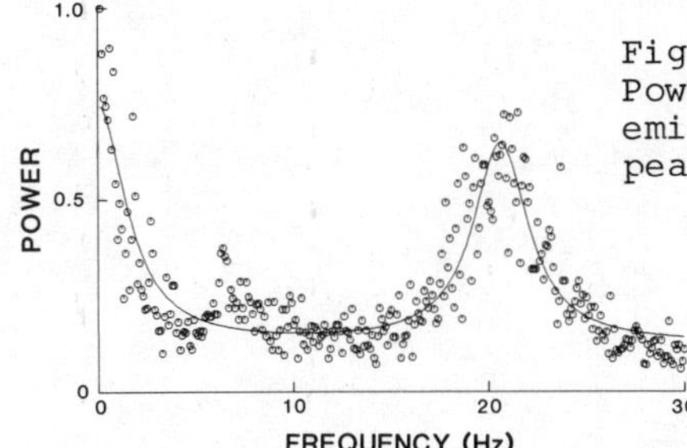

Figure 4.
Power spectrum of oto-acoustic emission signal envelope with extra peak at 21 Hz.

A spontaneous emission can be suppressed with an external tone. We have chosen the frequency of the suppressor 50 Hz. below that of the emission. Amplitude distributions for the filtered emission signal (through a bandwidth of 31.6 Hz) are given in figure 5. It is clear from this figure that adding an external tone of sufficient level changes the emission generator from an unstable active system into a stable system. A model to describe this phenomenon will be published elsewhere (Wit & Bialek, to be published).

If an external tone with the same frequency as that of the emission is supplied to the ear canal, phase-locking will occur. The higher the level of the external tone, the better the phase-lock (fig.6). The extent of phase-locking was measured by recording the zero-crossing distribution of the (filtered) emission signal with the Datalab system, which was triggered by the external tone. It was described with a model

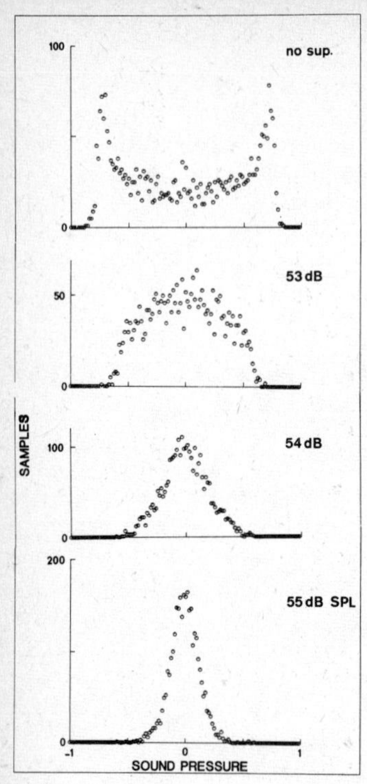

Figure 5.

Amplitude distribution for oto-acoustic emission signal for different levels of suppressor tone supplied to the ear canal. Horizontal scale is the same in all four panels. An external tone of 55 dB SPL suppresses the RMS-amplitude of the emission 13 dB below its unsuppressed value. This is still well above the background noise level (see fig.1).

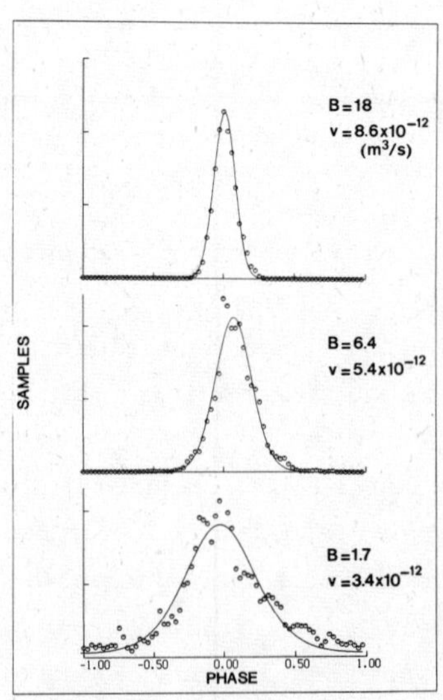

Figure 6.
Distribution for the phase difference (in π-units) between oto-acoustic emission signal and externally supplied tone of the same frequency, for different values of the volume velocity v in the ear canal. The solid lines are least-squares fits to $P(\phi) = A \exp[B \cos(\phi)]$.

(Bialek & Wit, 1984) in which parameter B describes the amount of phase-locking. If B is plotted as a function of pv/ω (with the dimension of an energy), we learn that very small energies are able to produce significant phase-locking (fig.7); v is the volume velocity created in the ear canal by the external tone, p is the sound pressure in the ear canal and ω is the frequency of the external tone (and the emission). Such small energies required to perturb the emission signal are in good agreement with a model that describes acoustic emissions from the human ear as quantum-limited oscillations (Bialek & Wit, 1984), which we would expect from the fact that the normal ear makes quantum-limited measurements (Bialek, 1983).

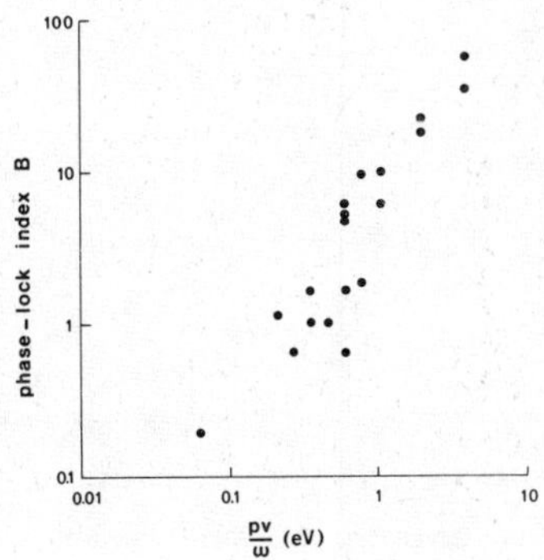

Figure 7.
Phase-lock index for oto-acoustic emission signal as a function of the energy supplied to the ear canal by an external tone (per oscillation divided by 2π).

2. ELECTRONIC MODEL

A simple electronic model for a limit-cycle oscillator can be constructed by adding an operational amplifier to an LCR-network, as shown in fig.8. The amplification in the feedback loop is variable in this model, while saturation of the oscillator can be obtained by limiting the power supply to the amplifier. To adjust the model, the amplification is enlarged until oscillation starts; then the power supply is reduced to minimize oscillation amplitude. In this way the features of the model are (qualitatively) similar to those of an oto-acoustic emission.

Figure 8.
Electronic limit-cycle oscillator as a model for the oto-acoustic emission generator.

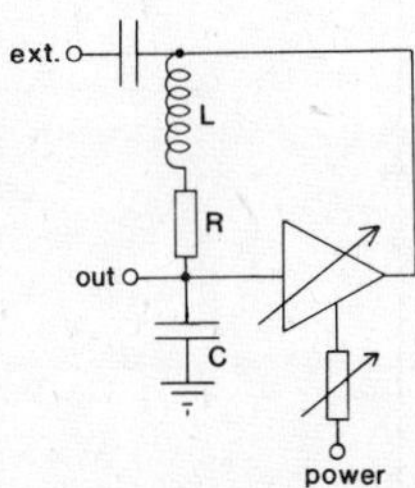

Suppression of the oscillation for instance can be obtained by supplying an external signal, with a frequency different from that of the spontaneous oscillation, to the circuit. Like for the emission, suppression changes the system from unstable into stable, as can be seen by comparing figures 5 and 9.

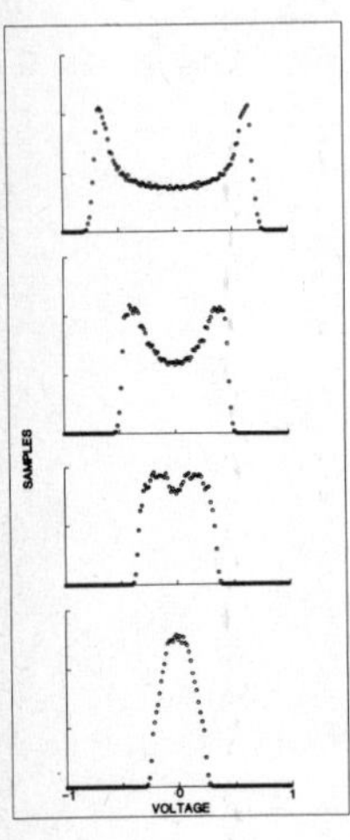

Figure 9.
Suppression of spontaneous oscillation amplitude in electronic model by externally supplied signal. (Horizontal scale is the same in all panels).

Phase-locking to an externally supplied signal with the same frequency as that of the oscillation is also possible. The amount of phase-lock depends on the level of the external signal (fig.10), just as this is the case for the oto-acoustic emission (fig.6).

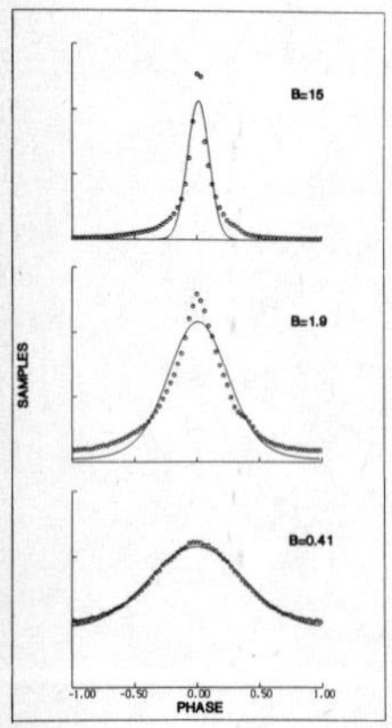

Figure 10.
Phase-lock of spontaneous oscillation in electronic model to externally supplied signal. Solid lines are fits to the same equation as in fig.6.

3. OSCILLATING HAIR CELLS

Oto-acoustic emissions are supposed to be related to the transduction process in cochlear hair cells. In this respect recent findings of oscillating hair cells (Crawford & Fettiplace, 1981; Ashmore, 1983; Lewis & Hudspeth, 1983) are of interest. We have analyzed the spontaneously fluctuating signal, obtained by in vivo intracellular measurement from a hair cell in the fish (Acerina) lateral line by Dr.Alfons Kroese (Dept.Biophysics, University of Groningen).
The average frequency spectrum (fig.11) of the signal (20 seconds duration) shows a narrow peak at 119 Hz., with a width of approximately 1.9 Hz. (Q_{3dB}=63).

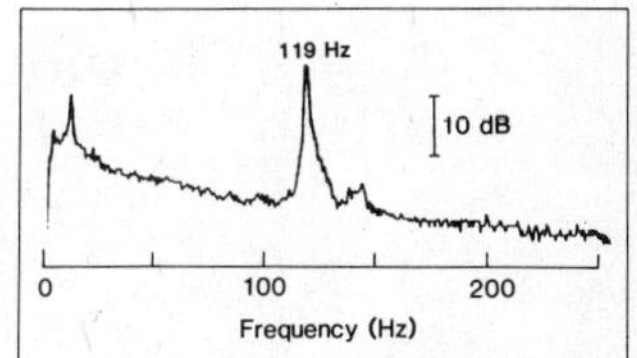

Figure 11.
Average frequency spectrum for spontaneous voltage fluctuation in a lateral line hair cell. These fluctuations occur in a narrow band around 119 Hz.

The amplitude distribution for this 119 Hz signal component was recorded with the Datalab system, in the same way as this was done for the oto-acoustic emission signal. The result, shown in fig.12,gives no evidence that the hair cell signal is generated by an unstable active system; it has the amplitude distribution for (sharply) filtered noise.

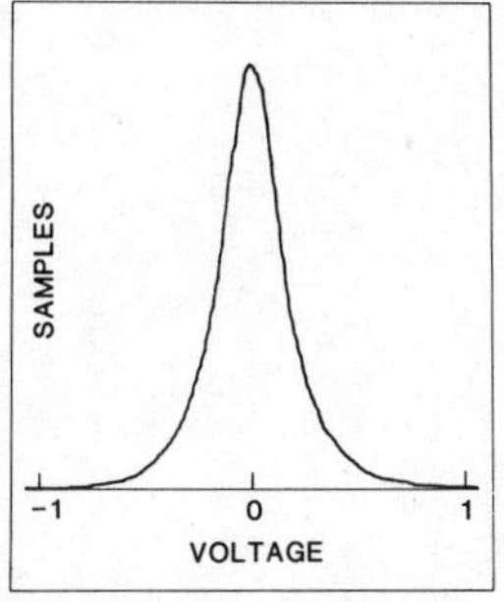

Figure 12.
Amplitude distribution for spontaneous voltage fluctuations in a lateral line hair cell (hor.scale has arbitrary units; the rms value of the voltage fluctuations is 1.3 mV).

This work was supported by the Heinsius Houbolt Fund.

REFERENCES

Ahsmore, J. F., "Frequency tuning in a frog vestibular organ." Nature 304, pp. 536-538, 1983.

Bialek, W. S., "Quantum effects in the dynamics of biological systems." Ph.D.Thesis, Lawrence Berkeley Laboratory, University of California, 1983.

Bialek, W., "Thermal and quantum noise in the inner ear". In Mechanics of Hearing, Ed. by E. de Boer and M. A. Viergever, Martinus Nijhoff Publishers, Delft University Press, 185-192, 1983.

Bialek, W. S., and Wit, H. P., "Quantum limits to oscillator stability: theory and experiments on acoustic emissions from the human ear." Phys.Lett. 104A, pp. 173-178, 1984.

Crawford, A. C., and Fettiplace, R., "An electrical tuning mechanism in turtle cochlear hair cells." J.Physiol. 312, pp. 377-412, 1981.

Lewis, R. S., and Hudspeth, A. J., "Voltage and ion-dependent conductances in solitary vertebrate hair cells." Nature 304, pp. 538-541, 1983.

Wit, H. P., Langevoort, J. C., and Ritsma, R. J., "Frequency spectra of cochlear acoustic emissions ("Kemp-echoes")."J.Acoust.Soc.Am. 70, pp. 437-445, 1981.

THE INFLUENCE OF TEMPERATURE ON FREQUENCY-TUNING MECHANISMS

J.P. Wilson,
Department of Communication and Neuroscience,
University of Keele,
Staffs ST5 5BG U.K.

ABSTRACT

Some hearing organs demonstrate frequency tuning that is strongly temperature-dependent whereas others do not. Previous experiments in mammals have shown little, if any, temperature-dependence. Spontaneous otoacoustic emissions (SOAEs) are very stable in some subjects and offer a sensitive indicator for temperature effects. SOAE frequency changes were investigated over the menstrual cycle, the diurnal cycle, and after irrigation of the ear canal. A consistent apparent negative correlation with temperature was found over the menstrual cycle. Over the diurnal cycle, however, different frequency components behaved differently with a negligible net temperature-dependence. Irrigation of the ear canal at 30°C indicated a slight negative temperature dependence, but further experiments on another subject gave opposite changes for 30°C but no change for 44°C. These latter effects may have been due to changes of middle-ear pressure, whereas the effects over the menstrual cycle may be due to changes of hormone level. The temperature dependence of human frequency tuning would appear to be less than 0.1%/°C, indicating a stiffness change of less than 0.2%/°C.

INTRODUCTION

If tuning of the hearing organ depends on the mechanical parameters of mass and stiffness, the facility for temperature dependence appears to be relatively limited. An octave change in tuning frequency would imply a fourfold change of stiffness. There are several species where tuning frequencies change by this order of magnitude for a 10°C shift (toad: amphibian papilla but not basilar papilla, Moffat and Capranica, 1976, gecko: Eatock and Manley, 1981, caiman: Smolders and Klinke, 1984, pigeon: Schermuly and Klinke, 1982). Such a large change suggests an electro-chemical basis for tuning rather than a mechanical one. In the caiman (Wilson et al, 1985) direct measurements of basilar membrane (BM) vibration showed no significant temperature-dependence of tuning, and a highly-significant difference from the neural results (Smolders and Klinke, 1984). Sharp BM tuning in mammals (Khanna and Leonard, 1982, Sellick et al, 1982) appears to be due to a vulnerable mechanical positive feedback mechanism. The tuning frequency may be determined solely by the passive, poorly-tuned BM whilst its Q-factor depends on the feedback ratio. It would appear, however, that in caiman there must be another reactive or tuned mechanism beyond the BM which is strongly temperature dependent. Mammals show very little temperature dependence (Emde and Klinke, 1977, Gummer and Klinke, 1983, Klinke and Smolders, 1977).

Even in animals where, owing to strong temperature dependence, it is tempting to invoke electro-chemical tuning, there is evidence of mechanical involvement. Both the caiman (Strack et al, 1981) and the frog (Palmer and Wilson, 1981) produce otoacoustic emissions, which are believed to be another manifestation of active mechanical feedback. Caution is required both in generalising between species and in concluding that a specific mechanism can be excluded. It was decided to look again at mammalian temperature effects using a particularly sensitive indicator, the SOAE frequency. The basis of this test is that SOAEs depend on cochlear-tuning mechanisms and that any factor that influences cochlear-tuning frequency should alter SOAE frequency in a similar manner. Zurek (1981) came to the conclusion, by using radiant heat, that temperature does not influence SOAE frequency, but this may not have been successful in raising cochlear temperature.

METHODS

Temperature changes were produced in three ways. Firstly, body-temperature changes occurring over the menstrual cycle in a female subject were utilised. Secondly, the diurnal changes of body temperature were employed. Thirdly, temperature of the cochlea was influenced directly by irrigation of the ear canal with warm or cool water as for testing of vestibular function.

The ear canal was irrigated for 30 sec. with thermostatically-controlled water with the nozzle temperature at 30°C or 44°C. Recordings of the SOAEs were made with a B & K 4165 half-inch microphone sealed to the ear canal with a 20-mm length of 5-mm diameter tubing passing through an E.A.R. earplug. A lab-built miniature low-noise amplifier was used with filter B (Wilson, 1980). A 1-kHz crystal-controlled sinewave was recorded with the SOAE to allow correction for mains-frequency fluctuations and recorder instability.

Components were analysed using a pair of quadrature lock-in amplifiers (Brookdeal 401) set at 100 msec integrating time (2.5-Hz bandwidth) and displayed on an X-Y 'scope. The reference signal was adjusted to obtain a stationary spot on the screen, and the oscillator and "1-kHz" signals were measured simultaneously on frequency counters.

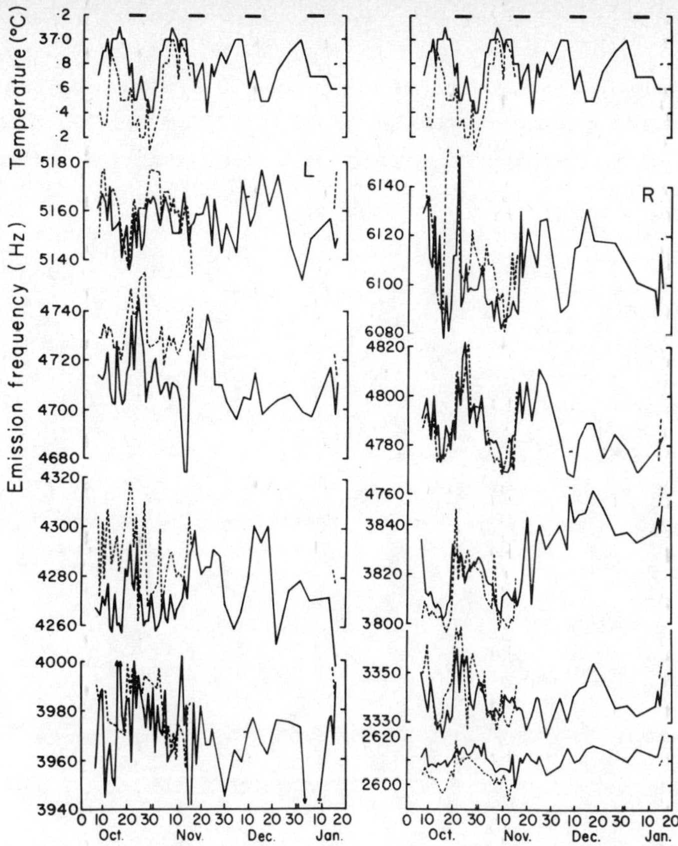

Fig. 1. Changes in SOAE frequency throughout four menstrual cycles (indicated by upper bars). The upper curves show the body-temperature variation and the lower curves show the frequencies of four emissions from the left ear (left) and five emissions from the right ear (right). Continuous lines show morning measurements (about 8 am) and broken lines and isolated points, evening (about 11 pm). All frequency components show a slight negative correlation with temperature.

RESULTS

The frequencies of the various components are shown in Fig. 1 as a function of time over a four-month period, for morning (continuous lines) and evening (broken lines). At the top are shown the corresponding temperature measurements with bars indicating menstruation, which corresponds to the falling-temperature phase. Inspection of the frequency curves revealed in each case a slight periodicity in antiphase with the temperature pattern, obscured by a considerable amount of noise. Temperature and frequency were compared directly using a standard curve fitting package. The results are shown in Table 1. All frequencies show weak negative correlations. The mean temperature coefficient was -0.42 ± 0.05%/°C. At this point the temperature hypothesis appeared plausible. However, the evening data (broken line, not included in the foregoing analysis) compromises the temperature model. Although the evening temperatures are consistently below the morning ones the evening frequencies are not consistently above the morning frequencies. It is interesting to note that in most cases there is a high degree of correspondence between morning and evening frequencies indicating that the variations are not simply random.

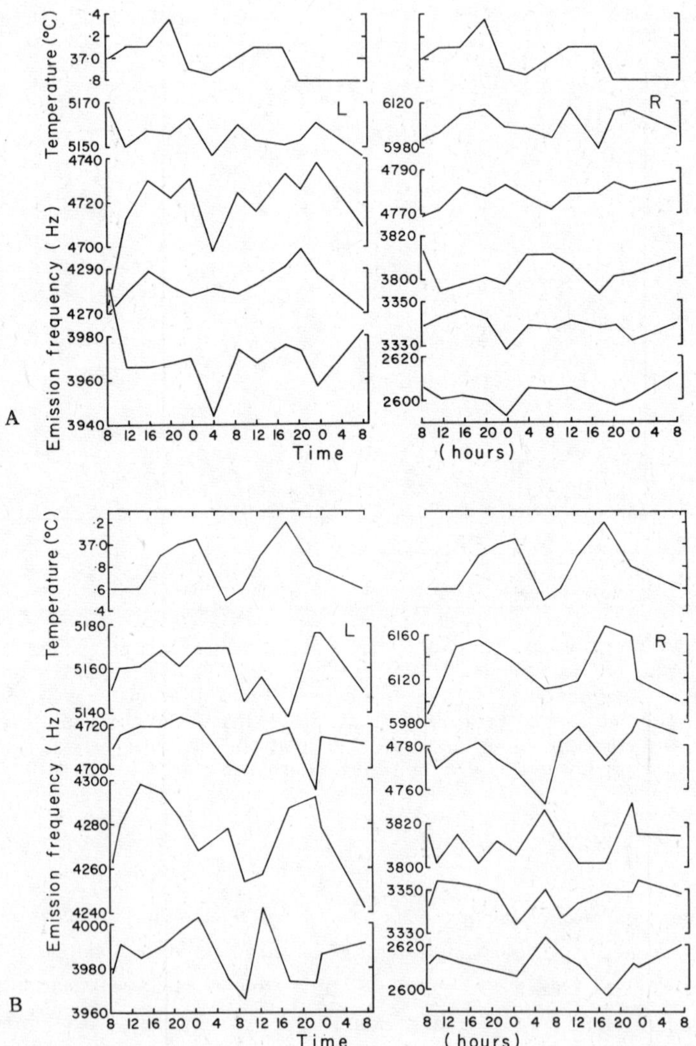

Fig. 2. Changes in SOAE frequency throughout the diurnal cycle for two separate periods of 48 hours. (A: 12-14 Nov., B: 14-16 Jan.) The upper curves show body temperature variation and the lower curves show the frequencies of four emissions from the left ear (left) and five emissions from the right ear (right). There is no consistent correlation with temperature.

In order to investigate this feature, and as a further test of the temperature hypothesis, an investigation of diurnal changes was made. Measurements were made about every four hours over two separate 48-hour periods.

Some frequency changes (Fig. 2) appear to be in phase with temperature changes, some show no correlation, and others show an opposite tendency. This would appear to be evidence against a strong temperature effect. Again the temperature hypothesis was tested by performing individual regressions of frequency versus temperature (Table 1). Overall the mean temperature coefficient was +0.05 ± 0.06 %/°C, much smaller and opposite to the monthly effect.

TABLE 1	LEFT				RIGHT				
Component (kHz)	3.97	4.27	4.71	5.15	2.61	3.34	3.82	4.79	6.10
MENSTRUAL									
Correlation (freq./temp.)	-0.05	-0.22	-0.30	-0.17	-0.54	-0.51	-0.50	-0.43	-0.23
Temperature coeff. (%/°C)	-0.19	-0.34	-0.39	-0.16	-0.41	-0.62	-0.66	-0.55	-0.35
" " " std. error	0.48	0.20	0.16	0.12	0.08	0.14	0.15	0.15	0.19
DIURNAL : A									
Correlation (freq./temp.)	0.10	0.20	0.04	0.08	-0.08	0.51	-0.32	-0.36	0.07
Temperature coeff. (%/°C)	0.20	0.02	0.09	0.06	-0.09	0.43	-0.32	-0.22	0.09
" " " std. error	0.61	0.34	0.68	0.24	0.33	0.22	0.30	0.20	0.37
DIURNAL : B									
Correlation (freq./temp.)	-0.13	0.33	0.13	-0.34	-0.42	0.57	-0.25	0.32	0.28
Temperature coeff. (%/°C)	-0.15	0.47	0.09	-0.26	-0.36	0.33	-0.20	0.22	0.51
" " " std. error	0.34	0.42	0.23	0.23	0.25	0.15	0.24	0.20	0.54

Finally temperature was influenced directly by ear irrigation at 30°C for 30 sec. in the same subject (Fig. 3). The results are suggestive of a negative temperature effect, but its magnitude is unknown because the cochlear temperature is unknown.

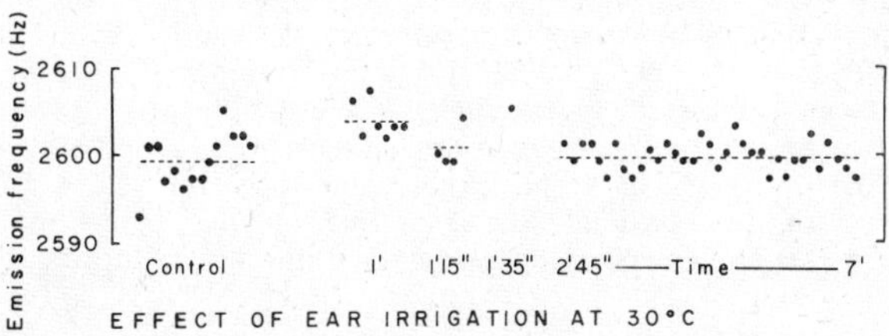

Fig. 3. Change in SOAE frequency following 30 sec. of irrigation of the ear canal at 30°C. The emission frequencies are shown immediately before irrigation (control) and at the times indicated after the end of irrigation. The recording is of the strongest component from the right ear of the same female subject as Figs. 1 and 2.

Finally measurements were made in a male subject who had particularly stable SOAEs in both ears, using irrigation. The results are shown in Fig. 4 for both the stimulated ear (A) at 44°C (above) and 30°C (below) and in the non-stimulated ear (B, 37°C). As expected, no influences were observed in the contralateral ear. In the ipsilateral ear, however, 30°C irrigation consistently reduced the two SOAE frequencies with recovery within two minutes. This indicates a positive temperature coefficient. For 44°C irrigation, however, no influence was observable. As the temperature changes at the cochlea would be quite small there should be symmetry between the two conditions and overall the results cannot be considered to support the temperature model.

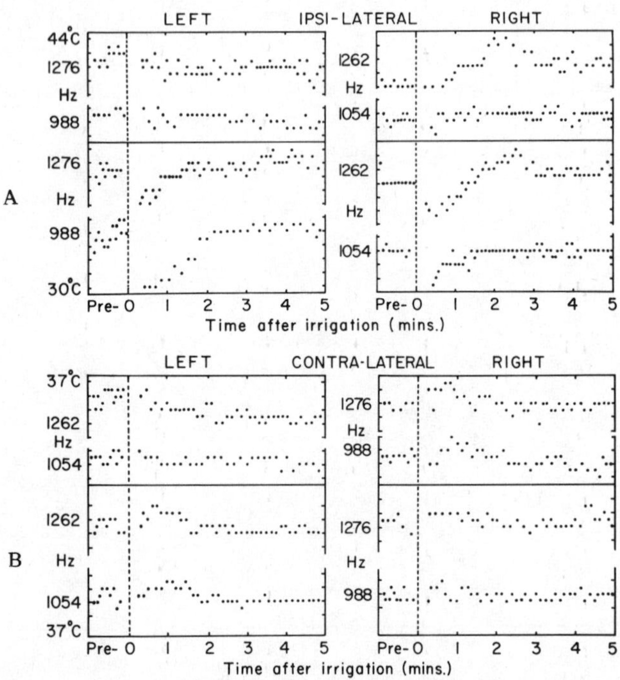

Fig. 4. Changes in SOAE frequency following irrigation of the ear canal in a male subject. The subject had two strong components in each ear at similar frequencies. Simultaneous recordings were made from the irrigated ear (A) and the non-irrigated ear (B). Each ear was irrigated with warm (44°C, A, upper plots) and cool (30°C, A, lower plots) water, in the sequence L-30, L-44, R-44, L-30. Emission frequencies (2 Hz/division) of each component are plotted as a function of the time before (pre-) and after the end of 30 sec. irrigation. Cooling appears to reduce emission frequency whereas warming appears to have no influence.

DISCUSSION

The temperature hypothesis is not supported by the opposite results obtained for the two subjects for 30°C irrigation, nor by the null results obtained at 44°C in the second subject. The subject volunteered that strong thermal and vestibular effects were observed in every case. One possible explanation is that the temperature changes induce middle-ear pressure changes which can influence OAE frequency (Kemp, 1979), or SOAE frequency (Wilson and Sutton, 1981). Although this usually produces a frequency increase, the opposite effect can also occur. On this hypothesis it would be necessary to assume that a positive pressure would release itself through the Eustachian tube whereas a negative pressure would not.

The diurnal influences appear to be the strongest evidence against a thermal effect and indicate that if it exists it is probably less than 0.1 %/°C. This suggests that mammalian frequency tuning is predominantly determined by mechanical processes.

The results obtained over the menstrual cycle indicate the possibility of hormonal effects, presumably influencing mechanical properties. Hormonal factors were believed to influence subjects' absolute pitch judgements in a study by Wynn (1973), although in this case a bi-menstrual rhythm was observed. A variation of about 0.5 - 1% was observed comparable to the present variation of about 0.7%. It is possible that the present subject shows one component of the bi-monthly rhythm much more strongly than the other. It is interesting to note that hormonal effects have also been observed in the tuning of fish electro-receptors (Bass & Hopkins, 1984), although this is complicated by concurrent changes in the fish's electrical discharges, which appear to entrain the tuning of electroreceptors. It has also been suggested that menstrual effects on auditory threshold may be due to changes in middle-ear pressure perhaps mediated by varying fluid retention. Such an effect could also influence SOAE frequency. Gibbs (1984), however, was unable to detect any systematic middle-ear pressure change. She investigated changes in threshold fine-structure but did not observe a consistent pattern, perhaps due to the size of frequency steps used. Another feature of the present data which argues against a middle-ear explanation for the menstrual effects is that all frequency components shifted in the same direction, whereas in an earlier experiment the same subject showed both increases and decreases with external ear pressure for different frequency components. A middle-ear-pressure explanation is, however, quite plausible for the diurnal changes. It is also possible to speculate that hormonal influences may act on the central nervous system but affect the cochlea indirectly via efferent pathways. This possibility would be supported by the report of Mountain (1980) that COCB stimulation influences the generation of the distortion product, $2f_1-2f_2$.

REFERENCES

Bass, A.H., and Hopkins, C.D. (1984): Shifts in frequency tuning of electroreceptors in androgen-treated mormyrid fish. J. Comp. Physiol. A. 155, 713-724.

Eatock, R.A., and Manley, G.A. (1981): Auditory nerve fibre activity in the Tokay gecko: II. Temperature effect on tuning. J. Comp. Physiol. 142, 219-226.

Emde, C. and Klinke, R. (1977): Does absolute pitch depend on an internal clock? In: Inner Ear Biology. Eds: M. Portmann and J.M. Aran, INSERM, Paris, 68, 145-146.

Gibbs, E.R. (1984): Hormonal factors in hearing. B.Med.Sc. project report, University of Nottingham.

Gummer, A.W. and Klinke, R. (1983): Influence of temperature on tuning of primary-like units in the guinea pig cochlear nucleus. Hearing Res. 12, 367-380.

Kemp, D.T. (1979): The evoked cochlear mechanical response and the auditory microstructure - evidence for a new element in cochlear mechanics. Scand. Audiol. Suppl. 9, 35-47.

Khanna, S.M. and Leonard, D.G.B. (1982): Basilar membrane tuning in the cat cochlea. Science 215, 305-306.

Klinke, R. and Smolders, J. (1977): Effect of temperature shift on tuning properties. In: Psychophysics and Physiology of Hearing. Eds. E.F. Evans & J.P. Wilson. Academic Press, Lond. pp. 109-112.

Moffat, A.J.M. and Capranica, R.R. (1976): Effects of temperature on the response properties of auditory nerve fibers in the American toad (**Bufo americanus**). J. Acoust. Soc. Am. 60, S80.

Mountain, D.C. (1980): Changes in endolymphatic potential and crossed olivocochlear bundle stimulation alter cochlear mechanics. Science 210, 71-71.

Palmer, A.R. and Wilson, J.P. (1982): Spontaneous and evoked acoustic emissions in the frog **Rana esculenta**. J. Physiol. 324, 66P.

Schermuly, L. and Klinke, R. (1982): Tuning properties of pigeon primary auditory afferents depend on temperature. Pflüg. Arch. Suppl. 394, R63.

Sellick, P.M., Patuzzi, R. and Johnstone, B.M. (1982): Measurement of basilar membrane motion in guinea pig using the Mössbauer technique. J. Acoust. Soc. Am. 72, 131-141.

Smolders, J. and Klinke, R. (1984): Effects of temperature on the properties of primary auditory fibres of the spectacled caiman, **Caiman crocodilus** (L.). J. Comp. Physiol. 155, 19-30.

Strack, G., Klinke, R. and Wilson, J.P. (1981): Evoked cochlear responses in **Caiman crocodilus**. Pflügers Arch. Suppl. 391, R43.

Wilson, J.P. (1980): Evidence for a cochlear origin for acoustic re-emissions, threshold fine-structure and tonal tinnitus. Hear. Res. 2, 233-252.

Wilson, J.P., Smolders, J.W.T., and Klinke, R. (1985): Mechanics of the basilar membrane in **Caiman crocodilus**. Hear. Res. 18, 1-14.

Wilson, J.P., and Sutton, G.J. (1981): Acoustic correlates of tonal tinnitus. In: **Tinnitus**, Eds. D. Evered and G. Lawrenson, Pitman Medical, Lond., 82-107.

Wynn, V.T. (1973): Absolute pitch in humans, its variation and possible connection with other known rhythmic phenomena. Prog. in Neurobiol. 1, 113-143.

Zurek, P.M. (1981): Spontaneous narrowband acoustic signals emitted by human ears. J. Acoust. Soc. Am. 69, 514-523.

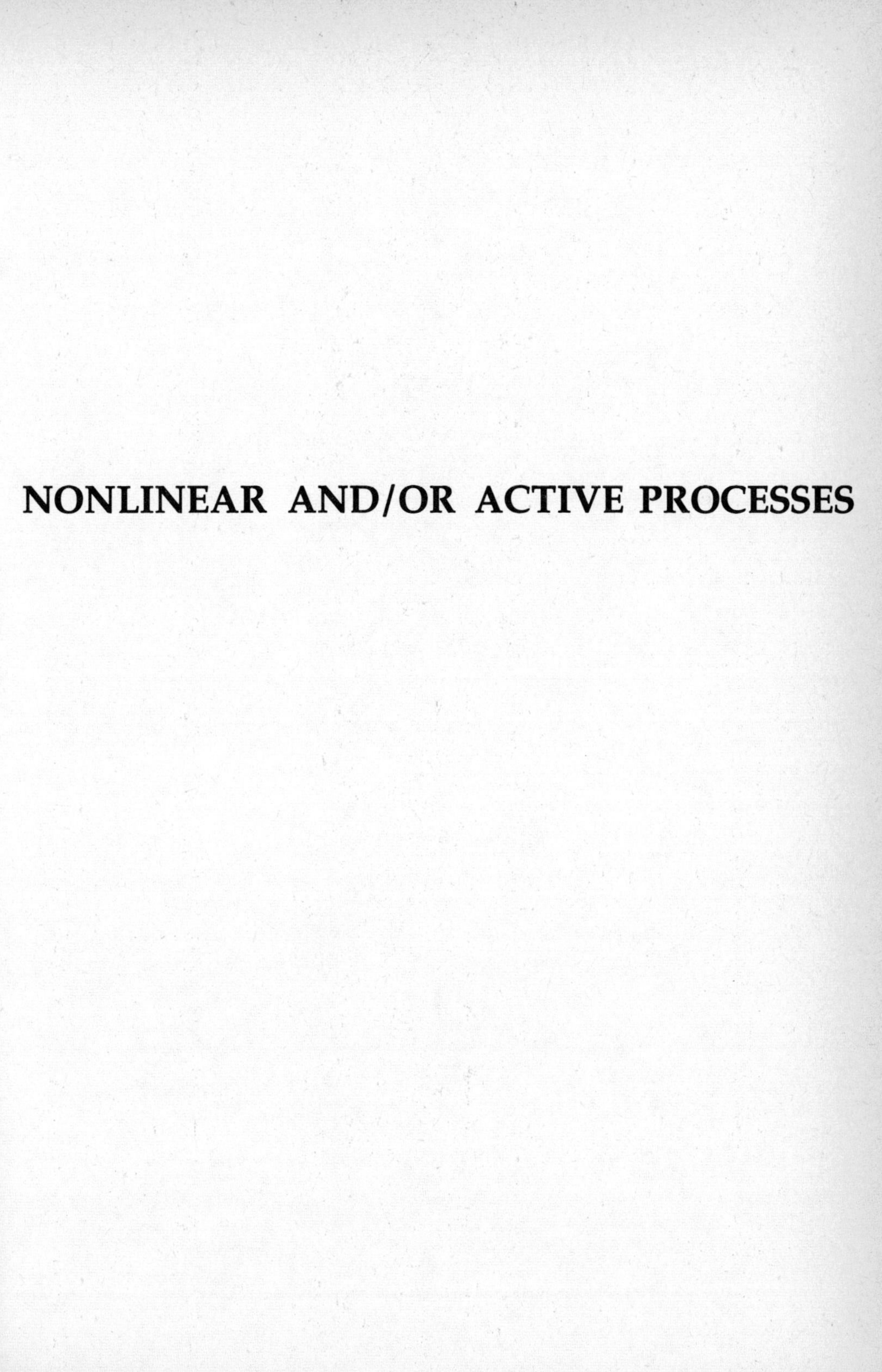

NONLINEAR AND/OR ACTIVE PROCESSES

A REVIEW OF NONLINEAR AND ACTIVE COCHLEAR MODELS

D.O. KIM
Washington University School of Medicine, Box 8101
St. Louis, MO 63110

ABSTRACT

This paper reviews nonlinear and active cochlear models with special attention to the question whether a "second filter" is needed for modeling two-tone suppression below the characteristic frequency (CF). The concept of a unidirectionally coupled "second filter" is inconsistent with experimental evidence that major cochlear mechanical nonlinear phenomena are affected by alterations of the organ of Corti. A possible way of modeling the below-CF suppression is suggested in terms of an indirect effect mediated by a baseline shift of the cochlear partition.

1. INTRODUCTION

There have been considerable modeling studies on cochlear mechanics over the past several years. Soon after Rhode's (1971) discovery of nonlinear behavior in basilar-membrane motion, several nonlinear models of the cochlea have been developed. These models (Hubbard and Geisler, 1972; Kim et al., 1973; Hall, 1974) were successful in reproducing Rhode's observations by incorporating a relatively simple hypothesis that the observed nonlinear behavior might arise from a nonlinearity in the damping term of the cochlear partition dynamics where the magnitude of the damping increased with increasing response. Besides Rhode's direct mechanical observations, the models of Kim et al. and Hall also reproduced several nonlinear phenomena observed in responses of single cochlear nerve fibers under two-tone (Sachs and Kiang, 1968; Goldstein and Kiang, 1968) or click-pair (Goblick and Pfeiffer, 1969) stimulations. Thus the model results suggested that the neurally observed nonlinear phenomena might be present in the mechanical response as well.

Particularly, Hall's (1974) model results made a specific prediction about the spatial distribution of two-tone distortion products for which there were no experimental data at that time. This model prediction provided an important motivation in a subsequent development of the experimental technique of the population studies of cochlear nerve fibers (Pfeiffer and Kim, 1975) and experimental verification of Hall's model prediction at the level of cochlear nerve fiber populations (Kim et al., 1980a). Although there was a period of controversy about whether two-tone distortion products were significantly present in basilar-membrane motion (e.g., Wilson and Johnstone, 1973), the author believes that the evidence available now is quite conclusive that basilar-membrane motion is significantly nonlinear (e.g., Sellick et al., 1982; Kemp, 1978; Kim, 1980).

For a number of reasons, many investigators have postulated in models of the cochlea the existence of a "second filter" distinct from the frequency selectivity associated with the motion of the cochlear partition (e.g., Evans and Wilson, 1973). Recent experimental data on basilar-membrane motion (Sellick et al., 1982; Khanna and Leonard, 1982) demonstrate that, when the physiological condition of the cochlea is optimal, the mechanical response is quite sensitive and very sharply tuned. Thus the need to postulate additional tuning for bringing the mechanical tuning in agreement with cochlear neural tuning now appears to have been greatly reduced.

However, there is one particular aspect of observed cochlear nonlinear phenomena whose explanation may still require a "second filter". That is, the response of a cochlear nerve fiber to an excitor tone at a frequency fe equal to the characteristic frequency (CF) can be suppressed by another tone at a frequency fs either below or above the CF; this phenomenon is called two-tone suppression (e.g., Sachs and Kiang, 1968). Nonlinear cochlear models of Kim et al. (1973) and Hall (1977) reproduce well two-tone suppression for fs above the CF. These models, however, have some difficulty of reproducing the two-tone suppression for fs below the CF.

The reason that nonlinear cochlear models can readily reproduce two-tone suppression behavior for fs above the CF is as follows. An fs tone above the CF excites a cochlear region basal to the characteristic place of the fe tone at the CF under observation. This leads to an increase in the damping of the cochlear partition motion of the model which in turn increases a compressive nonlinearity in the region around the fs place. A compressive nonlinearity inherently produces a suppression of a weaker signal by a stronger signal (Blachman, 1964). Thus the response to the fe tone is suppressed over a region encompassing the fs place and the fe (observation) place. At the same time, the fs tone above the CF does not in general excite the observation place because of the marked steepness of the apical slope of the cochlear spatial tuning.

The situation is different for fs below the CF primarily because the steepness of the cochlear spatial tuning is much less on the basal slope than on the apical slope especially at high stimulus levels as used in two-tone suppression studies. In the models, an fs tone below the CF can suppress the response to an fe tone at the CF only at high enough levels where the fs response component itself becomes large. Therefore, the total rms value of the displacement of a point on the basilar membrane is not suppressed in these models for fs below the CF. Hall (1977) remedied the problem by postulating a "second filter" which transforms basilar-membrane displacement into a receptoneural excitation signal. The "second filter" in Hall's model filters out the fs component after the fs signal has a chance of suppressing the fe component through a compressive nonlinear behavior arising from the basilar-membrane damping nonlinearity. The net outcome is that the total rms of the receptoneural excitation signal is suppressed by the fs even if it is below the CF. The situation here is similar to Pfeiffer's (1970) band-pass nonlinearity model in the sense that the "second filter" removes the large fs component after the fs component suppressed the fe component.

This modeling approach may appear to be reasonable on the surface. The author believes, however, that there is a conflict between Hall's hypothesis and a conclusion from several studies listed below that cochlear nonlinear behavior present in the mechanical response of the basilar membrane is physiologically vulnerable and is introduced into the mechanical response after being generated within the transduction processes of the hair cells; psychophysical studies showing effects of permanent or temporary threshold shifts on combination tone levels (Smoorenburg, 1972; 1980); electrophysiological (Dallos et al., 1980; Siegel et al., 1982) and otoacoustical (Anderson and Kemp, 1979; Kim, 1980; Mountain, 1980; Siegel and Kim, 1982) studies showing effects of irrevesible or reversible alteration of the organ of Corti on intermodulation distortion products or on cochlear "echoes"; electrophysiological

(Robertson, 1976; Schmiedt et al., 1980; Dallos et al., 1980;) and psychophysical (Wightman et al., 1977; Leshowitz and Lindstrom, 1977) studies showing effects of alteration of the organ of Corti on two-tone suppression behavior. According to this point of view, the compressive nonlinearity, which is key to creating a signal suppression, originates as an integral part of the transduction process rather than preceding the transduction process distinctly separated by a unidirectionally coupled "second filter".

This paper will review various aspects related to this apparent conflict regarding a "second filter" used in modeling two-tone suppression with fs below the CF. In addition, this paper will also briefly review active and nonlinear cochlear models.

2. IS A "SECOND FILTER" NECESSARY FOR MODELING TWO-TONE SUPPRESSION?

Figure 1 shows a scheme of classifying various cochlear models that have been proposed. The double arrows signify the model assumption that the coupling between the neighboring components is allowed to be bidirectional. In contrast, the single arrows signify the model assumption that the coupling is unidirectional. That is, the output of the left component "drives" the right component, but the left component is not affected by the one on the right. A bidirectional coupling is a general assumption, but a unidirectional coupling is a restricted assumption that the influence operating in the reverse direction is negligible. In all of the three types of the cochlear model, the cochlear hydronamics (CH) subsystem is assumed to be intrinsically linear. Nevertheless, the fluid pressure will exhibit prominent nonlinear behavior in Type B and Type C models because of bidirectional couplings between the CH subsystem and other subsystems which are nonlinear. Likewise, distortion products generated inside the cochlear part of a model will propagate into the ear canal (e.g., Matthews, 1983).

Examples of Type A cochlear model are those of Pfeiffer (1970) and Evans and Wilson (1973). Evans and Wilson's model postulates that cochlear partition (CP) motion behaves linearly and that all of the neurally observed nonlinear phenomena are generated in the components which are unidirectionally coupled to the CP component. This model is inconsistent with observations of mechanical and otoacoustic manifestations of cochlear nonlinear behavior.

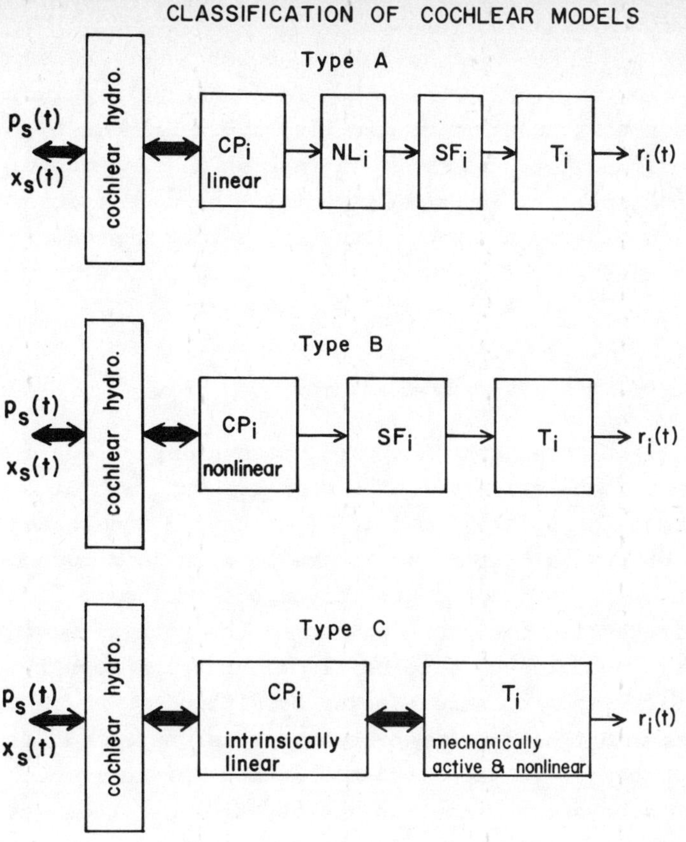

Figure 1. A classification scheme of various types of cochlear models. For each of the three types of models, the left most box represents the cochlear hydrodynamics subsystem. For the remaining boxes in each case, the subscript "i" indicates that the chain of boxes represents the component subsystems at a particular location along the length of the cochlea. ps(t): pressure at the stapes; xs(t): displacement of the stapes; CP: cochlear partition mechanics; NL: nonlinearity; SF: "second filter"; T: transduction process; r(t): spike discharge signal of a cochlear nerve fiber. The double arrows signify that the model allows bidirectional interaction between the neighboring components, whereas the single arrows signify the model assumption that the interaction between the neighboring components is unidirectional.

An example of Type B cochlear model is that of Hall (1977; 1981). Hall (1981) stated: "An essential feature for observing suppression of total rms response for fs less than fe is that a distinction must be made between the mechanism controlling the nonlinearity and the mechanism producing the response. There must be a stage of sharpening between the nonlinearity and the response." The unidirectional nature of the coupling between the CP and the transduction-process (T) subsystems is a serious flaw of Type A and Type B models.

3. ACTIVE AND NONLINEAR COCHLEAR MODELS

The author believes that the evidence available now is compelling that an adequate cochlear model should incorporate a bidirectional coupling between the cochlear-partition mechanics and the hair-cell transduction processes (Type C in Figure 1). In this type of model, the CP subsystem is assumed to be intrinsically linear; i.e., the intrinsic differential equations governing the mechanics of the cochlear partition is assumed to be linear under the condition where the transduction processes are blocked or clamped. A Type C model further assumes that all nonlinear behavior is generated within the T subsystem and then introduced into the mechanical subsystems (i.e., the CP and CH subsystems) from the T subsystem. To the best of the author's knowledge, there are no examples of a Type C cochlear model available now which succeed in reproducing the below-CF two-tone suppression behavior.

There are a number of recent studies attempting to develop cochlear models of Type C. Some cochlear models (e.g., Kim et al., 1980b; Neely and Kim, 1983; de Boer, 1983) have incorporated a hypothesis of an active energy-releasing mechanism of transduction processes but has not incorporated nonlinear behavior yet. There is increasing amount of support for the hypothesis that the outer hair cells of a mammalian cochlea give rise to an enhaced sensitivity and sharper tuning of cochlear-partition motion through an active and nonlinear bidirectional transduction mechanism (e.g., Mountain, 1980; Dallos et al., 1980; Siegel and Kim, 1982; Weiss, 1982; Neely and Kim, 1983; Davis, 1983; Kim, 1984; Brownell, 1985). Several investigators are developing cochlear models incorporating both active and nonlinear behavior of transduction processes which is coupled back to cochlear-partition motion (e.g., Koshigoe and Tubis, 1983; Mountain et al., 1983; Netten and Duifhuis, 1983; Diependal and Viergever, 1983).

These models which are rather in a beginning stage now have not been studied in detail yet and it remains to be determined whether a Type C cochlear model can reproduce the below-CF two-tone suppression.

Although the below-CF two-tone suppression may not be important by itself, this review has paid much attention to it because the author believes that it helps bring out important issues related to cochlear modeling. One may wonder whether it would be possible to combine the "second filter" effect of filtering out the fs component in the context of a Type C cochlear model. One may ask, "Can a Type C cochlear model exhibit sharper tuning of receptoneural response than basilar-membrane displacement?" The answer is yes. When more than one degree of freedom is incorporated for the CP subsystem by assigning a degree of freedom each to the basilar membrane and the tectorial membrane, the receptoneural tuning in such a model can be made sharper than the basilar membrane tuning (e.g., Zwislocki and Kletsky, 1979; Allen, 1980; Neely and Kim, 1985). The idea of combining a "second filter" with Type C model does not appear promising though according to the following reasoning. If indeed the CP subsystem is intrinsically linear and all nonlinear behavior originates from the T subsystem, then an fs tone should reach the T subsystem with large enough amplitude in order to activate a compressive nonlinearity thereby suppressing the response of a cochlear neuron evoked by an excitor tone at the CF and yet the fs tone itself should not be excitatory to the neuron. This seems to be difficult.

The discussion in this paper so far has implicitly assumed that the only means of a longitudinal coupling in the cochlea is through the fluid. One possible way for reproducing the below-CF suppression is a hypothesis of longitudinal coupling in the cochlear partition itself besides the coupling through the fluid. Jau and Geisler (1983) proposed a cochlear model where the local partition damping is increased by an integral of an exponentially weighted function of partition motion integrated over a distance. Their model exhibited an example of the below-CF suppression. It appears that their model requires quite large values of the integration distance and the longitudinal coupling space constant in order to produce the below-CF suppression. Physiological meaning of this is not clear.

Another possible mechanism (somewhat related to Jau and Geisler's model) which may help account for the below-CF suppression is as follows. Nonlinear behavior of cochlear-partition motion is believed to be asymmetric in the sense that even order distortion products such as (f2-f1) are generated (Hall, 1974; Kim et al., 1980a). When an fs tone below the CF is applied at a high level, such an asymmetry in cochlear-partition nonlinearity is inherently expected to produce a slow baseline shift of the partition which should exist during the period when the fs tone is kept on. This in turn may lead to an increase of a compressive nonlinearity and signal suppression at the observation place in view of the pronounced modulating effects of a very low frequency tone on the response sensitivity at a higher frequency CF observed neurally (LePage, 1981; Schmiedt, 1982; Sellick et al., 1982b) or psychophysically (Zwicker, 1977). In other words, the author suggests that an fs tone below the CF may give rise to a suppression of the response indirectly through a DC component of cochlear-partition motion (rather than by a direct action of the AC component of fs) in the case where the total response is suppressed. The author believes that further careful experimental and modeling examinations of such a possibility including effects of a below-CF suppressing tone on the spontaneous rate of a cochlear neuron should be helpful.

Acknowledgement: This work was supported by an NIH grant NS18426. The author thanks C.H. Ha, K. Divakar and J.B. Sunwoo for their general assistance.

REFERENCES

Allen, J.B.(1980). "Cochlear micromechanics - a method for transforming mechanical to neural tuning within the cochlea." J. Acoust. Soc. Am. 62, 930-939.

Anderson, S.D. and Kemp, D.T.(1979). "The evoked cochlear mechanical response in laboratory primates, a preliminary report." Arch. Otolaryn. 224, 47-54.

Blachman, N.M.(1964). "Band-pass nonlinearities." IRE Trans. Inform. Theory IT-10, 162-164.

Brownell, W.E., Bader, C.R., Bertrand, D. and Ribaupierre, Y. de (1985). "Evoked mechanical responses of isolated cochlear outer hair cells." Science 227, 194-196.

Dallos, P., Harris, D.M., Relkin, E. and Cheatham, M.A.(1980). "Two-tone suppression and intermodulation distortion in the cochlea: effect of outer hair cell lesion." In Psycho. Physiol. Behav. Stud. Hearing, Eds. van den Brink and Bilsen, Delft Univ. Press, Netherlands, pp242-249.

Davis, H.(1983). "An active process in cochlear mechanics." Hearing Res. 9, 79-90.

de Boer, E.(1983). "Power amplification in an active model of the cochlea - short-wave case." J. Acoust. Soc. Amer. 73, 577-579.

Diependal, R,J. and Viergever, M.A.(1983). "Nonlinear and active modeling of cochlear mechanics: a precarious affair" In Mechanics of Hearing, Eds. de Boer and Viergever, Martinus Nijhoff Pub., Delft Univ. Press, pp153-160.

Evans, E.F. and Wilson, J.P.(1973). "The frequency selectivity of the cochlea." In Basic Mechanisms in Hearing, Ed. Moller, Academic Press, New York, pp519-551.

Goblick, T.J. and Pfeiffer, R.R.(1969). "Time domain measurements of cochlear nonlinearities using combination of click stimuli." J. Acoust. Sco. Amer. 46, 924-938.

Goldstein, J.L. and Kiang, N.Y.S.(1968). "Neural correlates of the aural combination tone 2f1-f2." Proc. IEEE 56, 981-992.

Hall, J.L.(1974). "Two-tone distortion products in a nonlinear model of the basilar membrane." J. Acoust. Soc. Amer. 56, 1818-1828.

Hall, J.L.(1977). "Two-tone suppression in a nonlinear model of the basilar membrane." J. Acoust. Soc. Amer. 61, 802-810.

Hall, J.L.(1981). "Observations ona nonlinear model for motion of the basilar membrane." In Hearing Res. Theory Vol.1, Eds. Tobias and Schubert, Academic Press, New York, pp2-61.

Hubbard, A.E. and Geisler, C.D.(1972). "A hybrid computer model of the cochlear partition." J. Acoust. Soc. Amer. 51, 1895-1903.

Jau, Y.C. and Geisler, C.D.(1983). "Results from a cochlear model utilizing longitudinal coupling." Same reference as Diependal and Viergever, pp169-176.

Kemp, D.T.(1978). "Stimulated acoustic emissions from within the human auditory system." J. Acoust. Soc. Amer. 64, 1386-1391.

Khanna, S.M.and Leonard, D.G.B.(1982). "Basilar membrane tuning in the cat cochlea." Science 215, 305-306.

Kim, D.O., Molnar, C.E. and Pfeiffer, R.R.(1973). "A system of nonlinear differential equations modeling basilar-membrane motion." J. Acoust. Soc. Am. 54, 1516-1529.

Kim, D.O.(1980). "Cochlear mechanics: implications of electrophysiological and acoustical observations." Hearing Res. 2, 297-317.

Kim, D.O., Molnar, C.E. and Matthews, J.W.(1980a). "Cochlear mechanics: nonlinear behavior in two-tone responses as reflected in cochlear nerve fiber responses and in ear-canal sound pressure." J. Acoust. Soc. Am. 67, 1704-1721.

Kim, D.O., Neely, S.T., Molnar, C.E. and Matthews, J.W.(1980b). "An active cochlear model with negative damping in the partition: comparison with Rhode's ante- and post-mortem observations." Same reference as Dallos et al, pp7-14.

Kim, D.O.(1984). "Functional roles of the inner- and outer-hair-cell subsystems in the cochlea and brainstem." In Hearing Science: Recent Advances, Ed. Berlin, College-Hill Press, San Diego, CA, pp241-262.

Koshigoe, S. and Tubis, A.(1983). "A nonlinear feedback model for outer-hair-cell stereocilia and its implications for the respone of the auditory periphery." Same reference as Diependal and Viergever, pp127-134.

LePage, E.L.(1981). "The role of nonlinear mechanical processes in mammalian hearing." Ph.D. thesis, Univ. Western Australia, Nedlands, Australia.

Leshowitz, B.H. and Lindstrom, R.(1977). "Measurement of nonlinearities in listeners with sensorineural hearing loss." In Psycho. Physiol. Hearing, Eds. Evans and Wilson, Academic Press, New York, pp283-292.

Matthews, J.W.(1983). "Modeling reverse middle ear transmission of acoustic distortion signals." Same reference as Diependal and Viergever, pp11-18.

Mountain, D.C.(1980). "Changes in endolymphatic potential and crossed olivocochlear bundle stimulation alter cochlear mechanics." Science 210, 71-72.

Mountain, D.C., Hubbard, A.E. and McMullen, T.A.(1983). "Electromechanical processes in the cochlea." Same reference as Diependal and Viergever, pp119-126.

Neely, S.T. and Kim, D.O.(1983). "An active cochlear model showing sharp tuning and high sensitivity." Hearing Res. 9, 123-130.

Neely, S.T. and Kim, D.O.(1985). "A model for active elements in cochlear biomechanics." To be submitted for publication.

Netten, S.M. van and Duifhuis, H.(1983). "Modeling an active, nonlinear cochlea." Same reference as Diependal and Viergever, pp143-152.

Pfeiffer, R.R.(1970). "A model for two-tone inhibition of single cochlear nerve fibers." J. acoust. Soc. Am 48, 1373-1378.

Pfeiffer, R.R. and Kim, D.O.(1975). "Cochlear nerve fiber responses: distribution along the cochlear partition." J. Acoust. Soc. Amer. 58, 867-869.

Rhode, W.S.(1971). "Observations of the vibration of the basilar membrane in squirrel monkeys using the Mossbauer technique." J. Acoust. Soc. Am 49, 1218-1231.

Robertson, D.(1976). "Correspondence between sharp tuning and two-tone inhibition in primary auditory neurones." Nature 259, 477-478.

Sachs, M.B. and Kiang, N.Y.S.(1968). "Two-tone inhibition in auditory nerve fibers." J. acoust. Soc. Am. 43, 1120-1128.

Schmiedt, R.A., Zwislocki, J.J. and Hamernik, R.P.(1980). "Effects of hair cell lesion on responses of cochlear nerve fibers.I." J. Neurophysiol. 43, 1367-1389.

Schmiedt, R.A.(1982). "Effects of low-frequency biasing on auditory nerve fiber activity." J. Acoust. Soc. Am. 72, 142-150.

Sellick, P.M., Patuzzi, R. and Johnstone, B.M.(1982a). "Measurement of basilar membrane motion in the guinea pig using the MOssbauer technique." J. Acoust. Soc. Am. 72, 131-141.

Sellick, P.M., Patuzzi, R. and Johnstone, B.M.(1982b). "Modulation of responses of spiral ganglion cells in the guinea pig cochlea by low frequency sound." Hearing Res. 7, 199-221.

Siegel, J.H. and Kim, D.O.(1982). "Efferent neural control of cochlear mechanics? Olivocochlear bundle stimulation affects cochlear biomechanical nonlinearity." Hearing Res. 6, 171-182.

Siegel, J.H., Kim, D.O.and Molnar, C.E.(1982). "Effects of altering organ of Corti on cochlear distortion products f2-f1 and 2f1-f2." J. Neurophysiol. 47, 303-328.

Smoorenburg, G.F.(1972). "Combination tones and their origin." J. Acoust. Soc. Am. 52, 615-632.

Smoorenburg, G.F.(1980). "Effects of temporary threshold shift on combination tone generation and two-tone suppression." Hearing Res. 2, 347-355.

Weiss, T.F.(1982). "Bidirectional transduction in vertebrate hair cells: a mechanism for coupling mechanical and electrical processes." Hearing Res. 7, 353-360.

Wightman, F.L., McGee, T. and Kramer, M.(1977). " Factors influencing frequency selectivity in normal and hearing-impaired listeners." Same reference as Leshowitz and Lindstrom, pp295-306.

Wilson, J.P. and Johnstone, J.R.(1973). "Basilar-membrane correlates of the combination tone 2f1-f2." Nature 241, 206-207.

Zwicker, E.(1977). "Masking-period patterns produced by very-low-frequency maskers and their possible relation to basilar-membrane displacement." J. Acoust. Soc. Am. 61, 1031-1040.

Zwislocki, J.J. and Kletsky, E.J.(1980). "Micromechanics in the theory of cochlear mechanics." Hearing Res. 2, 505-512.

EVALUATING TRAVELING WAVE CHARACTERISTICS IN MAN BY AN ACTIVE NONLINEAR COCHLEA PREPROCESSING MODEL

Eberhard Zwicker and Georg Lumer
Institute of Electroacoustics, Technical University München,
8000 München 2, Arcisstr.21; F.R. Germany

Introduction

Since the discovery of v. Békésy (1943) that frequency-place transformation in the cochlea of vertebrates is accomplished by traveling waves of the displacement of the basilar membrane, many experiments have been undertaken to add to the knowledge about the vibration patterns taking place in the inner ear. For almost three decades, the data elaborated post mortem in birds and vertebrates looking through a microscope by v.Békésy have been used to describe cochlea wave patterns at any sound level. Introducing the indirect, but much more sensitive Mössbauer probe new data from living animals could be obtained. Rhode (1971) was the first pointing to the level dependence of basilar membrane displacement. More data at even lower levels and smaller displacement values have been published meanwhile (for references see for example Patuzzi et al., 1984). The new animal data showed very clearly that the inner ear and its normal activity are very vulnerable. From this point of view it seems to be unlikely that equivalent data from humans may become available. The level dependent, i.e. nonlinear data of the traveling wave characteristics of man are, however, of great interest not only for the people working in physiology and psychology but also in ENT-departments. Therefore indirect methodes leading to the data of interest may be used in exchange. As mentioned earlier (Zwicker, 1983) oto-acoustic emissions (OAE's) can be used as an effective tool to enlarge our knowledge about the inner ear's signal processing especially at low and very low levels at which normal methodes mostly fail. Meanwhile, a hardware model of the nonlinear preprocessing activity of the inner ear (Zwicker, 1985a) and its application to describe qualitatively the creation of different kinds of OAE's (Zwicker, 1985b) has been described. This model seems to be so effective that its parameters may be successivley changed in such a way that not only its qualitative results agree with data measured in human subjects but also the quantitative details. With regard to the traveling wave characteristics, the conspicuous frequency distance of neighboring spontaneous or simultaneous delayed oto-acoustic emissions in man which, expressed in critical band rate, shows a constant value of 0.4 to 0.5 Bark (Zwicker and Schloth, 1984; Dallmayr, 1985) seems to be such an interesting detail. Using the model we may be able to understand the reason for the existance of a minimal distance between neighboring OAE's. The parameters of the model may be changed so that the OAE-data can be simulated. In this case, the model's data may correlate to the characteristic data of the human cochlea we are searching for.

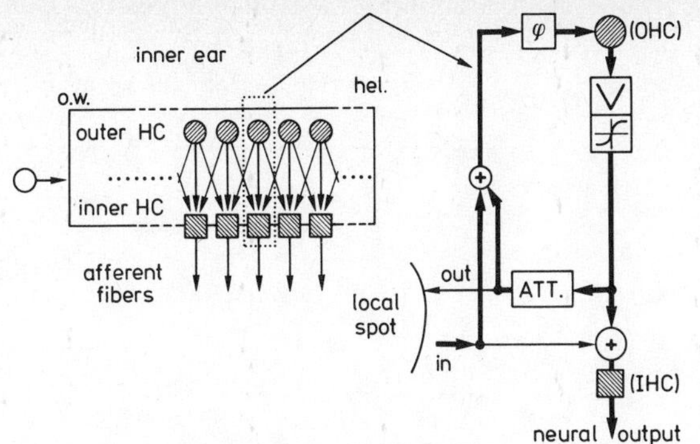

Fig.1: Simplified concept of the active nonlinear preprocessing in the cochlea (adopted from Zwicker, 1979)

In order to follow this goal, the analog hardware model and the computer version will be introduced and the influence of the number of sections, i.e. the degree of dissection discussed. OAE-data from human subjects will be compared with the emissions seen in the models. After describing the important reason for the above mentioned minimal distance, the data produced in the model are compared with measured simultaneous evoked OAE-data.

The Models

The structure of the active nonlinear cochlea preprocessing model was published a few years ago (Zwicker, 1979). It is drawn in Fig.1 in simplified form which, however, still contains the important elements. The model assumes that the outer hair cells do not transfer information towards higher neural levels but act as amplifiers with nonlinear saturating characteristic, still processing AC-values. The outer hair cells may realize this amplifier through mechanical-electrical-mechanical transformations in such a way that the vibration can produce feedback which, however, is effective mostly at medium and especially at low levels. This nonlinear feedback is responsible for the level dependent frequency selectivity and produces additionally a mixture of traveling waves and standing waves at very low levels (Zwicker, 1985a).

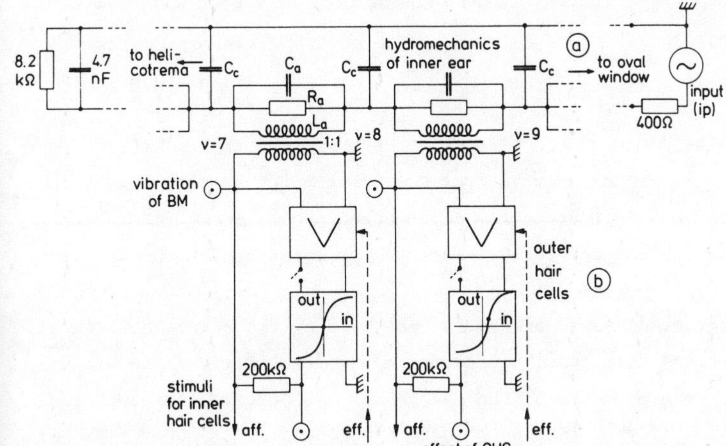

Fig.2: Schematic diagram of the hardware model (adopted from Zwicker, 1985a)

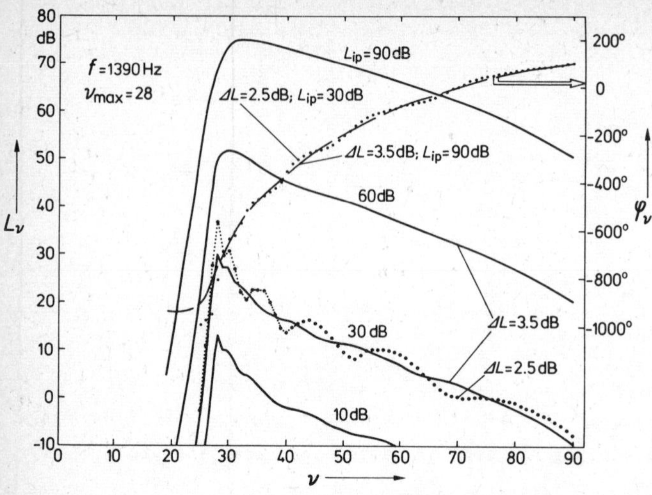

Fig.3: Level and phase response for different input levels L_{ip} as a function of the number of ν of the 90 sections of the hardware model. Note the level dependence and the influence of the feedback (expressed in ΔL)

Hardware Model

The parameters of the set up of the hardware model are described elsewhere (Zwicker, 1985a). Therefore, only the blockdiagram (Fig.2) and some amplitude and phase patterns along the basilar membrane, i.e. along the section number ν (Fig.3) are shown. The hardware model in its present form consists of 90 sections for the frequency range from about 1000 to 7000 Hz, i.e. about 8 sections per Bark or 6 sections per mm length along the basilar membrane. The most important effects produced by the model are - besides the level dependent sharpening of the frequency selectivity - the ripples (see Fig.3 for L_{ip}=30 dB and ΔL=2.5 dB) of the level response towards higher section numbers ν, i.e. towards the oval window. The minima of the ripples in the level response have their counterpart in pronounced steps in the phase response, indicating standing waves. The ripples occur at low levels only and increase in size for larger feedback. An example is given in Fig.3 for L_{ip}=30 dB with dots, which belong to the condition ΔL=2.5 dB instead of ΔL=3.5 dB for the solid line curves of Fig.3. As outlined in Table 1, ΔL decreases for larger feedback which corresponds to higher peaks in the level response.

Computer Model

We have been interested in two questions to be answered by the computer model, the description of steady state masking effects (not discussed here, see Lumer 1985) and the effect of varying the number of sections. Since simultaneous evoked oto-acoustic emissions are also a steady state effect, we searched for solutions in the frequency domain.

In the computer model, the electrical network is described by the equivalent signal flow chart in which, however, the signal direction is inverted (Fig.4). The blocks contain the impedances of the elements of the electrical network. The feedback loop with the saturation nonlinearity is incorporated in the impedance $\underline{Z}_{nl}$ of the parallel resonant circuit as a special negative conductance G_{nl} the value of which has to be

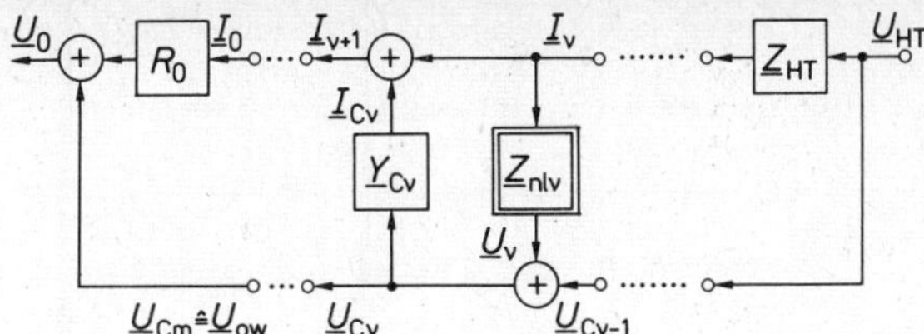

Fig.4: Signal flow chart of the computer model (adopted from Lumer, 1985)

found by an iteration process. G_{nl} is computed using the method of the "Harmonic Balance", where the distortion products arising from the nonlinearity are neglected. The computation of the level and phase pattern of a sinusoidal input signal as a function of the section number ν starts at the "helicotrema" with an arbitrary value $\underline{U}_{HT}$. In several iterations $\underline{U}_{HT}$ can be adjusted so that the desired voltage $\underline{U}_0$ is achieved. The dependences of the elements along the basilar membrane as indicated in a former paper (Zwicker,1985a, Fig.2) have been expressed in approximating equations as a function of critical band rate, z. Besides the distance Δz (expressed in Bark) between two neighboring sections the strength of the feedback is the important parameter. The more feedback, the higher the peak in the level-response pattern. In the hardware model, it is the gain reduction ΔL of the amplification in each single section from the point of ringing, which is responsible for the effect in question. In the computer model, the feedback factor k_f plays the important role. The relation between peak elevation and the two mentioned values ΔL and k_f are given in Table 1 for 10 and for 40 sections per Bark.

Table 1:						
peak elevation at low levels	0	17	25	35	48	dB
gain reduction ΔL (10 sect./Bark)	∞	5.5	3.5	2.5		dB
feedback factor k_f (40 sect./Bark)	0	0.707	0.825	0.925	0.995	

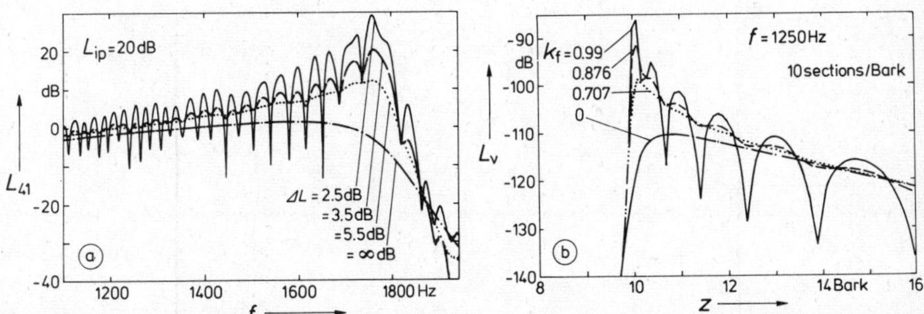

Fig.5: (a) Level-frequency pattern measured in the hardware model at section 41 for different feedback expressed in ΔL and (b) level-place pattern calculated in the computer model for different feedback factors k_f, also at low levels

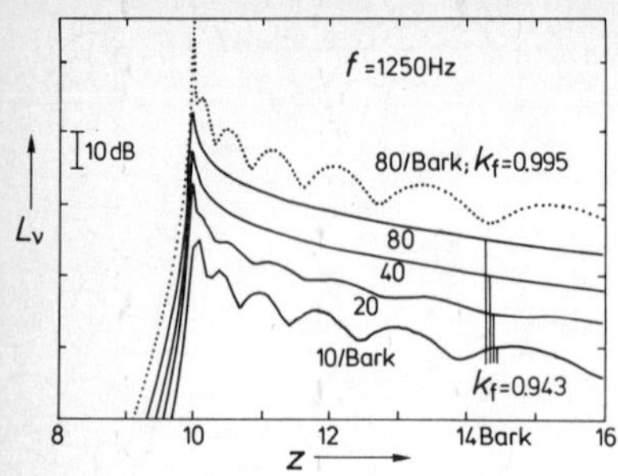

Fig.6: Level-place patterns calculated in the computer model for different dissection rate (expressed in sections per Bark) and low input levels. Feedback factor k_f is 0.943 for the lower curves (solid). 80 sections per Bark and strong feedback (0.995) are used for the dotted curve. Note the 10 dB shift between curves

For approximately the same number (10) of sections pro Bark, the computer model and the hardware model produce comparable data when the same amount of feedback is produced and nonlinearities have similar characteristics. Fig.5 gives an example of level-frequency patterns (a) measured in the hardware model and corresponding level-place patterns (b) calculated in the computer model. Parameter is in both cases the strength of the feedback.

Influence Of Number Of Sections Per Bark

The number of sections in the hardware model was chosen as small as possible but as large as to fulfill the condition that the phase changes not more than 90° from section to section (see Fig.3). This could be realized for small feedback but not for large feedback (ΔL=2.5 dB) and small levels. In this case, reflections from each section near its characteristic frequency can arise. These side-effects produced by the subdivision play a secondary role for qualitative effects described earlier (Zwicker, 1985b), become destructive, however, when searching for quantitative dependences of oto-acoustic emissions. The computer program offers an easy way to enlarge the number of sections, i.e. to refine the subdivision. Fig.6 shows the change of level-place patterns for changing the subdivision from 10 (as in the hardware model) to 20, 40 and 80 sections per Bark. The last number corresponds to a section distance of about 16µm, close to the distance between hair cells.The value k_f=0.943 is kept constant for the curves drawn as solid lines. From these curves it becomes clear that the ripple magni-

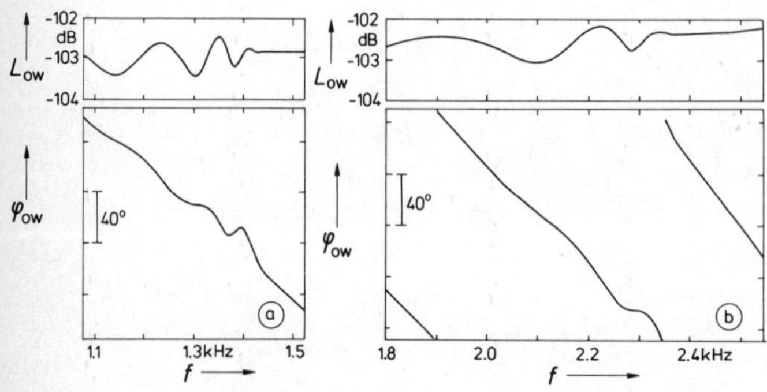

Fig.7: Calculated level- and phase-frequency response (low levels) in the neighborhood of 1.3 kHz (a) and 2.2 kHz (b) in case of narrow discontinuities as described in the text

tude deminishes for enlarging the subdivision. This dependence is reasonable and corresponds to the reduced change in impedance from section to section, producing less reflection, i.e. standing waves. The ripple magnitude enlarges again for stronger feedback as indicated in the upper dotted curve in Fig.6, for which k_f was enlarged to 0.995. This indicates that the ripples are generated by impedance changes from section to section, which can occur for coarse subdivision as well as for fine subdivision and strong feedback near ringing.

Simultaneous Emissions

Data equivalent to simultaneous evoked oto-acoustic emissions are obtained in the computer model by calculating the voltage $\underline{U}_{OW}$ corresponding to the input value of the inner ear (see Zwicker, 1985a, b). In order to produce neglectable ripple magnitude, 40 sections per Bark (320 in total) have been used for the calculations. A small discontinuity in the sequence of sections was introduced for seven cross capacities. The capacities increased for four consecutive sections by 10% each and then decreased to normal. Such a discontinuity seems to be more natural than an abrupt change.
The discontinuities were located around section numbers 110 or 240 corresponding to critical band rates of 10.75 Bark and 14 Bark, i.e. 1411 Hz and 2335 Hz characteristic frequency, respectivley. The results are outlined in Fig.7a and b as voltage level, L_{OW}, and phase φ_{OW} as a function of frequency. The level as well as the phase response shows a ripple towards frequencies lower to the frequency corresponding to the place of the discontinuity. The distance between neighboring minima or maxima increases somewhat towards lower frequencies. This distance increases, however, from Fig.7a to 7b, i.e. for increasing frequency related to the place of discontinuity.
The voltage and the phase ripples may be outlined in the same way as for simultaneous evoked emissions from human subjects. Using voltage and phase measured at high input levels as reference, the vector diagram of what is produced additionally at low levels can be drawn. Such a diagram based on the data of Fig.7b is outlined in Fig.8b. It may be compared with data measured in a human subject (Zwicker and Schloth, 1983) which are redrawn in Fig.8a. The similarity of the diagrams is very strong and reveals that the computer model produces the same effect that is produced in the inner ear at low levels for many subjects: standing waves which show up at low levels, which decrease in relative value for increasing level and which in many cases grow so much that rin-

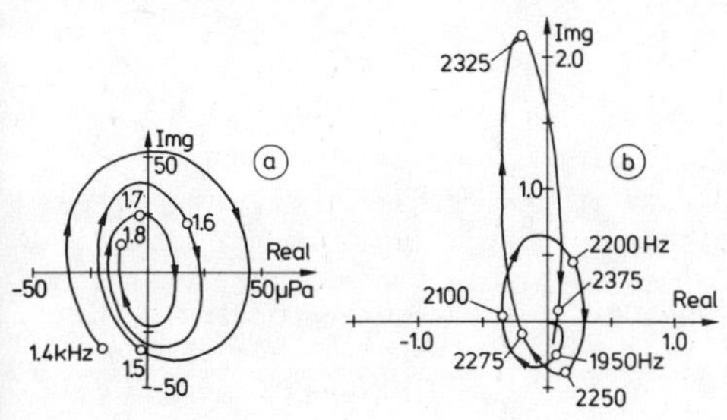

Fig.8: Vector diagrams of simultaneous evoked emissions; (a) measured in the closed ear canal of a human subject (adopted from Zwicker and Schloth, 1983) and (b) transformed from calculated data outlined in Fig.7b

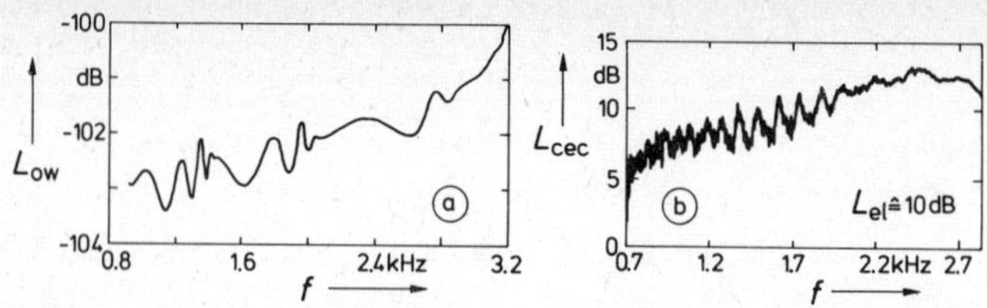

Fig.9: Level-frequency responses at low input levels (a) calculated introducing three discontinuities near the places corresponding to 1.4, 2.0, and 2.9 kHz and (b) measured in the closed ear canal of a human subject

ging occurs at very low levels, i.e. spontaneous emissions are generated.

In a few cases many SEOAE's can be observed in human subjects. The reason for so many ripples may be that not only one but a few discontinuities are present (Manley, 1984). Such a situation is simulated in the computer model. The result is indicated in Fig.9a and compared with the frequency dependence of the SPL measured in the closed ear canal of subject A.S. outlined in Fig.9b.

Using a different model (linear with 2nd filters), Sutton and Wilson (1983) report on the appearance of delayed emissions, but only in the case of irregularities.

Summary

Computer models have the advantage that the subdivision can be realized much finer than in a hardware model. The ripples occuring in both models for a subdivision of only 10 sections per Bark are interpreted as standing waves and due to discontinuities produced by poor dissection. For high resolution of 40 sections per Bark, the ripples diminish in the homogeneous model to very small size, show up again, however, for small discontinuities installed deliberately at limited range along the line. The frequency distance of the observed ripples increases with frequency in the model and in human subjects in similar fashion enlarging from about 50 Hz at low frequencies to about 250 Hz at 3 kHz. This frequency distance is directly related to the frequency distance needed to produce 180° phase difference at a certain place of the line. This frequency distance calculated in the model is plotted in Fig.10 (solid) as a function of frequency. A comparison with human data (Zwicker and Schloth, 1983, dashed; Dallmayr, 1985, dotted) indicates, that the assumed parameters of the cochlea and the basilar membrane producing the traveling wave patterns are very close to the reality.

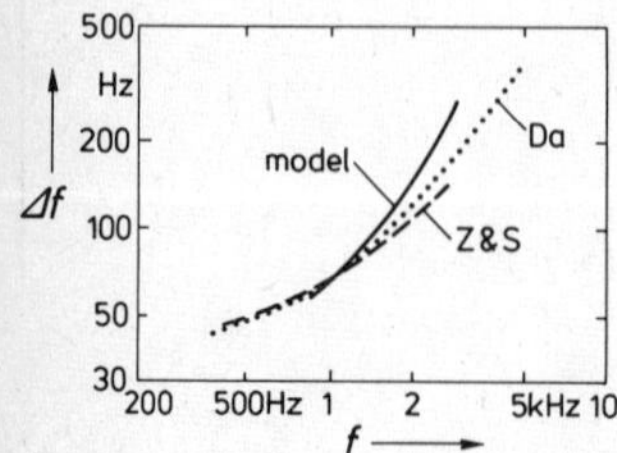

Fig.10: Measured data of the frequency distance Δf of neighboring spontaneous (dotted) and simultaneous evoked (dashed) oto-acoustic emissions. The solid line indicates the frequency distance corresponding to an 180°-phase difference at the steepest place of the phase-frequency pattern computed without feedback

The results indicate that more research is needed to answer the question in more detail and for more parameter variations. The nonlinear feedback model, however, seems to be a perfect tool to do so.

Acknowledgements

This research was carried out in Sonderforschungsbereich 204, Gehör, München, supported by the Deutsche Forschungsgemeinschaft. We are very thankful for many comments by Dr.-Ing. habil. H. Fastl on a former version of the manuscript.

Literature

von Békésy, G. (1943) Über die Resonanzkurve und die Abklingzeit der verschiedenen Stellen der Schneckentrennwand. Akust. Zeits., 1943, 8, 66-76.

Dallmayr, Ch. (1985) Spontane oto-akustische Emissionen, Statistik und Reaktion auf akustische Störtöne. Acustica, in press.

Lumer, G. (1985) Nachbildung nichtlinearer simultaner Verdeckungseffekte bei schmalbandigen Schallen mit einem Rechnermodell. Dissertation, TU München.

Manley, G.A. (1984) Frequency spacing of acoustic emissions: a possible explanation. In "Mechanisms of Hearing" (Eds. W.R.Webster and L.M.Aitkin) Melbournes Monash University press, 36-39.

Patuzzi, R., Sellick, P.M. and Johnstone, B.M. (1984) The modulation of the sensitivity of the mammalian cochlea by low frequency tones.III. Basilar membrane motion. Hearing Research 13, 19-27.

Rhode, W.S. (1971) Observations of the Vibration of the Basilar Membrane in Squirrel Monkeys using the Mössbauer technique. J.Acoust.Soc.Am. 49, 1218-1229.

Sutton, G.J. and Wilson, J.P. (1983) Modelling cochlear echoes: The influence of irregularities in frequency mapping on summed cochlear activity. In: Mechanics of Hearing, Eds. E. de Boer, M.A. Viergever, Delft University Press, Delft, pp. 83-90.

Zwicker, E. (1979) A Model Describing Nonlinearities in Hearing by Active Processes with Saturation at 40 dB. Biol. Cybernetics 35, 243-250.

Zwicker, E. (1983) On peripheral processing in human hearing. In: Hearing - Physiological Bases and Psychophysics (Eds. Klinke, R. and Hartmann, R.) Springer-Verlag, Berlin, 104-110.

Zwicker, E. (1985a) A hardware cochlea nonlinear preprocessing model with active feedback. Submitted to J.Acoust.Soc.Am..

Zwicker, E. (1985b) Oto-acoustic emissions in a nonlinear cochlea hardware model with feedback. Submitted to J.Acoust.Soc.Am..

Zwicker, E. and Schloth, E. (1984) Interrelation of different oto-acoustic emissions. J.Acoust.Soc.Am. 75, 1148-1154.

MODELING INTRACOCHLEAR AND EAR CANAL DISTORTION PRODUCT ($2f_1-f_2$)

John W. Matthews and Charles E. Molnar
Institute for Biomedical Computing
Washington University
St. Louis, MO 63110

ABSTRACT

Responses of a mechanical model of the peripheral auditory system are calculated for two-tone stimuli with primary frequencies chosen so that the frequency of the cubic difference tone ($2f_1-f_2$) is maintained constant for all pairs used. The comprehensive model explicitly includes linear characterizations of the earphone driver, acoustic coupler, and middle ear, and a passive cochlear model with nonlinear damping of the cochlear partition motion. The amplitude and phase plots of the ($2f_1-f_2$) distortion signal, determined both in the ear canal and at the ($2f_1-f_2$) characteristic place, show rapid variations with f_1, and quite different patterns depending on the stimulus level. These plots, as well as plots showing the spatial distribution of primary and distortion signals along the cochlea for six selected f_1 values, are interpretable as showing interactions between multiple propagating waves at the distortion product frequency.

1. INTRODUCTION

The generation of acoustic signals within the cochlea by nonlinear or active mechanisms can give rise to mechanical waves that propagate toward the stapes and are reflected back into the cochlea, requiring that the middle ear and external ear be included in a model since they affect the impedance matching conditions at the stapes. Results of our earlier studies (Matthews, 1980; 1983) indicated that there is complex variation with frequency of distortion amplitude and phase of signals in the ear canal and especially at the distortion characteristic place, suggesting the interference of forward- and backward- propagating distortion waves in the cochlea, but a much more detailed study with many more stimulus conditions appeared necessary to gain a less ambiguous picture. This study presents results obtained with three different levels of two-tone stimuli, and with a fine enough spacing of stimulus frequencies to resolve rapid changes in distortion magnitude and phase.

2. THE MODEL

Figure 1 shows a block diagram for the model. This model includes the effects of the stimulus delivery system, the middle ear, and the cochlea. The middle ear and acoustic coupler models are those developed by Matthews (1980, 1983) while the earphone driver model is that of Neely (1981a). These three elements of the model are linear; only the model for the cochlea is nonlinear.

Except for the nonlinear damping (see below), the parameters of the cochlear model are the same as used by Neely (1981b). Within the cochlear model, the fluid is represented by a two-dimensional, linear, incompressible, inviscid fluid. The basilar membrane, which separates the rectangular fluid chamber into two symmetric scalae, is locally reactive and is represented by mass, stiffness, and damping functions versus distance along the cochlea. The mass and stiffness functions are those of Neely (1981b) while the damping at each point is a nonlinear function of basilar membrane velocity:

$$R(x,t) = R_o[1 + \alpha v(x,t) + \beta v^2(x,t)]$$

where R_o is Neely's damping value of 200 dyne-sec/cm^3, and $\alpha = 850$ sec/cm and $\beta = 650{,}000$ (sec/cm)^2 after Matthews (1980). The cochlear model was solved by the finite difference method of Neely (1981b) with m = 4 points representing the scala height and n = 240 points representing the cochlear length. The model was solved in the time domain using a time step of approximately 3.15 microseconds. The computations were done in 32 bit floating point on a VAX 11/750. For further details of model assumptions and approximations see Matthews (1980) and Neely (1981a, 1981b).

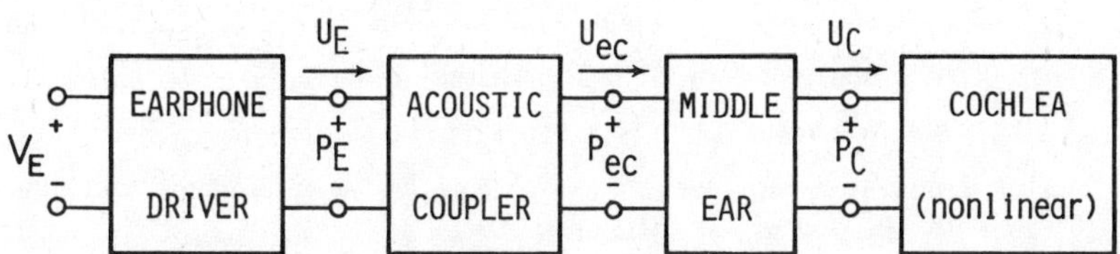

Figure 1. Block diagram of a "comprehensive" model of the peripheral auditory system of cat and a stimulus delivery system. Pressure (P), volume velocity (U), and voltage (V) are identified by subscripts at the earphone driver (E), ear canal near the eardrum (ec), and the most basal position of the cochlear spiral (C).

3. PARADIGM

The data presented here were obtained using an iso-($2f_1-f_2$) paradigm designed to closely mimic that used by Zwicker (1980). The stimuli used were frequency pairs whose frequencies were varied such that they were harmonically related and such that the "cubic difference tone" ($2f_1-f_2$) had a constant frequency of 1550 Hz. All stimulus frequencies used were harmonics of 38.75 Hz.

The magnitude and phase of the f_2 component of the voltage applied to the earphone driver V_E were adjusted such that the f_2 component of the ear canal pressure P_{ec} was 60 dB

SPL and zero cosine phase. Likewise, the magnitude and phase of the f_1 component of V_E were adjusted such that the f_1 component of P_{ec} was 60, 65 or 70 dB SPL and zero cosine phase. The actual ear canal pressures obtained varied slightly from the target values. The error in the magnitude ranged from -0.10 dB to +0.14 dB with a mean of -0.001 dB and standard deviation of 0.034 dB. The error in the phase ranged from -0.006π to $+0.004\pi$ with a mean of -0.0003π and standard deviation of 0.002π radians.

In order to obtain near steady state solutions, each stimulus was slowly turned on during 0.0315 seconds of simulated time using a raised cosine gating function. A full cycle of the fundamental frequency was then collected for each response variable of interest for later Fourier analysis. The fundamental frequency was either 38.75, 77.5, or 155 Hz. A total of 72 frequency pairs were used in this study, each requiring approximately three hours of CPU time.

4. RESULTS

Figures 2 and 3 summarize the results of the study with respect to the $(2f_1-f_2)$ distortion product. Figure 2 shows the magnitude and phase of the $(2f_1-f_2)$ component of the ear canal pressure plotted versus the frequency of f_1 for three levels of f_1. Figure 3 shows a measure of the $(2f_1-f_2)$ distortion product within the cochlea.

Figure 3 shows the variation of magnitude and phase of the displacement of the basilar membrane at the 1550 Hz characteristic place. In the vicinity of this place, the magnitude and phase of the $(2f_1-f_2)$ distortion product are distributed similarly to those of a single 1550 Hz tone (Hall, 1974; Matthews, 1980; Kim, et al., 1980). Therefore, the magnitude and phase of the basilar membrane displacement are plotted as the magnitude and phase of the ear canal pressure of a 1550 Hz tone that would produce the same displacement at the 1550 Hz place. In this study, this was accomplished by normalizing the $(2f_1-f_2)$ component of the displacement with respect to the displacement produced by a single 1550 Hz tone at 0 dB SPL and 0 phase. For single 1550 Hz tones, this method is accurate in estimating the ear canal pressure to within 0.18 dB and 0.002π radians for tones up to 40 dB SPL; at 35 dB SPL the error is 0.04 dB and 0.000π radians. We consider this to be sufficient accuracy for the data of figure 3.

The behavior shown in figure 3 is quite elaborate and some insight can be gained by examining the spatial distributions of various frequency components of the basilar membrane motion. Figure 4 shows the f_1, f_2, $(2f_1-f_2)$, and (f_2-f_1) components of the basilar membrane displacement plotted versus fraction of the cochlear length measured from the stapes for six of the stimulus conditions used in obtaining the 60 dB curves of figures 2 and 3. The stimulus conditions illustrated are indicated by f_1 frequency and by the lower case letters in each panel, which correspond to the letters shown on figures 2 and 3. The $(2f_1-f_2)$

component is shown with a bold line. The other components can be identified by the place of their apical peak. Left to right they are: f_2, f_1, and (f_2-f_1).

5. DISCUSSION

The elaborate behavior shown by the intracochlear distortion product resembles the psychophysical results obtained by Zwicker (1980) in both magnitude and phase. Much of this behavior can be interpreted in the model as the combining of two distortion product "waves". These waves originate in the cochlea at the locus where f_1 and f_2 are both strong. One wave propagates apically and the other propagates basally. The basal-going wave is partially reflected at the stapes boundary and the reflected portion becomes an apical-going wave. The two apical-going components combine, depending on their relative phase, and propagate to the 1550 Hz place. The null at f_1 = 2751 Hz in the 60 dB curve of figure 3 shows almost total cancellation; most of the 1550 Hz component that is observed at the 1550 Hz place can be accounted for by a secondary nonlinear interaction between f_1 and the (f_2-f_1) distortion signal (see figure 4 panel b).

Figure 3 shows an abrupt peak at f_1 = 3100 Hz for each stimulus level. This peak is somewhat of an artifact since it occurs for a stimulus where $(f_2-f_1) = (2f_1-f_2)$. Note that in figure 2, the ear canal distortion signal displays only a small dip for this stimulus condition. As can be seen in panels d, e, and f of figure 4, which show the intracochlear spatial distributions for the corresponding 60 dB stimulus condition and for the two nearest frequency pairs, the relative amplitudes of the $(2f_1-f_2)$ and (f_2-f_1) components at the stapes and in the apical region can be quite different, so that strong interference effects can be seen at one location and not at the other, even with the same stimulus conditions. The difference in the relative strength of forward- and backward-propagating waves of similar frequency may be accounted for with an interpretation like that of Zwicker (1980), since the contributions of sources distributed along the cochlea may combine quite differently into waves propagating in the two directions. Note also that the phase of $(2f_1-f_2)$ and (f_2-f_1) components of distortion have a different relationship to the phases of f_1 and f_2, even when they are of the same frequency.

Kim, et al. (1980; Kim, 1980) presented distortion products measured in cat ear canal pressure under several paradigms. With fixed stimulus frequencies of 2170 Hz and 2790 Hz, the levels of the primaries were varied together between 40 and 95 dB SPL. Under this paradigm, the level of the $(2f_1-f_2)$ distortion product in the ear canal pressure varied non-monotonically. We have not duplicated this paradigm precisely, but figure 2 shows non-monotonic growth of $(2f_1-f_2)$ level versus f_1 level for f_1 frequencies near 1750 Hz.

Kim showed that the (f_2-f_1) ear canal pressure under this paradigm was typically 10 to 20 dB less than that of $(2f_1-f_2)$. In our model results, the (f_2-f_1) level in the ear canal (not

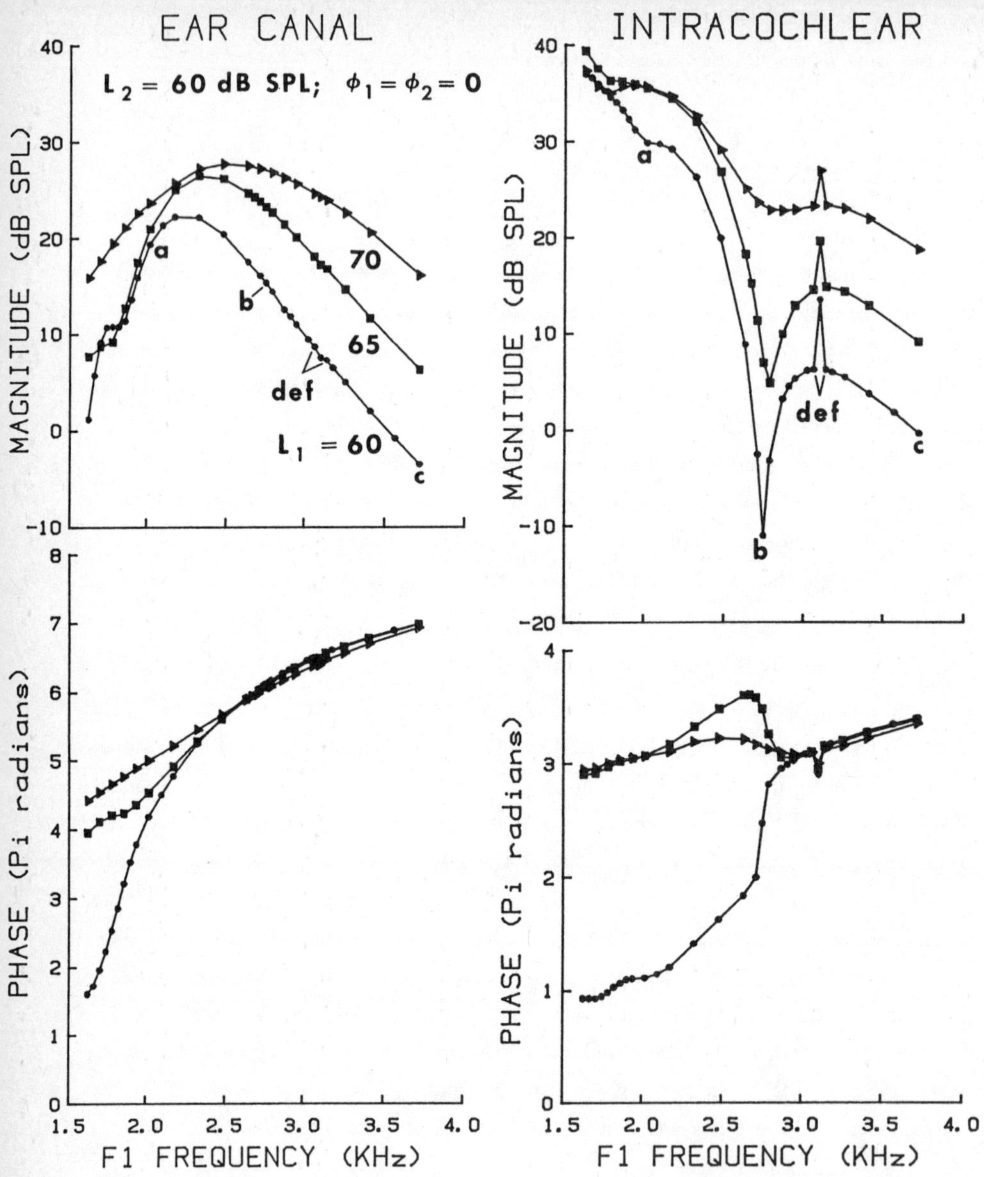

Figure 2 (left). Ear canal distortion product under an iso-$(2f_1-f_2)$ paradigm. The 1550 Hz component of the ear canal pressure is plotted versus the frequency of f_1 when the earphone driver voltage V_E contains only f_1 and f_2 varied in frequency such that $(2f_1-f_2) = 1550$ Hz.

Figure 3 (right). Intracochlear distortion under the same conditions as figure 2. A measure of the 1550 Hz component of the displacement of the basilar membrane at the 1550 Hz place is plotted versus f_1 frequency. The measure plotted is the level and phase of a single 1550 Hz tone that would produce the same displacement at the 1550 Hz place (see text).

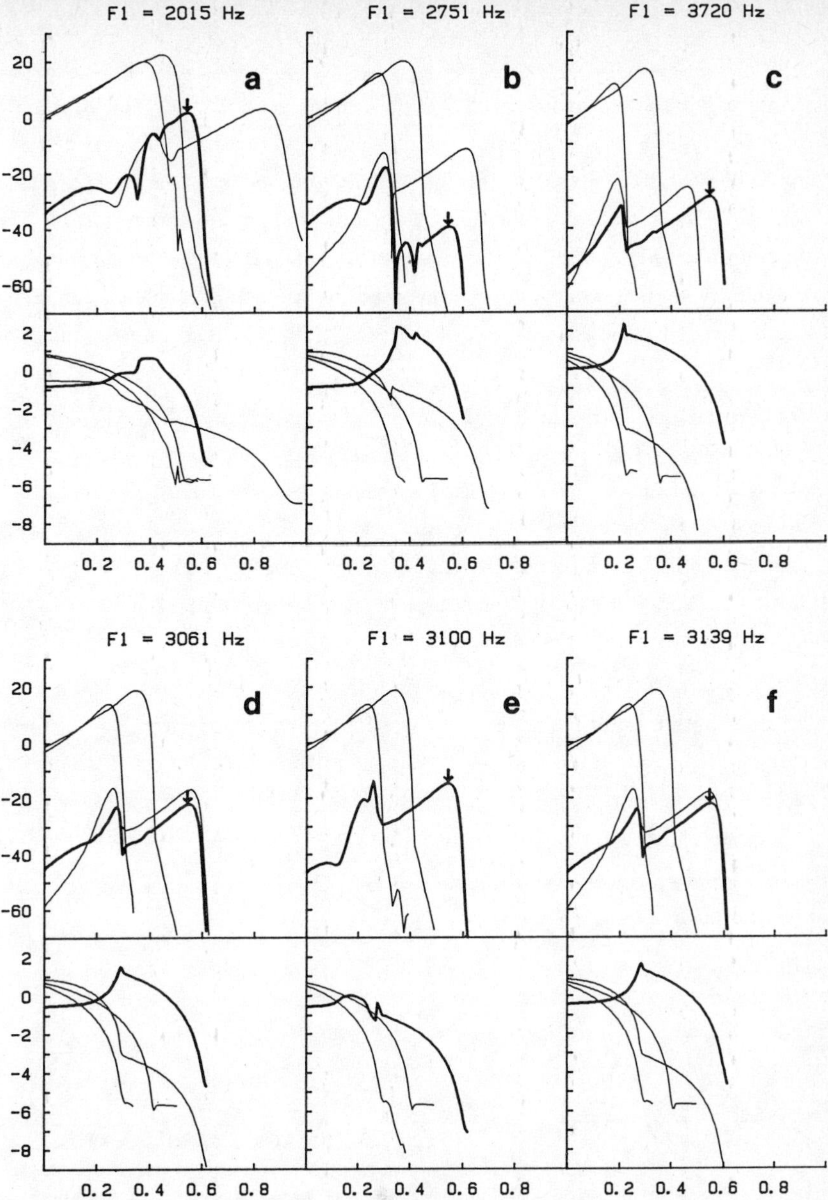

Figure 4. Intracochlear spatial distribution of several components of the basilar membrane displacement versus fraction of cochlear length for six different stimulus conditions. The bold line in each panel shows the $(2f_1-f_2) = 1550$ Hz component of the displacement of the basilar membrane. The other lines (by peak position from left to right) are f_2, f_1, and (f_2-f_1). The stimulus conditions are indicated by the f_1 frequency given above each panel. The letters in each panel correspond to the letters shown on figures 2 and 3. Displacement is shown in dB re 1 angstrom (upper) and phase in π radians (lower). The arrow in each panel indicates the position at which the $(2f_1-f_2)$ component of the displacement was determined for figure 3.

shown) was generally below the $(2f_1-f_2)$ level by a similar amount.

Kim (1980) also showed results for an iso-$(2f_1-f_2)$ paradigm similar to that presented here. The general shape and absolute level of his curves are well matched by the curves of figure 2. However, Kim's data show pronounced notches that he called the "lobing phenomenon." These notches are not reproduced by our model with the parameters and stimulus conditions that were used in this study, but there is an indication of a notch in both the 60 and 65 dB SPL curves of the model for f_1 near 1750 Hz. We speculate that this "notching" could be increased by altering the cochlear map and cochlear tuning, but that the conditions needed to produce cancellations yielding a deep notch may be quite particular.

The tuning and cochlear map for the model parameters used here do not correspond well to the cat cochlea. Incorporation of active mechanics into the present nonlinear model would allow us to obtain tuning more nearly resembling that of cat. However, our initial attempts at doing this have produced no stable solutions. We have not yet determined whether this instability is a computational problem, an artifact of the mathematical formulation of the model, a result of unfortunate parameter value choices, or whether the real cochlea itself is unstable.

Figure 5 shows a further illustration of the difficulty in interpreting the behavior of a system with distributed nonlinear sources. The stimulus conditions used to obtain figure 4a were repeated with one model parameter changed; the cubic term of the relationship between velocity and pressure at each point of the basilar membrane was eliminated ($\beta = 0$). One might expect this change to eliminate the $(2f_1-f_2)$ distortion product in the model. In fact the $(2f_1-f_2)$ distortion product observed in the ear canal actually increases from 19.4 to 21.6 dB SPL!

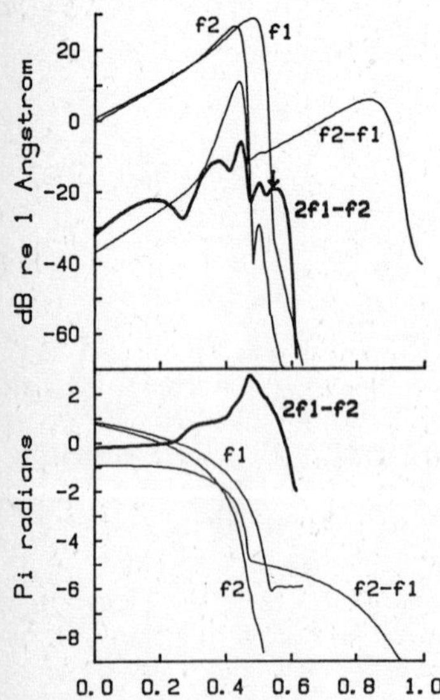

Figure 5. Spatial distributions of several components of the basilar membrane displacement versus fraction of cochlear length. Identical to figure 4a except that the nonlinear basilar membrane damping was changed such that $\beta = 0$. Note that the "cubic" distortion product $(2f_1-f_2)$ is present even though the basilar membrane damping has no cubic nonlinear term. The $(2f_1-f_2)$ distortion product is generated by secondary interactions such as between f_1 and (f_2-f_1).

The richness of the sources of distortion components and the complexity of the ways in which they can interact make it difficult to identify the aspects of these phenomena that are most important to hearing without considerable further study. We can hope that in the future relatively simple models and conceptualizations can provide a useful and accurate summary of the causes and effects of these phenomena, but at the present both experimental and modeling studies leave large areas of uncertainty to be explored.

ACKNOWLEDGEMENTS

We thank Walter Bosch and Don Ronken for assistance in preparation and revision of this paper and Jerome R. Cox, Jr. for helpful comments. This work was supported by NIH grants NS-21592, NS-07498, RR-00396, and RR-01379.

REFERENCES

Hall, J. L., "Two-tone distortion products in a nonlinear model of the basilar membrane," J. Acoust. Soc. Am. *56,* pp. 1818-1828, 1974.

Kim, D. O., Molnar, C. E., and Matthews, J. W., "Cochlear mechanics: Nonlinear behavior in two-tone responses as reflected in cochlear-nerve-fiber responses and in ear-canal sound pressure," J. Acoust. Soc. Am. *67,* pp. 1704-1721, 1980.

Kim, D. O., "Cochlear mechanics: Implications of electrophysiological and acoustical observations," Hearing Research *2,* pp. 297-317, 1980.

Lynch, T. J., III, Nedzelnitsky, V., and Peake, W. T., "Input impedance of the cochlea of cat," J. Acoust. Soc. Am. *72,* pp. 108-130, 1982.

Matthews, J. W., "Mechanical Modeling of Nonlinear Phenomena Observed in the Peripheral Auditory System," D.Sc. dissertation, Washington University, St. Louis, Missouri, USA, 1980.

Matthews, J. W., "Modeling reverse middle ear transmission of acoustic distortion signals," in *Mechanics of Hearing,* ed. E. de Boer and M. A. Viergever, Delft University Press, Delft, The Netherlands, pp. 11-18, 1983.

Neely, S. T., "Fourth-order Partition Mechanics for a Two-dimensional Cochlear Model," D.Sc. dissertation, Washington University, St. Louis, Missouri, USA, 1981a.

Neely, S. T., Finite difference solution of a two-dimensional mathematical model of the cochlea," J. Acoust. Soc. Am. *69,* pp. 1386-1393, 1981b.

Zwicker, E., "Cubic difference tone level and phase dependence on frequency difference and level of primaries," in *Psychophysical, Physiological and Behavioral Studies in Hearing,"* ed. G. van den Brink and F. A. Bilson, Delft University Press, Delft, The Netherlands, pp. 268-271, 1980.

INTERACTIONS AMONG MULTIPLE SPONTANEOUS OTOACOUSTIC EMISSIONS

Kenneth Jones and Arnold Tubis
Department of Physics
Purdue University
West Lafayette, IN 47907

Glenis R. Long
Department of Audiology and Speech Sciences
Purdue University
West Lafayette, IN 47907

Edward M. Burns
Department of Speech and Hearing Sciences
University of Washington
Seattle, WA 98195

Elizabeth A. Strickland
House Ear Institute
Los Angeles, CA 90057

ABSTRACT

Evidence has recently been obtained (Burns et al., 1984) for several interactions among spontaneous otoacoustic emissions (SOAEs) including intermodulation distortion products, mutual suppression, and noncontiguous-linked SOAEs which apparently share energy between two quasi-stable states. In this paper, we give an updated record of our findings on intermodulation distortion products and linked emissions, and give evidence that the former tend to occur when a distortion product frequency is close to that of a cochlear resonance. Computer simulations of the interactions among van der Pol oscillators, which represent nonlinear active elements in a simplified cochlear model, appear to qualitatively account for some of the observed features of SOAE interactions.

1. INTRODUCTION

Human ears (about 40% of those tested) spontaneously emit narrowband acoustic signals which can be recorded with a sensitive microphone in the ear canal. These signals called spontaneous otoacoustic emission (SOAEs) are usually found at levels less than 20 dB SPL. Although their existence was predicted as early as 1948 by Gold (1948) as a consequence of a hypothetical active filtering process in the cochlea, and they had been reported in the clinical literature, SOAEs were first investigated systematically by Kemp (1979, 1981), Wilson (1980), Wilson and Sutton (1981), and Zurek (1981).

At the present time, a model for SOAEs and related evoked otoacoustic emissions (EOAEs) (Kemp, 1978, Kemp and Chum, 1980; Wilson, 1980), which is firmly grounded in known cochlear anatomy and physiology, does not exist. Kemp (1980) developed a schematic model of the auditory periphery which incorporates retrograde cochlear wave generation and resulting quasi-standing-wave resonance, which qualitatively accounted for a number of observed properties of emissions, and their correlation with the microstructure of behavioral thresholds (e.g., Zwicker and Schloth, 1984; and Long 1984). Significant retrograde wave generation in the cochlea may arise from irregularities in the spatial variation of (passive or active) cochlear response, and/or nonlinear cochlear interactions.

As was pointed out by Kemp (1981) and J. B. Allen (private communication), the existence of SOAEs is not necessarily evidence of active biomechanical cochlear response since these emissions may be due to resonant enhancement of narrow bands of noise. Evidence that at least some SOAEs may arise from active nonlinear feedback forces in the cochlea includes: 1) the physiological vulnerability of emissions to oxotoxic drugs (e.g., McFadden and Plattsmier, 1984); 2) recent measurements by Bialek and Wit (1984) of the statistical properties of the ear-canal-incremental pressure in a frequency band containing a prominent SOAE, the results of which were compatible with a cochlear limit-cycle oscillation plus a small amount of noise, rather than with a resonantly-enhanced band of noise; 3) the observed frequency locking (entrainment) of SOAEs by external tones (e.g., Wilson and Sutton, 1981) which is a well known feature of externally driven limit-cycle oscillators (e.g., Hayashi, 1964); and 4) the discovery (Burns et al., 1984) of interactions among SOAEs which are suggested by models of interacting limit-cycle oscillators (e.g., Pavlidis, 1973).

In this paper, we give an updated summary of our data on two classes of interactions among SOAEs (Burns et al., 1984) - intermodulation distortion products, and alternate multifrequency SOAE states with the transition from one to the other occuring spontaneously or in response to external tones. We also give a theoretical interpretation of these SOAE interactions in the context of an analysis of interacting van der Pol oscillators in a simplified cochlear model.

2. METHODS

SOAE measurements were obtained, and external tones presented, using Knowles transducers (models EA-1842 and BR-1888, respectively). The transducers were connected via plastic tubing to a Grason-Stadler oto-admittance earpiece which was placed in the ear canal of the subject using an appropriate tip size. An FFT analysis (1.25 Hz line spacing, spectral averaging of 32 samples) was applied to the output signal from the EA-1842 transducer. A careful search was made for artifacts in the measurement and signal presentation system using (among other calibration methods) a Zwislocki coupler in a KEMAR. When two external tones were presented via separate EA-1888 transducers at levels up to 80 dB SPL, intermodulation distortion products measured in the KEMAR's ear canal were at least 65 dB below the levels of the external tones.

3. RESULTS

To date, we have found in 8 ears of 7 subjects, 11 satellite SOAEs which are third order intermodulation distortion products of primary SOAEs. These satellite or cubic difference tone (CDT) SOAEs have frequencies, $2f_1 - f_2$ ($f_2 > f_1$), where f_1 and f_2 are the primary SOAE frequencies. We have also found in 3 ears of 3 subjects, 3 cases of alternate SOAE states. Both types of SOAE interactions are illustrated in Fig. 1.

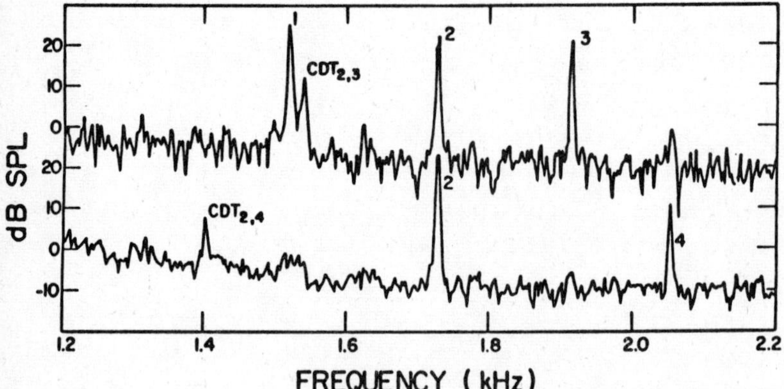

Fig. 1. The lower trace shows one of the alternate states in the left ear of subject 11, over the range, 1200-2200 Hz. Primary SOAEs 2 and 4 (1724 Hz and 2050 Hz, respectively) produce the satellite $CDT_{2,4}$ SOAE ($2f_2 - f_4$ = 1398 Hz). The upper trace shows the other alternate SOAE state with primary SOAEs 1,2 and 3 (1520 Hz, 1724 Hz, and 1914 Hz, respectively)producing the satellite $CDT_{2,3}$ ($2f_2 - f_3$ = 1534 Hz).

The CDT frequencies track day to day shifts in the primary SOAE frequencies (up to 10 Hz) within the resolution (1.25 Hz) of our frequency analysis. Also, as was demonstrated previously (Burns et al., 1984), external tone suppression of CDT SOAEs occurs only if external tone levels are sufficiently high enough to suppress one or more of the primary SOAEs. These results indicate that CDT SOAEs are not simply the fortuitous occurence of primary SOAEs at the appropriate frequencies. A listing of CDT satellite emissions is given in Table I.

The other cases of alternate SOAE states were found in subject 9(LE), with SOAEs of frequencies 1330 and 1595 Hz alternating with those of frequencies 1410 and 1700 Hz; and subject 13(RE) with SOAEs of frequencies 1100 and 1380 usually either absent or present as a pair. Clark et al. (1984) have reported a set of alternate SOAE states in the chinchilla after noise exposure, with one of the SOAEs alternately occuring at 4600 Hz and 5680 Hz.

TABLE 1

AVERAGE FREQUENCIES AND LEVELS OF PRIMARY AND CDT SOAEs

Subject	f_1(Hz)	f_2(Hz)	CDT(Hz)	f_2/f_1	I_1(dB SPL)	I_2(dB SPL)	I_{CDT}(db SPL)
4(RE)	1298	1400	1197*	1.08	14.0	3.5	-3.0
8(LE)	1465	1642	1288	1.12	5.0	0.5	-5.0
8(RE)	1469	1689	1248	1.15	9.0	6.0	-5.0
9(LE)	1331	1596	1065*	1.20	20.0	26.5	5.0
9(LE)	1225	1331	1120*	1.09	11.0	17.5	1.0
9(LE)	1596	1702	1489*	1.07	25.5	11.5	7.0
10(LE)	1541	1638	1442*	1.06	10.5	3.5	-2.0
11(LE)	1725	2051	1399	1.19	25.5	20.5	11.0
11(LE)	1725	1914	1536	1.11	21.5	20.0	12.1
F18(RE)	1745	2171	1319	1.24	23.6	13.6	17.5
13(RE)	1283	1383	1183*	1.08	18.1	13.2	9.0

In Table I, an asterisk indicates that the CDT SOAE is within a few Hz of a minima of the behavioral threshold microstructure. Subject 8 does not have minima at the CDT frequencies, and the threshold microstructures of subjects 11 and F18 have not yet been measured. Additional insight concerning the connection between SOAE CDTs and threshold microstructure is obtained by entraining the f_2 SOAE with an external tone of frequency f_2'. This results in the replacement of the CDT SOAE with another one of frequency $2f_1 - f_2'$. The results of varying f_2' are indicated in Figs. 2(a,b) for subject 13(RE). The amplitude ratio of the CDT and f_2' emissions is maximal for $f_2' = 1377$ Hz, with $2f_1 - f_2' = 1189$ Hz (the actual behavioral threshold minimum). These results suggest that a CDT SOAE is most likely to occur when there is a cochlear resonant enhancement at or near the CDT frequency. This is compatible with the fact (Wilson, 1980a) that distortion products in the ear canal produced by two external tones usually occur at frequencies corresponding to EOAEs or SOAEs.

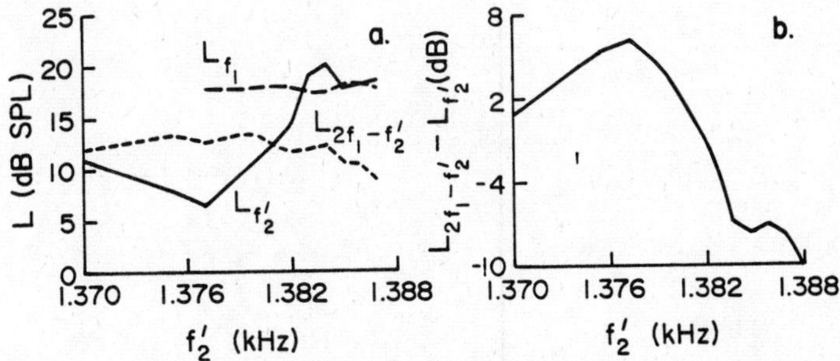

Fig. 2.(a) Levels of the entrained emission of frequency f_2', the emission of frequency f_1, and the CDT emission of frequency, $2f_1 - f_2'$, vs. f_2'. (b) The ratio of the $2f_1-f_2'$ and f_2' amplitudes vs. f_2'. The subject is 13(RE).

4. INTERPRETIVE MODEL OF SOAE INTERACTIONS

As a first attempt to model our results on SOAE interactions, we formally transform a one-dimensional nonlinear cochlear model into a system of interacting van der Pol oscillators (e.g., Pavlidis, 1973).

Consider an inviscid one-dimensional nonlinear cochlear model with: length = L; x = distance from oval window; S = scala cross-sectional areas; ρ = scala fluid mass density; β = basilar membrane (BM) width; $p_{sv}(x,t)$, $p_{st}(x,t)$ = scala vestibuli and scala tympani fluid pressures, respectively; $p_d = p_{st} - p_{sv}$; $\xi_{BM}(x,t)$ = BM displacement; $\dot{\xi}_{BM}(x,t)$ = BM velocity; $\ddot{\xi}_{BM}(x,t)$ = BM acceleration, K(x) = BM stiffness function; M = BM areal mass density; and $R(x,t) = R_1(x) + R_2(x)\,\xi_{BM}(x,t)^2$ = BM resistance function. $R_1(x) < 0$ corresponds to a region of active cochlear response. The BM dynamical equation is

$$M\,\ddot{\xi}_{BM} + K\,\xi_{BM} + R\,\dot{\xi}_{BM} = p_d \quad . \tag{1}$$

The differential equation for p_d,

$$\frac{\partial^2}{\partial x^2} p_d - \gamma\, p_d = -\gamma(K\,\xi_{BM} + R\,\dot{\xi}_{BM}) \quad ,$$

$$\gamma = 2\rho\beta/MS \quad , \tag{2}$$

may be converted by a Green function technique (e.g., Furst and Goldstein, 1982) to the integral relation,

$$p_d(x,t) = -\int_0^L dx'G(x|x')\gamma\,\{K(x)\xi_{BM}(x,t) + R(x,t)\dot{\xi}_{BM}(x,t)\} + G(x|0)\,\frac{\partial p_d}{\partial x}(x,t)\Big|_{x=0} \quad , \tag{3}$$

where

$$G(x|x') = \frac{-\cosh\sqrt{\gamma}\,x_< \sinh\sqrt{\gamma}\,(L - x_>)}{\sqrt{\gamma}\cosh\sqrt{\gamma}\,L} \quad , \tag{4}$$

with $x_>$ ($x_<$) the greater (lesser) of x and x', and

$$\frac{\partial p_d}{\partial x}(x,t)\Big|_{x=0} = 2\rho\cdot\dot{u}_s(t) \quad , \tag{5}$$

where $u_s(t)$ is the stapes velocity.

Eqs. (1),(3), and (4), in the spatially discretized form,

$$\ddot{\xi}_i + \omega_i^2\xi_i + (R_i/M)\dot{\xi}_i = \sum_j A_{ij}(\omega_j^2\xi_j + (R_j/M)\dot{\xi}_j) + B_i\dot{u}_s \quad , \tag{6}$$

where $\xi_i(t) = \xi_{BM}(x_i,t)$, $\omega_i^2 = K(x_i)/M$, $R_i = R_1(x_i) + R_2(x_i)\xi_i^2$, $A_{ij} = -\gamma\Delta G(x_i|x_j)$,

$B_i = 2\rho G(0|x_i)u_s$, and $\Delta x = x_i - x_{i=1}$, represent the dynamics of a system of interacting van der Pol oscillators. An equivalent alternative version of Eqs.(6) was used by Diependaal and Viergever (1983). These equations are solved using a six-point modified predictor-corrector method, and the frequency spectra of the $\xi_i(t)$ are computed from a East Fourier Transform (FFT) technique. Ear canal signals may be simulated by coupling the model cochlea to models of the middle and outer ears (e.g., Matthews, 1983).

The results of extensive computer simulations of the full cochlear model will be presented elsewhere. Preliminary calculations indicate that some of the essential features of CDT SOAEs may be simply interpreted in terms of a drastic truncation of Eqs.(6).

We assume that $R_1(x_i) < 0$ only at sites i and k so that in the absence of the fluid coupling ($A_{ik} = 0$) of the van der Pol oscillators, ξ_i and ξ_k would undergo limit-cycle oscillations of angular frequencies ω_i and ω_k ($> \omega_i$), and amplitudes, $2\sqrt{-R_1(x_i)/R_2(x_i)}$ and $2\sqrt{-R_1(x_k)/R_2(x_k)}$, respectively. Because of the nonlinearity of the model, each site along the BM will be driven at frequencies $\approx \omega_i$ and ω_k, as well as at distortion product frequencies, $2\omega_i - \omega_k, \ldots$. If $R_1(x_\ell)$, corresponding to $\omega_\ell \approx 2\omega_i - \omega_k$, is anomalously small, a CDT SOAE of frequency, $2\omega_i - \omega_k$, is expected to result. To illustrate this, consider just the coupling of the i,k, and ℓ terms in Eqs.(6), with $u_s = 0$. Typical results are shown in Fig. 3. The level of the f_ℓ SOAE decreases with increasing deviation of f_ℓ from the resonance condition, as is to be expected. It is assumed that ξ_o in Fig. 3 is a rough measure of the ear canal pressure that would be obtained in a more complete model calculation.

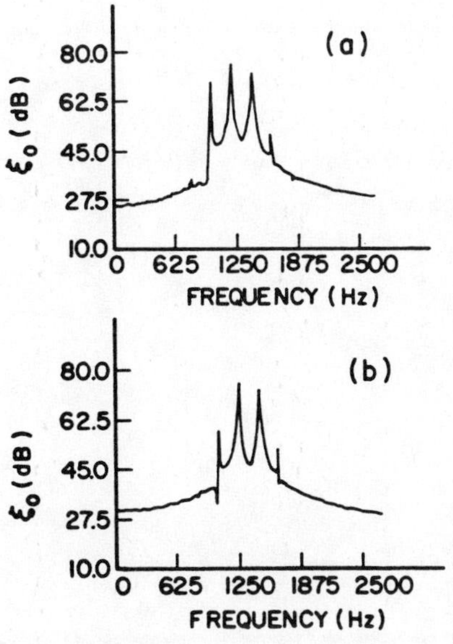

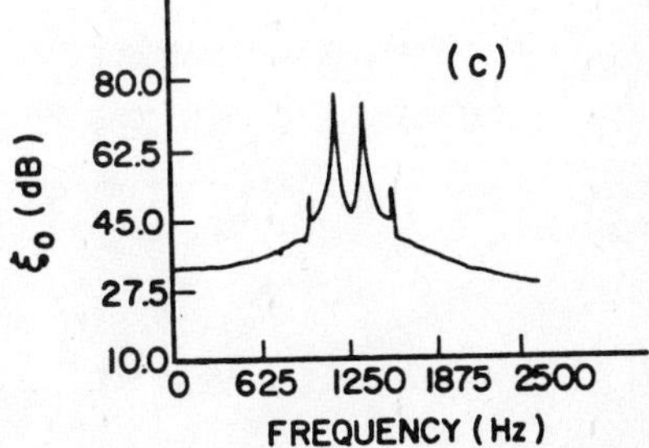

Fig. 3. Fourier amplitudes (in arbitrary dB units) of BM displacement, ξ_o, near the stapes based on coupled i,k, and ℓ terms in Eq. 6, with $f_i = 1200$ Hz, $f_k = 1400$ Hz, and $f_\ell = 1000$ Hz (a); 990 Hz (b); 980 Hz (c). ξ_o is assumed to be proportional to $\omega_i^2\xi_i + \omega_k^2\xi_k + \omega_\ell^2\xi_\ell$. Other parameter values used are: $L = 3.5$ cm; $S_2 = 0.04$ cm^2; $\beta = 0.1$ cm; $\rho = 1$ g/cm^3; $m = 0.15$ g/cm^2; $K(x) = 10^9 \cdot \exp(-3x)$ g/cm s^2; $R_1(x_i) = R_1(x_k) = -255$ g/cm^{2}s; $R_1(x_\ell) = 2.25$ g/cm^{2}s; $R_2(x_i) = R_2(x_k) = 15$; $R_2(x_\ell) = 2.25$. x is in cm units and the ξ_i are in units of 10^{-4} cm.

An equivalent effect, more closely related to the experimental data shown in Fig. 2, is illustrated in Fig. 4. An external tone of 1410 Hz entrains the 1400 Hz SOAE and drives the f_ℓ ($\equiv$ 1000 Hz) oscillator 10 Hz off resonance via the CDT of frequency, 2x 1200-1410 = 990 Hz. The 990 Hz amplitude is about 21 dB below that of the 1410 Hz one, whereas in Fig. 3(a), the 1000 Hz amplitude is only about 4 dB below that of the 1400 Hz one.

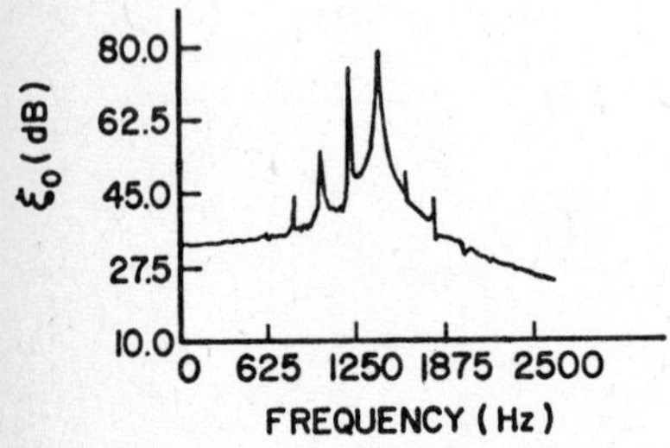

Fig. 4. Fourier amplitudes (in arbitrary dB units) of BM displacement, ξ_o, near the stapes. The parameter values are the same as those used in Fig. 3(a), with the addition of driving terms, $B_i\dot{u}_s = B_k\dot{u}_s = B_\ell\dot{u}_s = 10^8 \cos(2\pi\ ft)$, f = 1410 Hz.

5. DISCUSSION

CDT SOAE satellite emissions appear to be the results of typical types of cochlear nonlinearities suggested by other physiological and psychophysical measures. Experimental and model studies indicate that the CDT SOAEs tend to occur when a distortion product frequency is close to that of a cochlear resonance. An interesting possibility now being explored is the nonlinear-active coupling of a family of quasi-standing-wave modes associated with a single site of cochlear retrograde wave generation (Kemp, 1980). These modes have roughly equal spacing in frequency so that conditions such as $f_\ell \approx 2f_i - f_k$, would be highly probable.

Simulations of interacting nonlinear oscillator systems of the type considered in this paper may reveal alternate multifrequency limit-cycle states (e.g., Pavlidis, 1973), which may be correspondents of the alternate SOAE states which we have found.

Interactions among SOAEs provide a rich area for probing experimentally, and in model simulations, the nature of the nonlinear active cochlear response.

ACKNOWLEDGEMENT

This work was supported by NINCDS Grants NS 22095-1 (Purdue University) and NS 19805 (University of Washington).

REFERENCES

Bialek, W. and Witt, H., "Quantum limits to oscillator stability: theory and experiments on acoustic emissions from the human ear." Phys. Lett. 10A, pp. 173-177,1984.

Burns, E.M., Strickland, E.A., Tubis, A., and Jones, K., "Interactions among spontaneous otoacoustic emissions. I. Distortion products and linked emissions." Hearing Res. 16, pp. 271-277, 1984.

Clark, W.W., Kim, D.O., Zurek, P.M., and Bohne, B.A., "Spontaneous otoacoustic emissions in chinchilla ears; Correlation with histopathology and suppression by external tones." Hear Res. 16, pp. 299-314, 1984.

Diependaal, R.J. and Viergever, M.A., "Nonlinear and active modelling of cochlear mechanics: a precarious affair," in Mechanics of Hearing, Ed's E. de Boer and

M.A. Viergever, Martinus Nijhoff, The Hague, The Netherlands, pp. 153-160, 1983.

Furst, M. and Goldstein, J.L., "A cochlear nonlinear transmission-line model compatible with combination tone psychophysics." J. Acoust. Soc. Am. 72, pp. 717-726, 1982.

Gold, T., "Hearing II. The physical basis of the action of the cochlea." Proc. R. Soc. Lond. Ser. B. 135, pp. 492-498, 1948.

Hayashi, C., Nonlinear Oscillations in Physical Systems, McGraw-Hill, New York, pp. 287-291, 304-308, 1964.

Kemp, D.T., "Stimulated acoustic emissions from within the human auditory system." J. Acoust. Soc. Am. 64, pp. 1386-1391, 1978.

Kemp, D.T., "Evidence of mechanical nonlinearity and frequency selective wave amplification in the cochlea." Arc. Oto-Rhino-Laryngol. 224, pp. 37-45, 1979.

Kemp, D.T., "Towards a model for the origin of cochlear echoes." Hearing Res. 2, pp. 533-548, 1980.

Kemp, D.T., "Physiologically active cochlear micromechanics-one source of tinnitus," in Tinnitus, Ed's. D. Evered and G. Lawrenson, Pitman, London, pp. 54-76, 1981.

Kemp, D.T. and Chum, R., "Observations on the generator mechanism of stimulus frequency acoustic emissions-two tone suppression," in Psychological, Physiological and Behavioral Studies in Hearing, Ed's. G. van den Brink and F.A. Bilsen, Delft Univ. Press, Delft, The Netherlands, pp. 34-41, 1980.

Long, G.R., "The microstructure of quiet and masked thresholds." Hearing Res. 15, pp. 73-87, 1984.

Matthews, J.W., "Modeling reverse middle ear transmission in acoustic distortion signals," in Mechanics of Hearing, Ed's E. de Boer and M.A. Viergever, Martinus Nijhoff, The Hague, The Netherlands, pp. 11-18, 1983.

McFadden, D. and Plattsmier, H.S., "Aspirin abolishes spontaneous otoacoustic emissions." J. Acoust. Soc. Am. 76, pp. 443-448, 1984.

Pavlidis, T., Biological oscillators: Their Mathematical Analysis, Academic Press, New York, pp. 138-186, 1973.

Wilson, J.P., "Evidence for a cochlear origin for acoustic re-emissions, threshold fine structure, and tonal tinnitus." Hearing Res. 2, pp. 233-252, 1980.

Wilson, J.P., "The Combination tone $2f_1 - f_2$ in psychophysics and ear canal recording," in Psychophysical, Physiological and Behavioral Studies in Hearing, Ed's. G. van den Brink and F.A. Bilson, Delft Univ. Press, Delft, The Netherlands, pp. 43-50, 1980a.

Wilson, J.P. and Sutton, G., "Acoustic correlates of tonal tinnitus," in Tinnitus, Ed's. D. Evered and G. Lawrenson, Pitman, London, pp. 54-76, 1981.

Zwicker, E. and Schloth, E., "Interrelation of different oto-acoustic emissions." J. Acoust. Soc. Am. 74, pp. 1148-1154, 1984.

Zurek, P.M., "Spontaneous narrowband acoustic signals emitted by human ears." J. Acoust. Soc. Am. 69, pp. 514-523, 1981.

BASILAR MEMBRANE MOTION IN GUINEA PIG COCHLEA EXHIBITS FREQUENCY-DEPENDENT DC OFFSET.

E. L. LePage[+] and A. E. Hubbard[*]
Boston University,
Departments of Otolaryngology, Biomedical Engineering[+]
and Electrical, Computer and Systems Engineering[*],
110 Cummington Street, Boston, Massachusetts, 02215.

ABSTRACT

Direct mechanical measurements of the basilar membrane in the guinea pig cochlea have revealed that a dc component of basilar membrane motion exists and shows a systematic tendency to reverse polarity close to the best frequency (BF) of the ac response. The dc component is seen in time records as a change in the mean position of the basilar membrane with the onset of the tone burst. The offset is physiologically vulnerable and is substantially reduced as the preparation deteriorates. A computer model of the basilar membrane, employing nonlinear stiffness and damping elements, shows a pattern of rectification with polarity reversal not unlike the experimental data.

INTRODUCTION

A possible consequence of nonlinear mechanical behavior of the basilar membrane (Rhode, 1971; LePage and Johnstone, 1980; Sellick et al. 1983) is rectification, resulting in a dc component of the vibratory response. The Mössbauer measurements of Robles and Rhode (1974) suggest that such a component might exist. However, the Mössbauer technique is velocity-sensitive so an improvement might be to use a linear, displacement-sensitive technique, capable of faithful representation of the time waveform of the mechanical response. The capacitive probe technique is an example, which has the advantage that for a fixed working distance it transduces the displacement of the basilar membrane under small-signal conditions. In addition to showing the nonlinear mechanical behavior in guinea pig, the capacitive probe technique has also provided indirect evidence for a low frequency component of basilar membrane, associated with the nonlinear behavior in the region of the characteristic frequency (LePage, 1981). The aim of the present study was to test the existence of a physiologically vulnerable dc component, by measuring the shift in the mean position of the basilar membrane with the onset of a tone burst.

METHODS

Guinea pigs were anesthetised, paralyzed using Flaxedil to prevent middle ear muscle contraction and artificially respired. The N_1 compound action potential was used to obtain a threshold audiogram periodically throughout the procedure. The waveform of the basilar membrane response was recorded at a fixed location approximately 4 mm from the base of the guinea pig cochlea. Digitally synthesized tone-bursts with 1 ms rise and fall times were delivered over a fixed 28 ms period, in which 12 ms bursts occurred 3 ms from the start of the period. The stimulus set was chosen to be ten frequencies with greatest density at the expected characteristic frequency (CF) of the place measured. The bursts were delivered at constant precorrected sound levels ranging from 45 dB to 80 dB SPL. The filtered output signal of the capacitive probe was sampled at 13.5 μs intervals. The responses to the tone bursts were averaged for 64 repetitions of the burst. The total recording time for the stimulus set was less than two minutes. The frequency dependence of the dc component was obtained by averaging all the sampled values during the tone burst and subtracting the average of the pre-toneburst baseline values. The ac peak value was obtained by computing the first discrete Fourier component for the duration of the tone burst. These methods afford a further 57 dB signal-to-noise improvement, or an effective noise level of about 0.1 μV.

EXPERIMENTAL RESULTS

Time records of the response of the basilar membrane to the tone bursts showed a range of nonlinear behavior, including an offset of the ac response from the prestimulus mean value, coincident with the onset of the burst. This is shown in Fig. 1 for two fresh preparations with low thresholds before and after the measurement, and with not more than 35 dB loss of sensitivity during the measurement, due to draining scala tympani. Fig. 1A shows the capacitive probe response of the basilar membrane for tones of 55 dB SPL and two frequencies, the lower panel at the estimated CF and the upper panel a half octave below CF. Similarly, Fig. 1B shows responses in a different preparation at CF and one quarter octave below CF for 70 dB SPL. There are clear differences in the noise level of the two recordings due to different probe placements, nevertheless the net asymmetry coincident with the onset of the tone bursts is characteristic of the data.

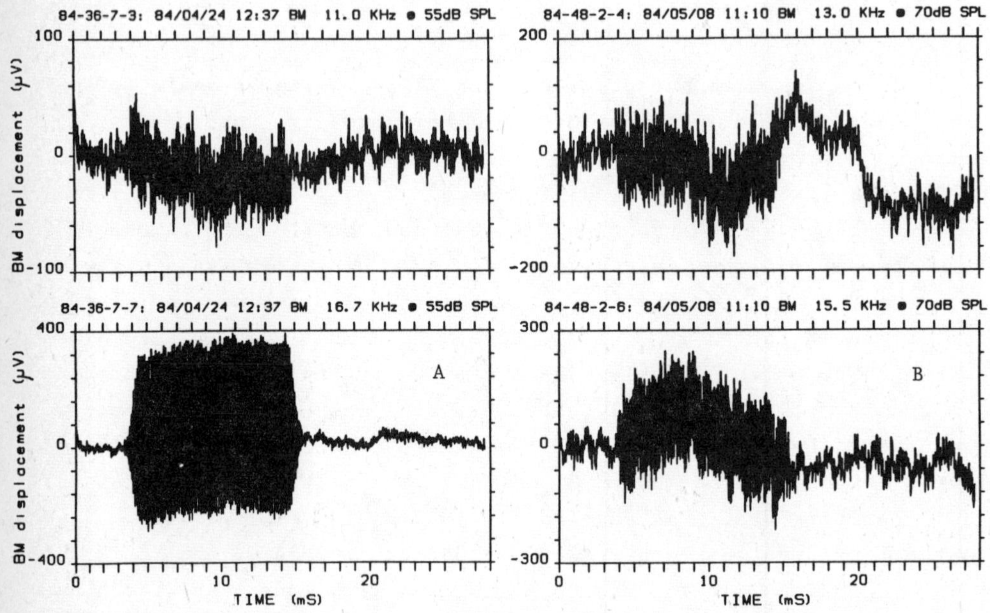

Figure 1. Time waveforms for two preparations are shown in panel pairs A and B. The top panel in each case is for a frequency one half and one quarter of an octave, respectively, below the expected CF. The lower panels are responses at CF.

The ac and dc components are plotted versus frequency in Fig. 2A and Fig. 2B respectively, for three animals. The ac response characteristics, obtained for a constant sound level of 70 dB, exhibit a slope rising with frequency towards the CF, notably with a null in the response a half octave below the CF. Simultaneous with these ac amplitude variations, Fig. 2B shows there is a net, negative-going dc displacement of the basilar membrane towards scala tympani (ST) which is maximal about an octave below the best frequency of the ac response. As the frequency is raised, the dc component exhibits a rapid polarity reversal, coincident with the null in the ac response within a half octave of BF. At the BF, there appears to be either a net positive offset towards scala vestibuli (SV) or simply a loss of the ST component. Examination of the computed levels in Fig. 2A and Fig. 2B shows that, under these conditions of measurement, at an octave below CF, the dc response can be comparable in size to the peak value of the ac response. At the CF however, the dc response may be at most a third to a half of the peak value of the ac response.

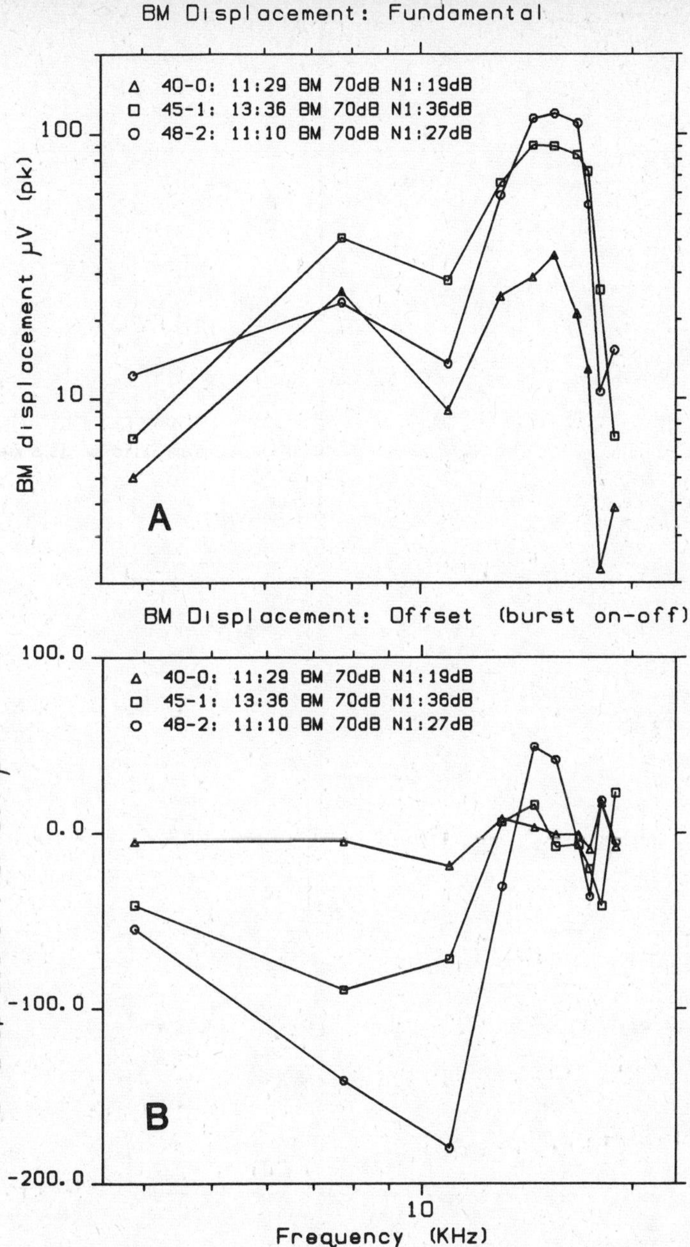

Figure 2. Panels A and B for three fresh preparations represent, respectively, the ac and dc components of the capacitive probe response to the motion of the basilar membrane. Each point is an average value for the whole period of the tone burst, representing a 57 dB noise reduction over the averaged time data. The best frequency of the ac response is aligned with the expected CF for the place measured. Polarity reversals in the dc component are characteristic of all fresh preparations. Static calibration is 2 nm/mV.

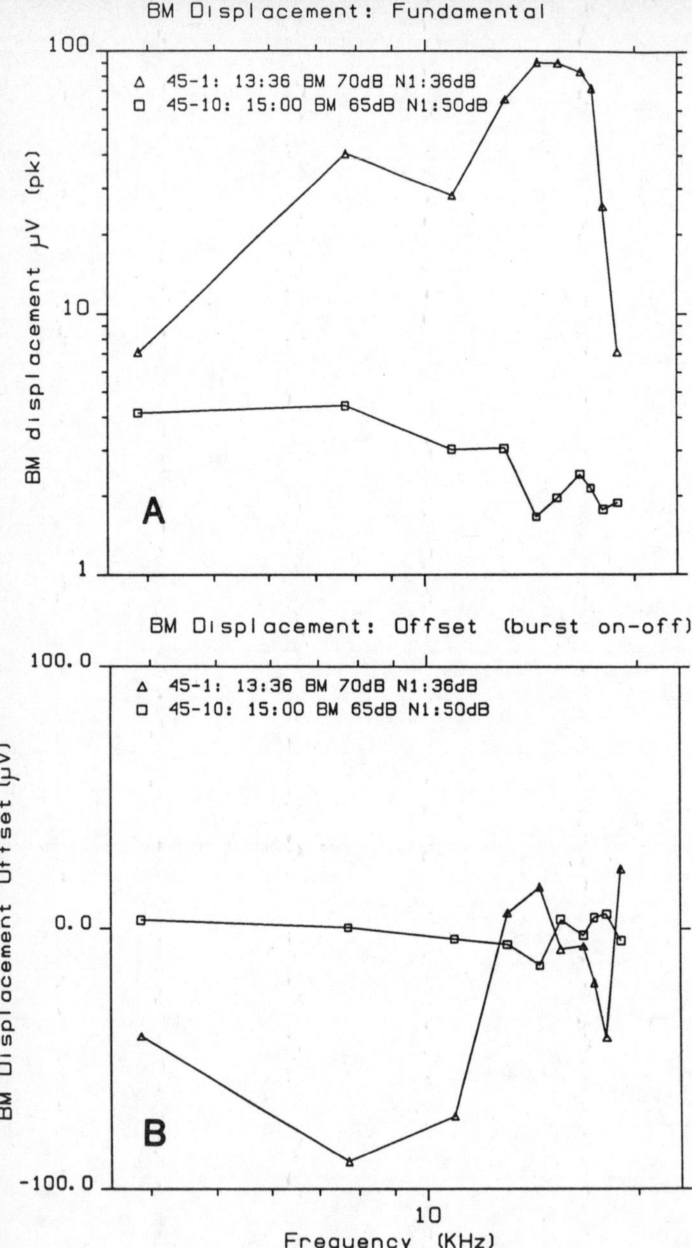

Figure 3. Comparison of two records for one guinea pig preparation taken early in the experiment (triangles) and 84 minutes later (squares) showing a substantial and permanent loss of both the ac and dc components of the motion.

Although all records are from low threshold animals, the baseline shift is not developed to the same extent in each record due to the variable effects of draining scala tympani. Based on a sensitivity calibration of the capacitive probe (approximately 2 nm/mV), these results suggest that the maximum dc offset observed at 70 dB is of the order of a nanometer.

The observed offset appears physiologically vulnerable similar to that of the nonlinear responses observed in a previous study (LePage and Johnstone, 1980). In the present study, it appears that the amplitude of both ac and dc components of the basilar membrane response is markedly and permanently reduced as is seen in Fig. 3. Panels A and B show the ac and dc components respectively for the first recording of the series and a recording later in one preparation.

MODEL RESULTS

The model used is a one-dimensional formulation of cochlear hydrodynamics (Zwislocki, 1950). The model, which uses 300 sections, is basically a tapered transmission line with tuned shunt branches. Computation of the time response is carried out in a fashion such that any model parameter can be a function of any other computed quantity, past or present. In this way, the model is capable of incorporating a wide range of nonlinearities. Both nonlinear compliance and damping are used to produce the results shown in Fig. 4. Each compliance is a linear function of the weighted summation of the instantaneous displacements at locations +/- 0.58 mm distant from any given central point. The nonlinear damping used at each point is a function of that point's instantaneous velocity squared, when the velocity is positive, and constant otherwise.

Basalward from the peak of the traveling wave, a dc shift is produced by asymmetry in the stiffness nonlinearity. At the peak, a dc shift in the opposite direction is produced by the damping nonlinearity. In the frequency domain, such spatial patterns produce a negative dc shift for frequencies less than CF. The positive dc peak occurs precisely at CF.

DISCUSSION

The observed mean position shift ceases to be measurable as the N_1 threshold rises and this suggests that the dc component is related to the vulnerable nonlinear behavior seen in previous studies. The fact that the component may be observed in a fresh preparation suggests that the effect is not due to draining ST required by the technique. To

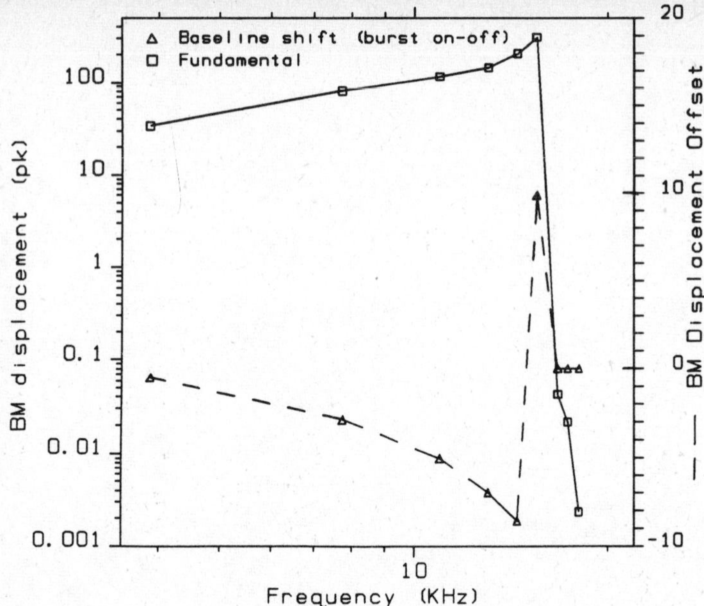

Figure 4. Result of a computer model showing a small bipolar dc component of a similar pattern to the experimental data. The positive peak is due to the dominance of nonlinear damping elements at the CF.

the contrary, draining has been shown to produce both a loss of neural sensitivity (Robertson, 1974) and potentiates temporary threshold shifts due to loud stimuli (Alder, 1978), so it is more likely that draining acts to produce the rapid loss of nonlinear behavior. This may account for why Wilson and Johnstone (1975), who used high level tones, found no dc component.

The polarity reversal may be related to cochlear dc potentials and hair cell electromechanical behavior. This view is supported by an earlier capacitive probe study employing kanamycin-treated guinea pigs, shown to lack outer hair cells, in which no vulnerable nonlinearity was seen (LePage, et al., 1981).

Recent evidence suggests that the sensitivity of the cochlea is influenced by low frequency tones which introduce a mechanical bias of the cochlear partition (Patuzzi et al. 1984). The capacitive probe results are not inconsistent with the idea that draining itself introduces a mechanical bias on the cochlear partition which is responsible for sensitivity loss at the CF and a smaller positive dc component.

The model produces a dc component which changes polarity near the peak of the traveling wave for any particular location. The model currently shows a far smaller dc/ac response ratio than does the data.

Broad minima in the dc component at about half the CF are not produced. The model qualitatively encompasses physiological deterioration as a loss of nonlinearity, combined with a substantial increase in damping.

ACKNOWLEDGEMENTS

E.LeP. thanks C.E. Molnar and J.W. Matthews for support in carrying out the experiments. We thank D.C. Mountain, T.C. McMullen and H.F. Voigt for helpful comments. This work was supported by NIH grants NS07498 and NS16589.

REFERENCES

Alder, V.A. (1978). Neural correlates of auditory temporary threshold shift. Ph.D. Thesis, The University of Western Australia.

LePage, E.L. (1981). The role of nonlinear mechanical processes in mammalian hearing. Ph.D. Thesis, The University of Western Australia.

LePage, E.L. (1985). Mammalian hearing: Sound evokes bipolar shift in mean position of the basilar membrane. (Submitted).

LePage, E.L. and Johnstone, B.M., (1980). Nonlinear mechanical behaviour of the basilar membrane in the basal turn of the guinea pig cochlea. Hearing Res., 2, 183-189.

LePage, E.L., Johnstone, B.M. and Robertson, D. (1980). Basilar membrane mechanics in th guinea pig cochlea - Comparison of normals with Kanamycin-treated animals. J.Acoust.Soc.Am. 67, S46.

Patuzzi, R., Sellick, P.M. and Johnstone, B.M. (1984). The modulation of the sensitivity of the mammalian cochlea by low frequency tones. I-III. Hearing Res. 13, 1-27.

Rhode, W.S., (1971). Observations of the vibration of the basilar membrane in squirrel monkeys using the Mössbauer technique, J.Acoust.Soc.Am., 49, 1218-1231.

Robertson, D., (1974). Cochlear neurons: Frequency selectivity altered by perilymph removal, Science, 186, 153-155.

Robles, L. and Rhode, W.S., (1974). Nonlinear effects in the transient response of the basilar membrane, in Facts and Models in Hearing, E. Zwicker and E. Terhardt, eds. Springer, Berlin, 1974, 287-298.

Sellick, P.M., Patuzzi, R. and Johnstone, B.M., (1982). Measurement of basilar membrane motion in the guinea pig using the Mössbauer technique. J.Acoust.Soc.Am., 72, 131-141.

Wilson, J.P. and Johnstone, J.R., (1975). Basilar membrane and middle-ear vibration in guinea pig measured by capacitive probe, J.Acoust.Soc.Am., 57, 705-723.

Zwislocki, J. J. (1950). Theory of acoustical action of the cochlea. J.Acoust.Soc.Am., 22, 778-784.

LINEAR AND NON-LINEAR EFFECTS IN A PHYSICAL MODEL OF THE COCHLEA

Salvo D'Angelo and Marcello Masili
Dipartimento Ingegneria Aeronautica e Spaziale - Politecnico di Torino
Corso Duca degli Abruzzi 24 - 10129 - Torino - Italia

Riccardo Malvano
Centro di Studio per la Dinamica dei Fluidi - C.N.R.
Corso Duca degli Abruzzi 24 - 10129 - Torino - Italia

ABSTRACT

The response of a physical three chamber model of the cochlea is described in the present paper. The model is rectilinear, geometrically scaled 50:1 and supplied with train-gauge transducers to measure axial and cross components of oscillating fluid near the Corti organ.A continuous motion of fluid in the scala media was observed and a complete picture of the stream line distribution was obtained using fluorescent powders and black light.Some peculiar aspects of this continuous flow can be closely correlated with basilar membrane (B.M.) response.

1. INTRODUCTION

Experimental data obtained measuring the B.M. vibration *in vivo* at moderate stimulus levels using the Mössbauer technique(6), suggested new mathematical and experimental models of the cochlea and helped test their validity.Physical models do not require the geometrical and mechanical simplifications proposed in current mathematical inner ear models and enable to analyse the peculiarities of the velocity field which can not be otherwise detected.The continuous motion of the fluid from the sulcus to the scala media(3) is an example of non linear behaviours discovered using physical models of the cochlea. Fig.1 shows the authors' model cross section compared to the real cochlea.The model (50:1) contains the constituent elements of the cochlea, including the B.M.,Reissner's and the tectorial membranes and the Corti organ.Thanks to the transparency of the ductus and sulcus, the oscillatory motion was visualized using stroboscopic light and the continuous flow using black light and fluorescent particles suspended in the fluid.

2. SIMILARITY AND EXPERIMENTAL CONDITIONS

By normalizing the Navier-Stokes equation and the dynamic equation of the cochlear partition, and by taking into account endolymph and perilymph kinematic viscosities a set of dimensionless coefficients is obtained(2). The dimensionless coefficients are:

$$\nu/L^2\omega \qquad \rho\omega^2\Sigma \qquad \omega^2 M\Sigma \qquad u/\omega L \qquad \nu/\nu_e$$

where L is the characteristic length of the cochlea; ω the radian frequency and u the velocity amplitude of the input signal; ν perilymph kinematic viscosity; ν_e endolymph kinematic viscosity; ρ perilymph density; M the generalized mass and Σ the compliance of a unit length strip of partition(4). The first and the second are two indipendent dimensionless coefficients in the same five magnitudes as Békésy's(1). The third one is the ratio between inertia and elastic component of the partition in the ω frequency oscillating motion. When Békésy's conditions are fulfilled and the input signal amplitude is low enough so as to satisfy the fourth coefficient then similitude conditions are satisfied. This is so if the shape of the internal structures of the scala media is mantained and the structures themselves are reproduced with a material whose density is like that of the surrounding liquid. Lastly the value that ν/ν_e coefficient has in the real cochlea can be mantained in the model. The existence of a membrane simulating Reissner's membrane makes it possible to vary the scala media viscosity in relation to the other two scales. Experiments(8) and theoretical considerations(4) both suggest anisotropy is an important characteristic of the real membrane. This anisotropy was simulated by stretching a thin rubber latex membrane (1.8 N/mm elastic modulus) crosswise with a stress varying from base to apex. Two different kinds of tectorial membranes were used: an elastic and a rigid one. The most important parameter in determining the scaling ratios between model and real cochlea is the B.M. compliance distribution (fig.2). According to Lighthill(4), Peterson & Bogert's(5) compliance distribution is the most probable one, even if phase velocity data for low-frequency waves seem to suggest the estimated compliance values are too low. A better agreement with real cochlea might be obtained with a compliance distribution 3÷5 times greater than P.& B.'s distribution. Lastly we must point out that configurations a,b,c,d were equipped with

strain-gauge transducers to measure the phase lag.
See table one for a complete list of the scaling ratios and of the model configurations adopted in the different tests.

3. EXPERIMENTAL RESULTS AND CONCLUSIONS

The aim of the present work was to give a complete picture of the continuous flows within the scala media using visualization methods. This is why the main feature of config.e is a complete transparency from the top and from the sides of the model.It was obtained by adopting a rigid and perfectly transparent tectorial membrane and removing the strain-gauge transducers used in the previous configurations. A rigid tectorial membrane could seem rather in disagreement with the real cochlea. Other configurations showed that B.M. response did not vary both with a flexible tectorial membrane (config.a) and with a rigid one (config.b). Furthermore in config.e we adopted an only partially filled scala vestibuli to reduce the possible interferences of the fast wave (4). Theoretical and experimental results are as for a fully filled scala vestibuli (7) (2). Fig.3 shows the slope of the response curves at fixed point vs. frequency (-30dB/oct) obtained in the config.e, which are slighter than the ones obtained in configs.a-b-c-d (-40÷-60dB/oct). However they were all plainly insufficient if compared to the -100÷-300dB/oct obtained in the *in vivo* experiments. This may be due to the longitudinal stiffness of all the B.Ms. of the model. Fig.4 shows the distribution of the maxima along the longitudinal axis. The shift towards the apex of the maxima in config.e with respect to config.a is in agreement with the distribution of B.M. compliance. Fig.5 shows the phase lag vs. distance obtained in configs.a-b-c-d for four frequencies, with the model supplied with strain-gauge transducers to measure this phase lag. The tests to obtain the picture of the continuous flows in the scala media were carried out with a pure tone or with a two tone stimulation. Fig.7 shows B.M. response at 40Hz, 120Hz and the two tone combination 120Hz+40Hz. The same fig. shows the stream lines at the same frequencies.
Notice the following features:

1) The location B.M. response maxima does not change when two tones are combined to generate a two-tone signal.
2) The continuous flow organizes itself in circulation cells.
3) In the area of the maximum a cell with a point located over the Corti organ before the maximum with zero continuous velocity is always present. Its location does not depend on the amplitude of the input signal but only on its frequency. Moreover it does not move forewards or backwards when its frequency is combined with another pure tone.
4) A strong outflow from the sulcus, following B.M. maximum defines the right boundary of this cell. A less intense inflow defines the left one. The positions of the left and right sides of the cell depend on the intensity of the input signal and on the presence of another second frequency.
5) B.M. response in config.e (at frequencies > 60Hz) shows a peculiar shape (fig.7) consisting in a noticeable new increase of displacement after the strong decrease following the maximum. It is probably due to the longitudinal stiffness. This unexpected *second maximum* produces a circulation cell of the same kind of the cell before described.
6) cells rotating in the opposite sense can be correlated to the minima.
7) fig.6 shows the value of the longitudinal component of the velocity in the scala media, measured near B.M. maximum as a function of the input signal amplitude.In the config.e (viscosity lower than in the other configs.) the velocity of the continuous motion increases linearly with the input signal up to a level corresponding to an acoustic signal of approximately 140dB in the real cochlea.

It is certainly difficult to attribute a physiological function to this continuous flow. A sharply concentrated zero point of the continuous velocity, for every frequency and strictly correlated to the location of the B.M. maximum could act as a better locator of the frequency along the B.M. compared to the maximum displacement itself.

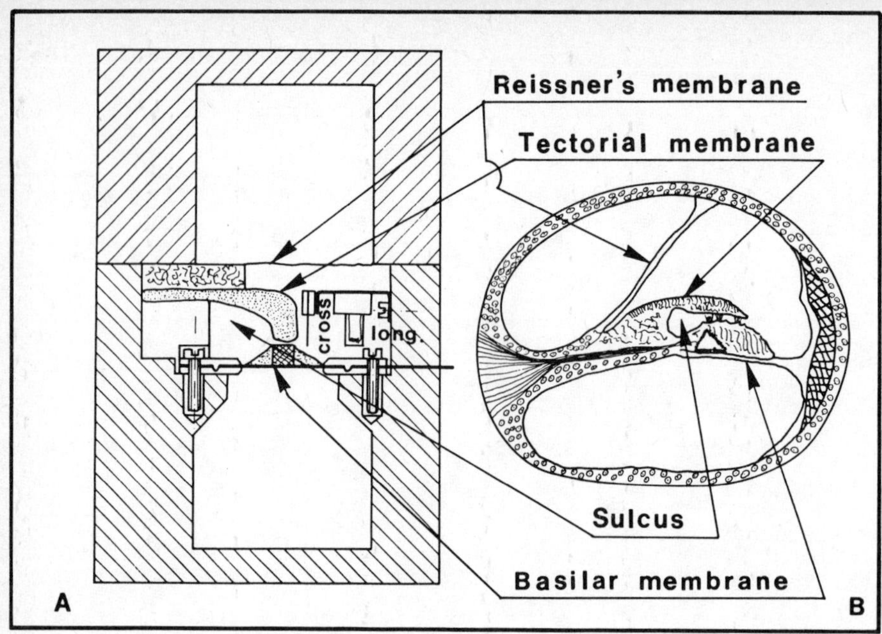

fig.1 Model(A) and cochlea(B) cross-section.

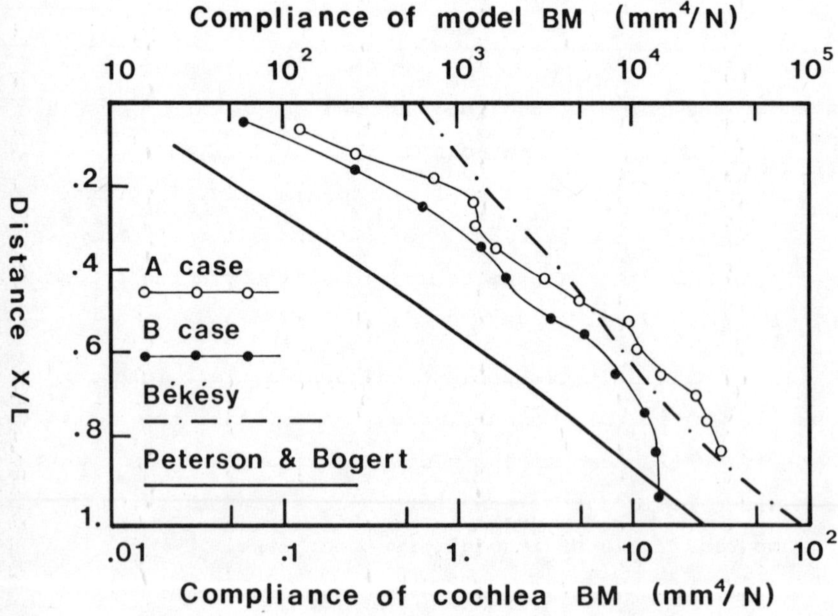

fig.2 Compliance distribution of the basilar membrane
upper axis - model compliace
lower axis - real cochlea compliance

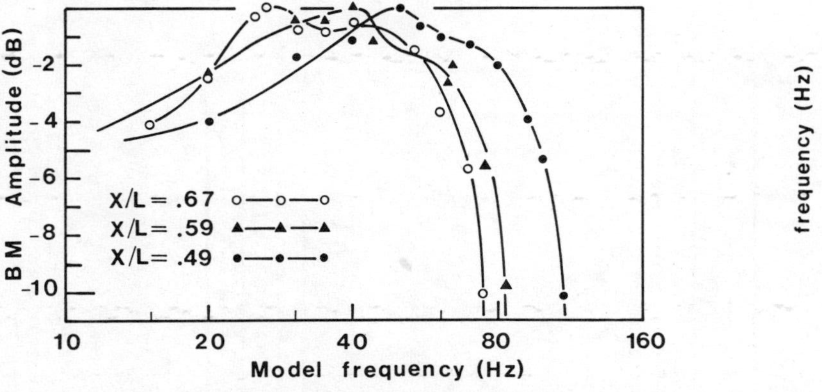

fig.3 Model basilar membrane response at fixed points versus frequency.

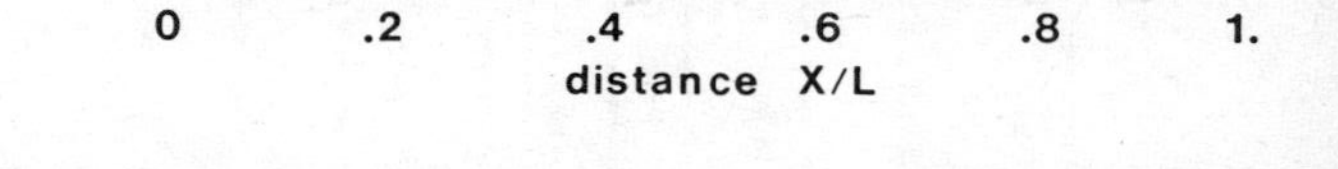

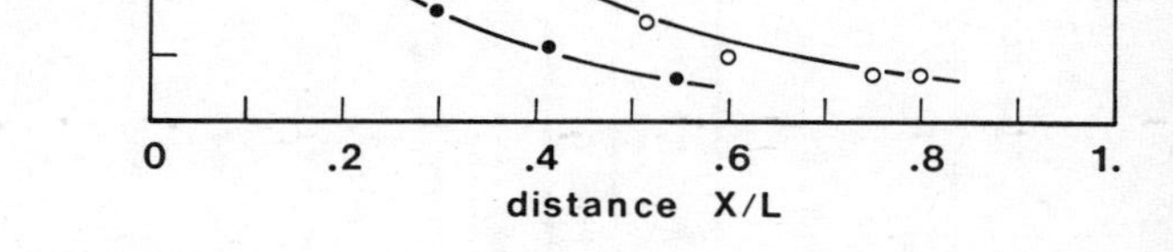

fig.4 Distribution of amplitude maxima of the B.M. response along the long. axis of the model vs. frequency.

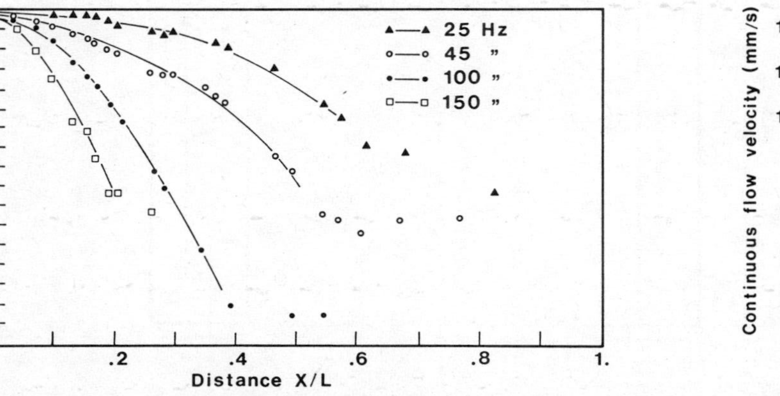

fig.5 Phase between the input signal and the transducers response.

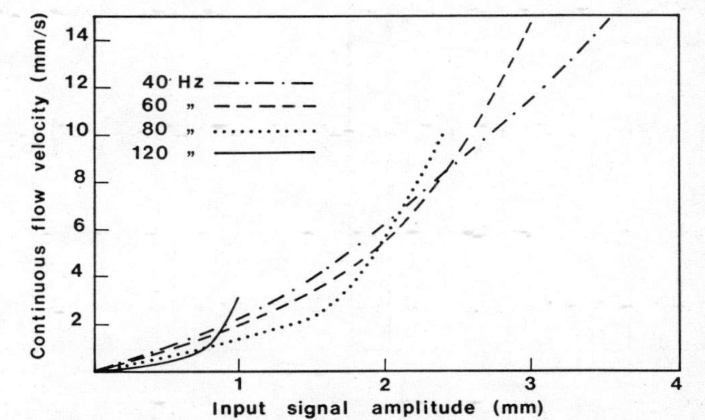

fig.6 Velocity of the continuous flow versus the input signal amplitude.

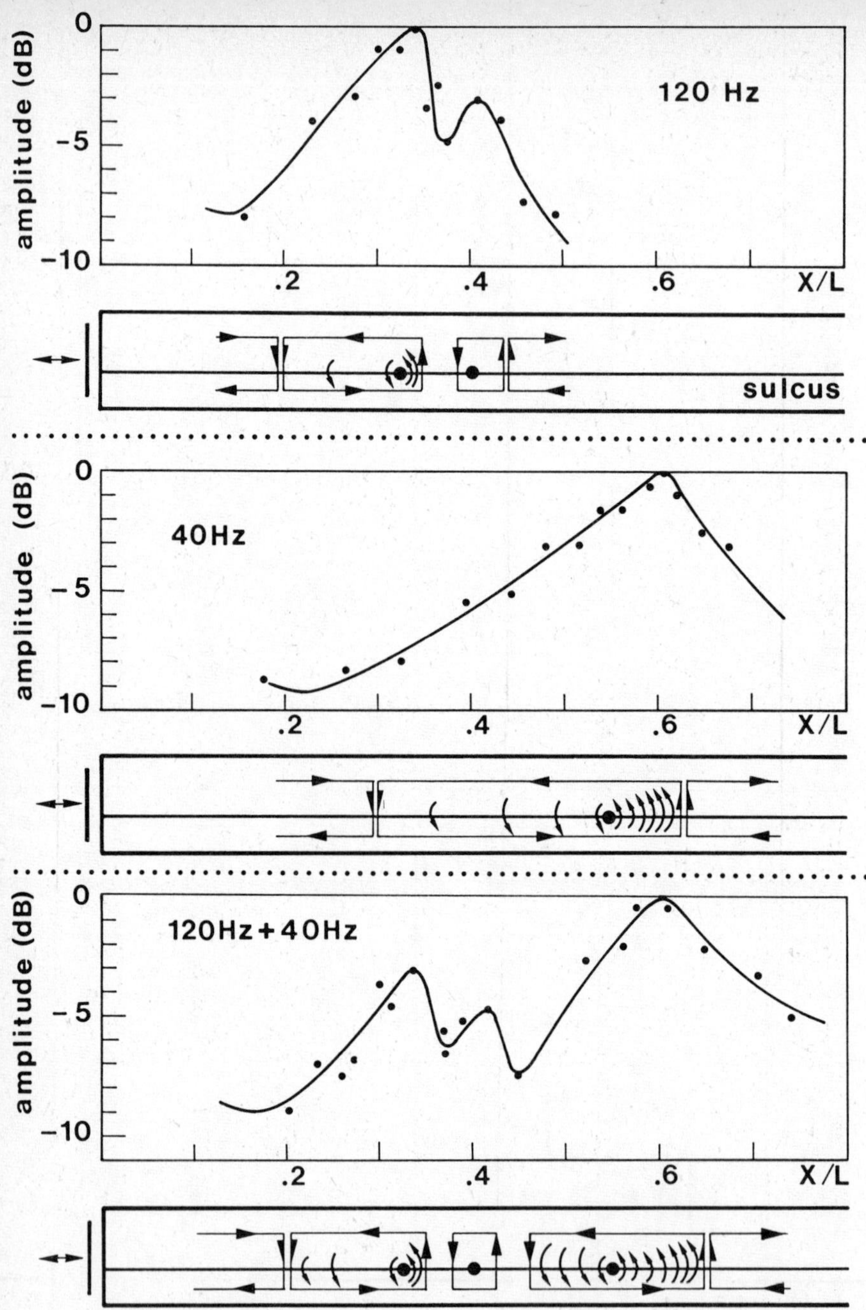

fig.7 Basilar Membrane response vs. distance at fixed frequency and schematic picture of the continuous flow at the same frequncy or combination of frequencies.

MODEL CONFIGURATIONS

	a	b	c	d	e
geometric ratio	50	50	50	50	50
density ratio	1	1	1	1	1
supposed compliance ratio	$2.5\text{x}10^3$	$2.5\text{x}10^3$	$2.5\text{x}10^3$	$2.5\text{x}10^3$	$1.2\text{x}10^3$
estimated frequency ratio	1/50	1/50	1/50	1/50	1/35
model perilymph viscosity (mm /s)	130	130	130	130	25
endolymph viscosity / perilymph viscosity	1	1	1	10	1
tectorial membrane	elastic	rigid	elastic	elastic	rigid
filling level in the scala vestibuli	full	full	partial	full	partial
distribution of the compliance (fig.2)	A-case	A-case	A-case	A-case	B-case

Table 1.

REFERENCES

(1) Békésy, G. von. "Experiments in Hearing." Mc Graw Hill, 1960.

(2) Cancelli, C., D'Angelo, S., Malvano, R., Masili, M. "Experimental Results in a Physical Model of the Cochlea." J. Fluid Mech. 153, pp. 361-388, 1985.

(3) Helle, R. "Enlarged Hydromechanical Cochlea Model with Basilar Membrane and Tectorial Membrane." in "Facts and Models in Hearing." ed. Zwicker & Terhardt, pp. 77-85, Springer, 1974.

(4) Lighthill, J. "Energy Flow in the Cochlea." J. Fluid Mech. 106, pp. 149-213, 1981.

(5) Peterson, L. C. & Bogert, B. P. "A Dynamical Theory of the Coclea." J. Acoust. Soc. Am. 22, pp. 369-381, 1950.

(6) Rhode, W. S. "Observations of the Vibration of the Basilar Membrane in Squirrel Monkeys Using the Mössbauer Technique." J. Acoust. Soc. Am. 49, pp. 1218-1231, 1971.

(7) Steele, C. R. & Zais, J. G. "Effects of Opening and Coiling in Cochlear Models." J. Acoust. Soc. Am., 1985.

(8) Voldřich, L. "Mechanical Properties of Basilar Membrane." Acta Otolaryngol, 86, pp. 331-335, 1978.

MODELLING THE COCHLEAR PARTITION WITH COUPLED VAN DER POL OSCILLATORS

H. Duifhuis[1,2], H.W. Hoogstraten[3], S.M. van Netten[1], R.J. Diependaal[4], and W. Bialek[1,5]
[1]Biophysics Dept., Rijksuniversiteit Groningen, the Netherlands
[2]also: Institute for Perception Research, Eindhoven, the Netherlands
[3]Dept.of Mathematics, Rijksuniversiteit Groningen, the Netherlands
[4]Dept.of Mathematics, Delft University of Technology, the Netherlands
[5]present address: Theoretical Physics Dept., U.C.Santa Barbara, USA

ABSTRACT

Within the context of interest in analyzing 'active' and nonlinear processes in the cochlea we have been studying a model cochlea in which the local membrane impedance is described by a Van der Pol-oscillator. The behaviour of the undriven and sinusoidally driven discretized model is examined numerically. The undriven model describes the behaviour of a discrete number of coupled oscillators, which, if uncoupled, would have limit cycles gradually differing in frequency. In the coupled case the limit cycle behaviour is less predictable: it appears to exhibit quasi stochastic properties. In the driven model a sufficiently strong stimulus causes entrainment to the stimulus, and odd order harmonics appear. In the range where the driven response is small compared with the average limit cycle, the response is almost linear. The strict Van der Pol-damping function, which is parabolic in velocity, produces strong saturation. A generalized Van der Pol-damping term, which causes small-amplitude instability and large-amplitude stability, produces much the same general behaviour, but the intensity response can be modelled more realistically.

1. INTRODUCTION

The motivation to study a model cochlea embodying 'active' and nonlinear properties has been given before (Van Netten and Duifhuis, 1983; Diependaal and Viergever, 1983). In short, we want to take the nonlinear data, the emission data, and the data suggesting a common or closely related origin seriously.

In order to minimize complexity we study the 1-dimensional, long-wave cochlea model, introducing small-amplitude negative damping and large-amplitude positive damping.

Van Netten and Duifhuis (1983) and Diependaal and Viergever (1983) have presented preliminary results of the model cochlea with a damping factor of the form $(-d_1 + d_2 y_t^2)$ as in the original Van der Pol equation. The former authors presented results from a harmonic analysis study, which only gave response components at the stimulus frequency; the latter presented results from a numerical analysis. The agreement between the results was not convincing. The results of this paper clear up some points and delineate new issues.

2. SOLVING THE COUPLED VAN DER POL-OSCILLATOR MODEL

The mathematical formulation of the coupled Van der Pol-oscillator model is:

$$(my_{tt} + dy_t + sy)_{xx} - \frac{2\rho}{h} y_{tt} = 0 \tag{1}$$

where m,d, and s are basilar membrane (BM) mass, damping, and stiffness per unit area, ρ is fluid density, h scala height, and y the BM-amplitude. As indicated above, d takes the form

$$d = - d_1 + d_2 y_t^2, \quad \text{with } d_1 \text{ and } d_2 \text{ positive.} \tag{2}$$

The parameters d_1, d_2 and s are supposed to depend on x. The problem is completed with the appropriate initial and boundary conditions (i.c. and b.c.). The common conditions for the linear case, viz.

$$\text{i.c.: } (t = 0) \quad y = 0, \quad y_t = 0 \tag{3a}$$

$$\text{b.c.: } (x = 0) \quad (my_{tt} + dy_t + sy)_x = f(t); \qquad (x = \ell) \quad (my_{tt} + dy_t + sy) = 0 \tag{3b}$$

appear to be inadequate, but that is where we started.

After the Delft symposium HWH and WB joined the project. HWH proposed an elegant method and FORTRAN code to solve Eq. 1. The first step which helps to assess the relevance of the different terms, is to scale all quantities and make Eq. 1 dimensionless. Using x_o = 35 mm, y_o = 1 nm, and t_o = 1 ms, one obtains

$$(y_{tt} + (-D_1 + D_2 y_t^2) y_t + Sy)_{xx} - \alpha^2 y_{tt} = 0. \tag{1a}$$

With $m = 0.5$ kg/m^2, $s = s_o \exp(-\lambda x)$ with $s_o = 10^{10}$ Pa/m and $\lambda = 300$ m^{-1}, $d_1 = \varepsilon\sqrt{ms}$ with $\varepsilon = 0.05$, and $d_2 = d_1(t_o/y_o)^2$, one has $D_1=D_2=5\sqrt{2}\exp(-5.25x)$ $S = 2\cdot 10^4 \exp(-10.5x)$ and $\alpha = 70$.

Putting $\psi = y_{tt} + D_1(y_t^2 - 1)y_t + Sy$, and $g = \psi - y_{tt}$, and finally writing $y_t = v$, yields the ordinary differential equations:

$$\psi_{xx} - \alpha^2\psi = -\alpha^2 g \tag{4a}$$

and

$$\begin{aligned} y_t &= v \\ v_t &= \psi - g, \text{ with } g = D_1(v^2-1)v + Sy \end{aligned} \tag{4b}$$

which have to be solved simultaneously. Note that ψ is the normalized pressure. To solve these equations numerically, the x-interval is broken up into a (large) number of subintervals of equal length and Eq. 4a is discretized accordingly. At fixed times this leads to a tri-diagonal set of linear algebraic equations for the ψ-values at all mesh points which can be solved straightforwardly by Gauss elimination. To advance the solution in time at each mesh point, we applied a 4th order Runge-Kutta method to Eqs. 4b. Since the results obtained in this way appeared to depend very strongly upon the x-discretisation, it should be stressed at this point that the numerical results presented in this paper should not be considered as valid approximate solutions to Eq. 1a. They represent the behaviour of a large set of Van der Pol-oscillators (Eq. 4b) with a coupling term ψ satisfying the *discretized* version of Eq. 4a.

So far we used a 401-point discretization of x and a time step of 2.5 µs. The time step may seem small. However, the highest resonance frequency to be encountered is 20 kHz. Appropriate representation of distortion requires that the overtones are adequately represented. The current choice describes 4 odd-order overtones at 20 kHz. Now we run into a storage problem: computing y or v on 401 points at 2.5 µs steps produces 0.64 Mbytes for each ms. Thus, for computations over longer time intervals it is impossible to store all information. (Because of potential aliasing problems, straight downsampling is not the appropriate approach to data reduction, even if one is only interested in, say, the mid-frequency range.)

3. INTERLUDE: FIRST RESULTS AND DISCUSSION

An undriven single Van der Pol-oscillator, $\ddot{v} + \varepsilon(v^2-1)\dot{v} + v = 0$, with $\varepsilon \ll 1$, will oscillate at its limit cycle frequency and velocity. The latter is determined by d_2/d_1. For the parameter values given in Sec. 2 the limit cycle peak velocity is $v = 2/\sqrt{3} \simeq 1.155$ nm/ms. However, the trivial solution $v = \text{constant} = 0$ is also a solution of the Van der Pol equation, albeit an unstable solution. Starting e.g., at 0.1 of the limit cycle amplitude it will take about $5/\varepsilon$ cycles to get to 0.9 of this value. This implies that the initial conditions of Eq. 3a will be inefficient in the driven situation, and even incorrect in the undriven case. Since we do not yet know how the system behaves, we postpone the question of the adequate initial condition. Just any condition $v \neq 0$ will do for the time being.

Although it was quite clear how a single Van der Pol-oscillator would behave, the behaviour of the coupled oscillators appeared unpredictable. One expects some influence of coupling on the behaviour of the individual oscillators. Since they all have slightly differing resonance frequencies they can not synchronize permanently. What happens is shown in Figs. 1 and 2. Figure 1 shows BM-velocities of two adjacent oscillators (11 and 12) over a 10-ms interval. It is apparent that the oscillators are neither completely uncorrelated, nor com-

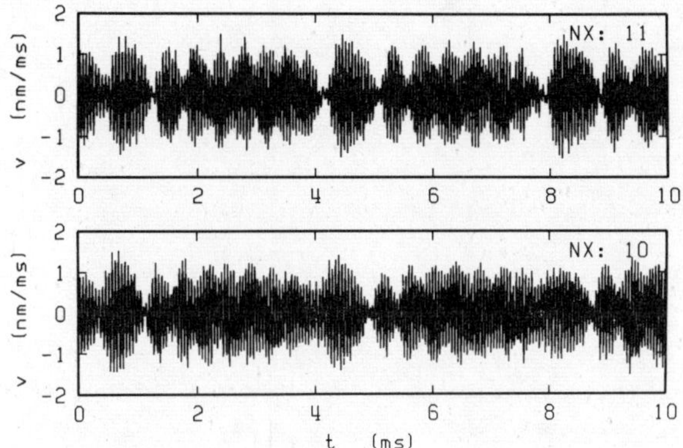

Figure 1. Responses in the adjacent channels 10 and 11 (0 base; 401 apex). No external stimulus. The two resonance frequencies are near 20 kHz. Uncoupled, the oscillators would have stable amplitudes of 1.155 nm/ms.

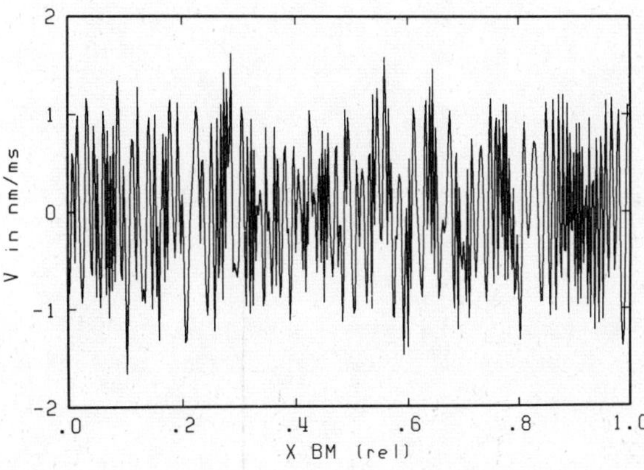

Figure 2. Response profile across the BM at an instant where correlation with the initial condition has disappeared. The instantaneous velocities of the 401 oscillators are connected by straight lines. No external stimulus.

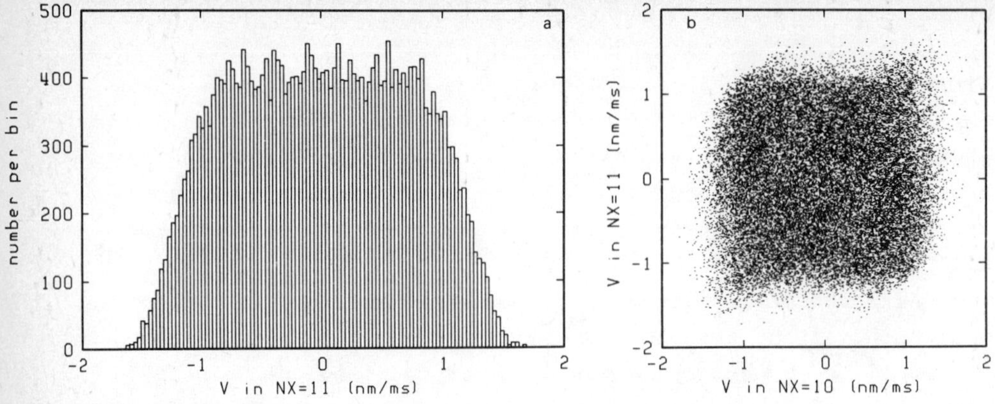

Figure 3a. Distribution of the velocity amplitude in channel 11, sampled at 2.5-μs intervals. N = 31200. No external stimulus. 3b: Dot display showing (the lack of) correlation between the responses in the two adjacent channels 10 and 11 (cf Fig. 1). Same time record as in 3a.

pletely correlated. Looking across the 401 different mesh points at the BM (Fig. 2) one sees a rather 'chaotic' response pattern, which changes completely from one moment to the other. In fact the rate of change at each point is inversely proportional to its resonance frequency.

Some statistics of the responses in the time domain are given in Fig. 3. Panel a shows the amplitude distribution measured over 78 ms. Note that the distribution is in between that for filtered noise (Gaussian) and that for an oscillator ($1/\{\pi\sqrt{1-v^2}\}$). The distribution is remarkably flat between + and - the limit cycle velocity. Panel b gives a sample by sample dot display of velocities in two adjacent channels. There seems to be little long term correlation. Short samples, however, do show correlation (cf. Fig. 1).

The implications for the adequate initial condition are as follows. When starting from some coherent i.c. one would expect to need several tens of cycles of the lowest frequency involved to obtain an uncorrelated pattern. Having established such a pattern once, it can be used as a typical i.c. for other computations.

Next we reconsider the boundary conditions. At the apex it is clear that the last oscillator is not kept at amplitude 0. The last oscillator is coupled to its neighbour at one side, at the other side it meets with its acoustic radiation impedance. This implies a mass load termination. In fact this is equivalent to adding a final section, using a somewhat different mass coupling, which section then is fixed at zero amplitude.

At the stapes nonlinearity invalidates the usual linear b.c.

$$\left(\frac{\partial p}{\partial x}\right)_{x=0} = f(t) \tag{5}$$

at least for simple harmonic stimuli. This problem is discussed by Matthews (1980). We propose, as the simplest physical approximation, to model the middle ear as an ideal transformer (ratio T) which meets the radiation impedance at the eardrum. For an external pressure source p_e we get the boundary condition

$$\frac{\partial p}{\partial x} = \frac{2\rho}{S_{sc}TR_a}\left\{-\frac{dp_e}{dt} + \frac{1}{T}\cdot\frac{\partial p}{\partial t}\right\} \quad \text{at } x=0 \tag{6}$$

where S_{sc} is the scala area, T the transformer ratio ($T > 1$), and R_a the radiation impedance at the eardrum. Boundary condition Eq. 5 causes severe problems as soon as the driven component at x=0 becomes of the order of magnitude of the limit cycle, but setting it equal to zero for the undriven case did not give trouble.

4. RESPONSE TO A SINGLE TONE

So far we have analyzed the response to a 1-kHz tone at two levels where b.c. Eq. 5 still applies, at approx. -10 and 30 dB SPL. The lower level stimulus hardly produces a recognizable response in the time pattern or in the BM-profile. Spectral analysis of the time records, however, does show a response component, which, except at about x = 0.6 is well below the limit cycle level, as shown by the lower heavy line in Fig. 4. Beyond x = 0.6 the response is not traceable. We found no indication of overtones. The response to the 30-dB tone, however, is quite different. After a few milliseconds a clear response builds up

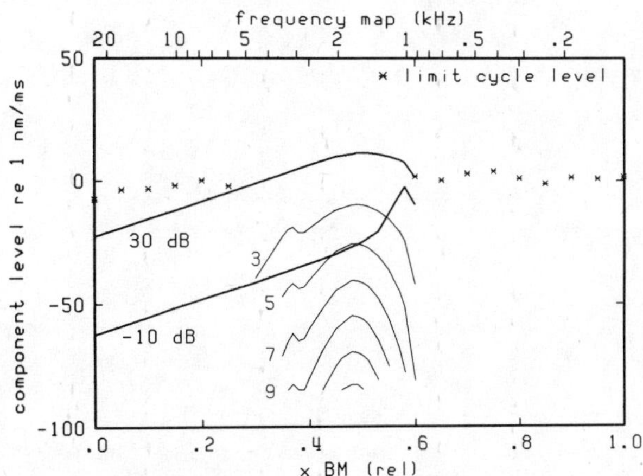

Figure 4. Response levels for a -10 dB and a 30 dB stimulus. For the -10 dB 1-kHz tone only a 'linear' 1-kHz response component is found, and each channel exhibits (modulated) limit cycle oscillations. The 30-dB tone causes entrainment in the 0.3 to 0.6 x-range. In this range we observe odd order distortion, as indicated by the thin lines, the first four of which are marked with their harmonic numbers.

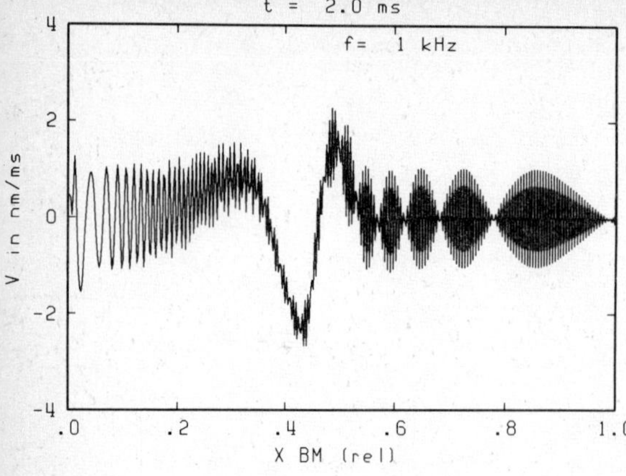

Figure 5. The BM-response to a 1-kHz 30 dB tone, 2 ms after the (smooth) tone onset. In the obvious response area one finds a reduction of the limit cycle. The coherent pattern in the apical part is due to the i.c., which in this case was an alternating +1,-1 velocity profile.

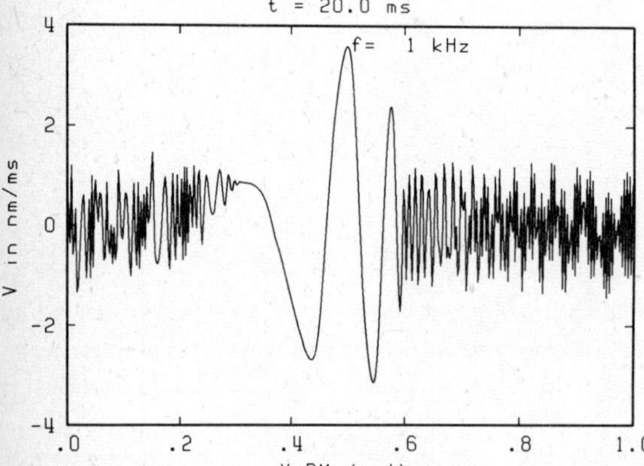

Figure 6. As Figure 5, but at 20 ms after tone onset.

in the range from about $x = 0.3$ to 0.6 (Figs. 5,6). The oscillators in this range are entrained by the stimulus. Within a few milliseconds the limit cycle oscillations are completely suppressed. Spectral analysis shows that within the range of entrainment odd-order distortion is quite pronounced (thin lines in Fig. 4). Even order distortion is absent. Note that the response maximum (heavy line) is shifted basically from the 1-kHz resonance site, and that a secondary maximum is visible in the distortion products near the 3-kHz resonance site. Due to the strong nonlinearity the response exceeds the limit cycle level only by some 12 dB for the 30-dB tone.

5. DISCUSSION

First a remark about the feasibility of the model as a representation of the cochlea. At this point we present our analysis as a contribution to efforts studying nonlinear and active processes in the cochlea, rather than as the ultimate ear model.

The limit cycle velocity of v=1.155 nm/ms is considered to be near threshold. Hence the scaling is such that the 'chaotic' behaviour may be considered threshold noise. This scaling implies that the intensity response follows a cube root, right from threshold. Changing the form of Eq. 2 to something like $(\sinh\beta\upsilon/\beta\upsilon - 2/\cosh\upsilon)d_o$ gives the possibility to modify the intensity response, but leaving general features the same. The value for d_2, which determines the scale, is 10^8 larger than in our 1983 papers. For the harmonic analysis this implies a scaling factor. For the numerical analysis the effect is rather more pronounced because it puts the limit cycle at 80 instead of 0 dB SPL.

The quasi-random undriven velocity profiles imply the occurences of antiphasic responses of adjacent points. On the cochlear scale, with a 10-μm grid, opposite deflections of the order of one nm do not appear to violate physical possibilities.

In this study we used smoothly varying cochlear parameters. One logical extension is to introduce irregularities. Another necessary one is the analysis of complex stimuli (suppression, combination tones).

The model cochlea consisting of coupled Van der Pol-oscillators produces several unexpected effects, a result to be expected when studying complex nonlinear systems. The gap between data and analysis is not yet bridged. An interesting prediction, however, is that hair cell receptor potentials as well as spontaneous activity in afferent nerve fibers would reflect tuning in their autocorrelations.

ACKNOWLEDGEMENTS

The authors thank E. de Boer, M.A. Viergever, and P.I.M. Johannesma for contributions in a number of discussions. This study received support from the Netherlands Organization for the advancement of Pure Research ZWO (SMvN and RJD) and from an US-NSF-NATO postdoctoral fellowship (WB).

REFERENCES

Diependaal, R.J. and Viergever, M.A.,"Nonlinear and active modelling of cochlear mechanics: A precarious affair," in Mechanics of Hearing (Eds. E.de Boer and M.A.Viergever) Delft Univ.Press,pp 153-160, 1983.

Matthews, J.W. Mechanical modeling of nonlinear phenomena observed in the peripheral auditory system, PhD-thesis,Washington Univ.,StL.1980.

Van Netten, S.M. and Duifhuis, H.,"Modelling an active, nonlinear cochlea," in Mechanics of Hearing, (E.de Boer and M.A.Viergever eds) Delft Univ. Press, pp. 143- 151, 1983.

NEW EFFECTS OF COCHLEAR NONLINEARITY IN TEMPORAL PATTERNS OF AUDITORY NERVE FIBER RESPONSES TO HARMONIC COMPLEXES

J.W.Horst*, E.Javel[+], and G.R.Farley[+]
*Institute of Audiology, Univ.Hospital Groningen
The Netherlands.
[+]Boys Town National Institute, Omaha, NE 68131

ABSTRACT

Discharge patterns of single auditory nerve fibers in adult cats were recorded in response to multi-component harmonic complexes. Stimuli possessed a variable number of equal-intensity harmonics of a common low-frequency fundamental. Data were analyzed by examining Fourier transforms of period histograms locked to the period of the stimulus fundamental. At low intensities transfer functions agreed with those defined by more traditional measures such as tuning curves. However, responses generally showed nonlinear level dependence at higher intensities, particularly above 60 dB SPL. In this, response spectra changed from resembling a bandpass filter to resembling a band-reject filter, i.e. responses were synchronized to components at the edges of the signal spectrum and to combination tones near the edges. The nonlinear effects became more pronounced as the number of components in the signal passband increased. Comparison of data from fibers with a wide range of thresholds indicated that the nonlinear behavior is mainly related to absolute stimulus level. This suggestes that the site of the nonlinearity is located prior to the spike generator.

1. INTRODUCTION

We have been investigating relationships between temporal processing of complex stimuli in single cat auditory nerve fibers and human psychophysical frequency-discrimination data. In the course of our physiological studies, we observed some unexpected nonlinear phenomena (Horst et al., 1985a, 1985b). The stimuli we used had relatively simple spectra. That is, they possessed flat spectral envelopes, were one octave wide, and consisted of successive harmonics of a common low-frequency fundamental. We observed that fiber responses changed considerably with stimulus intensity, and that they displayed particularly strong responses to the spectral edges of the stimuli at high intensities. These data may be interpreted in terms of enhancement of spectral contrast, similar to Houtgast's psychophysical data (1974). This paper further explores some of the properties of high-intensity "edge enhancement".

2. METHODS

Responses from single auditory nerve fibers were recorded from anesthetized adult cats. Details of the surgical and recording procedures are given in Horst et al. (1985a). Before stimulation with complex signals, a tuning curve was taken for short-duration pure tones. This yielded the fiber's characteristic frequency (CF) and approximate discharge rate threshold. Complex tones were multicomponent stimuli defined by a center frequency F and a spectral spacing factor N. The ratio F/N defined the fundamental frequency F_o of the stimulus. Signals consisted of successive harmonics of F_o. Spectra were limited to one octave in width and were centered geometrically about a fiber's CF, with F=CF. Component amplitudes were equalized during synthesis to produce stimuli that were acoustically flat. Stimulus intensities were expressed in terms of the intensity of each component, expressed in dB SPL (re 20 μPa). Responses of single auditory nerve fibers were analyzed by obtaining Fourier transforms of period histograms synchronized to the period of the waveform fundamentals.

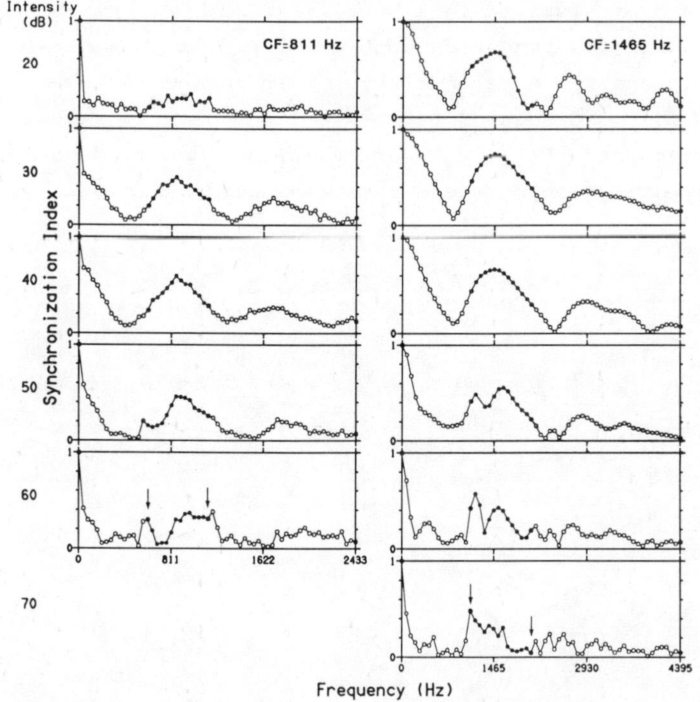

Figure 1. Fourier transforms of period histograms in response to a complex stimulus with flat spectral envelope and cosine-phase spectrum as a function of the intensity of each component. Data for a high- and for a low-spontaneous-rate fiber. The magnitudes of spectral components are made equal to synchronization indices by normalizing with respect to the DC-component. Filled symbols indicate frequencies present in the acoustic stimulus, and open symbols indicate frequencies absent from the stimulus. Arrows in bottom panels indicate the lower and upper frequency limits of the stimulus spectra.

3. RESULTS AND DISCUSSION

Fig.1 presents data from two different fibers having similar thresholds at CF. The responses shown in the left-hand column came from a fiber with a high spontaneous rate (45 spikes/s), and the responses in the right-hand column came from a fiber with a low spontaneous rate (0.1 spikes/s). At low intensities, the distribution of synchronization indices resembles the fiber's transfer function as defined by the pure-tone tuning curve (not shown). As stimulus intensity was increased, the shape of the transfer function changed gradually. At 60-70 dB, responses at the spectral edges dominated, and the relation to the tuning was obscured. Additional responses to frequencies not present in the stimuli can also be observed at these intensities. These can be presumed to have been generated by combination tones.

In a sample of 105 fibers from five cats, we observed strong intensity-dependent nonlinearities in the responses of 84% of the fibers. Three types of intensity dependencies could be distinguished. These were (1) shifts of CF, (2) narrowing of bandwidths of response spectra, and (3) enhancement of spectral edges. This last type was the most prevalent nonlinearity, and it occurred in 83% of the fibers who exhibited nonlinear behavior with increasing stimulus intensity.

In addition to intensity, a second factor that influenced edge enhancement was the number and spacing of spectral components present in the stimulus. By changing $N=F/F_o$, we could vary the number of components in the stimulus while keeping the stimulus bandwidth fixed at one

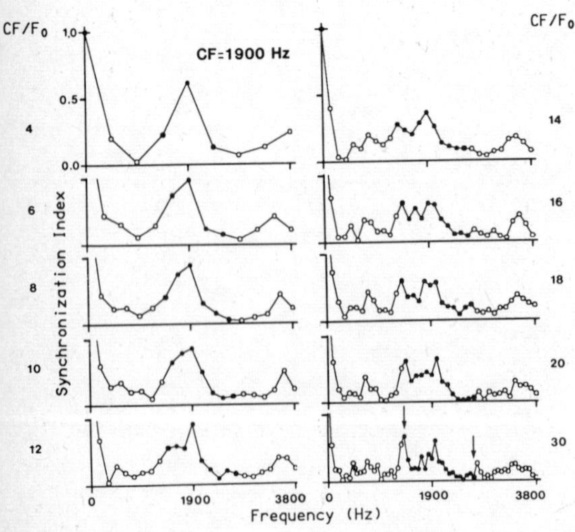

Figure 2.
Fourier transforms of period histograms at high (75 dB SPL) intensities as a function of the spacing factor $N = CF/F_o$. Usage of symbols as in Fig.1.

octave. At low intensities (where fiber responses are linear), increasing N simply provided a more accurate sample of the fiber's transfer function. At higher intensities increasing N generally produced greater edge enhancement. An example of this is shown in Fig.2. Here, the relationship between the tuning curve and the response spectrum is retained relatively well for N=10. When N is increased above 10, however, there is a rapid change in the shape of the response spectrum, so that for N=16 the low frequency edge is clearly enhanced. For yet higher values of N a clear peak is present at the lower edge of the response spectrum. Responses to combination tones can also be seen near this low frequency edge.

We also examined the abruptness of the transition between the linear response obtained at low intensities and the nonlinear response obtained at high intensities. We did this by presenting sets of stimuli in which we concurrently varied both N and stimulus intensity. An example is given in Fig.3. When data are compared for different values of N, it appears that responses at one value of N cannot be readily predicted from responses at other values. For example, the responses obtained at 75 dB SPL for Ns of 4 or 8 show a clear enhancement of the low frequency edge, while the response for N=16 is much flatter. On the other hand, responses obtained for higher values of N show distinct peaks at both high- and low-frequency edges.

Comparing the data in Fig.3 row by row, that is, across intensity, it is clear that no two rows are alike. Thus, these data provide additional evidence for the intrinsically nonlinear nature of single-fiber responses. Similar results were obtained when other aspects of the amplitude and phase spectrum were varied (Horst et al.,1985b).

The transduction from mechanical motion of the basilar membrane to hair cell responses and single-fiber spike trains involves several nonlinear processes. Among these are the stochastic nature of spike generation on only one stimulus phase. These processes clearly influence the relationship between the stimulus spectrum and the response spectrum. These influences may be reduced by compiling compound period histrograms. Examples are given in Fig.4a, for the same fiber that provided the data shown in Fig.3. The spectra in Fig.4a, which are based on compound period histograms, show reductions in some of the effects of rectification distortion apparent in the Fourier transforms of simple period histograms. This is especially clear in the reduced magnitudes of components below 0.5 CF and around 2CF (compare Fig.4a with column 4 of Fig.3). However, the nonlinear behavior around CF and the enhancement of stimulus spectral edges is still clearly present.

Previous studies of single fiber transfer functions were mainly

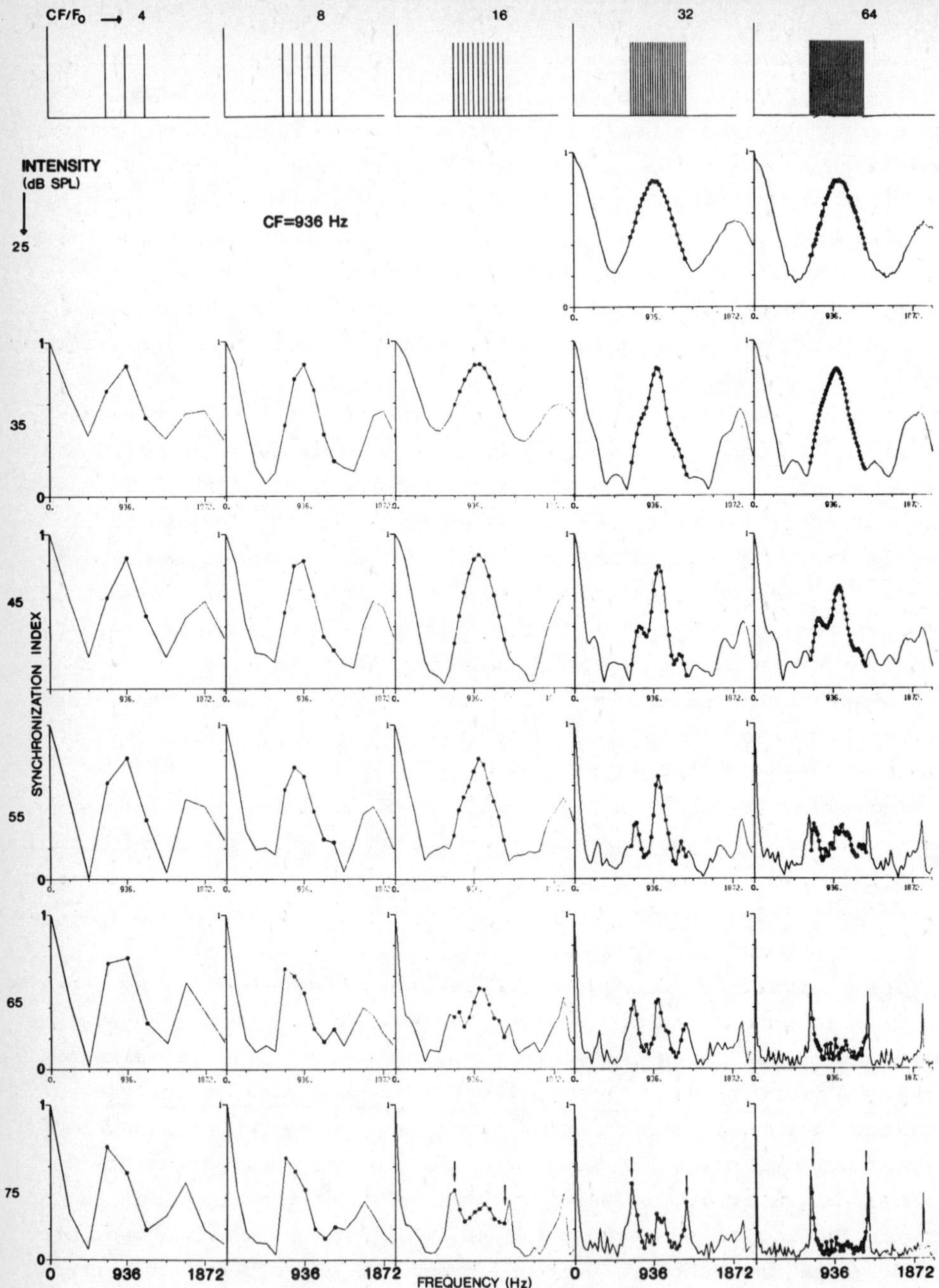

Figure 3. Fourier transforms as a function of stimulus intensity and spacing factor. Meaning of filled symbols and little arrows as in Fig.1. The stimulus spectra are shown at the top.

carried out using noise stimuli. These studies (e.g. de Boer and Kuyper, 1968; Møller, 1977) did not report the dramatic edge enhancements seen in the present data. For comparison with these studies, we also obtained data for periodic stimuli having random phase relationships among spectral components. It was generally the case for these stimuli that edge enhancement was much reduced in fiber responses at all intensities. An example is shown in Fig.4b, which shows Fourier spectra of compound period histograms of data from the same fiber whose responses are shown in Figs.3 and 4a. Clearly, the bandpass or tuning-curve-like nature of the transfer function is better retained at higher intensities for random-phase stimuli than for in-phase stimuli.

Also notable in Fig.4b is the observation that the random-phase data show less smooth response spectra, even though we collected more spikes for these (i.e. the spectra in Fig.4b were based on 4000 or more spikes, compared with approximately 2000 spikes for the spectra in Fig.4a). The random-phase stimulus may be regarded as intermediate between completely deterministic and purely random stimuli, and the observed effects are in good agreement with the notion that stimulus randomization tends to linearize the system's response.

The enhancement of the spectral edges observable at high intensities in Figs.1-4a strongly suggests the involvement of both suppression and combination tones of the type $2F_1-F_2$ and $2F_2-F_1$. As the number of stimulus components increases, the number of contributions to a given combination tone will also increase. These contributions will tend to cancel when they are added in random phase, and they will be augmented when they are added in phase. The generation of combination tones is usually attributed to interactions at the basilar membrane level and/or in the hair cells themselves (Goldstein, 1967; Kim et al., 1980). The same generation mechanisms have been proposed for suppression (Rhode, 1977; Sellick and Russell, 1979). These arguments imply that the generators of suppression, combination tones and, by extension, the edge enhancement phenomenon are all caused by nonlinear mechanisms operating peripheral to the spike generator.

Additional evidence concerning the locus of origin of nonlinearities in response to complex tones was obtained by comparing responses from fibers with different absolute thresholds. For a sample of 24 fibers in to cats, we compared the absolute thresholds of the fibers with the intensities producing clear edge enhancement at N=16. The correlation coefficient was 0.24, which was not significant (p=.1). The poor correlation implies that the occurrence of edge enhancement is mainly determined by absolute stimulus intensities and not the intensity above the the fiber's threshold.

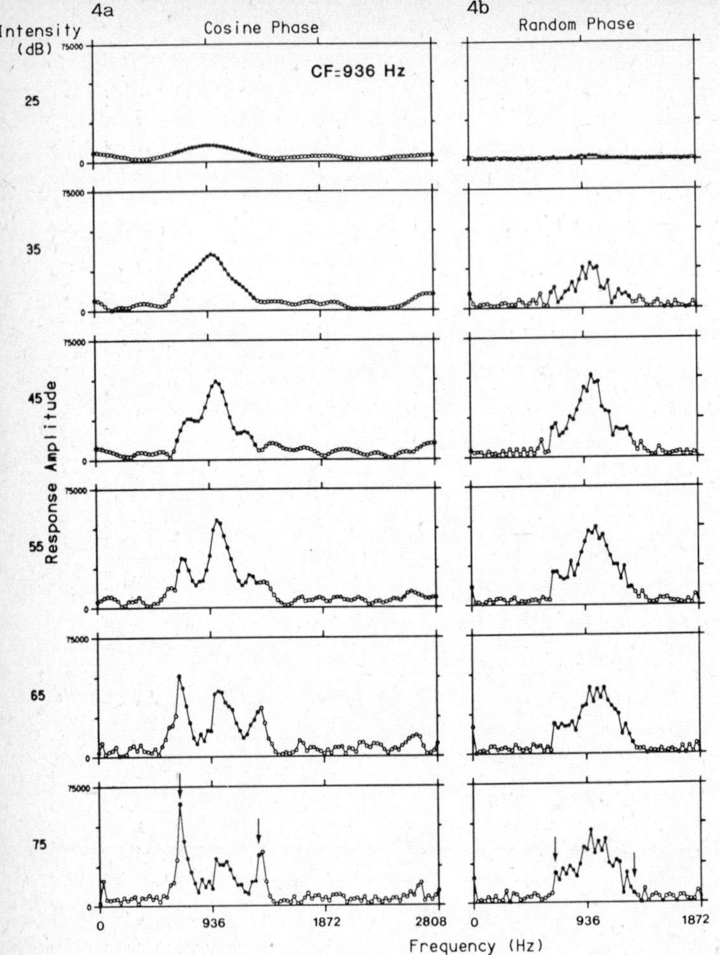

Figure 4. Fourier transforms of compound period histograms at N=32 and at several intensities for the same fibers as in Fig.3. Usage of symbols as in Fig.1.
a. Cosine-phase stimulus spectrum.
b. Random-phase stimulus spectrum.

It was usually the case in our data that edge enhancement was accompanied by a systematic change in the temporal envelope of the period histogram. That is, what was a single-peaked response at low intensities became a multi-peaked response at high intensities. In these cases, increasing stimulus intensity resulted in a redistribution of spikes over the entire stimulus period. This is generally not the case for responses to click trains (e.g.Pfeiffer and Kim, 1972), which indicates the tendency of single-fiber responses to retain the shape of the stimulating waveform. This in turn indicates that the observed edge enhancement is present before the spike generating mechanism. The edge enhancement, then, may be explained by the existence of a compressive nonlinearity. Because the low-amplitude waveform peaks in the band-limited stimulus are mainly tuned to the spectral edges of the stimulus, the compressive nonlinearity would influence the fiber's response considerably as intensity is increased.

The data in this paper present nonlinear phenomena which have not been reported previously. The phenomena likely reflect nonlinear aspects of basilar membrane motion and hair cell stimulation, and they are highly relevant to the understanding and modelling of cochlear mechanical processes.

AKNOWLEDGEMENTS

This work was supported by grant NS-14880 award to EJ by NINCDS and by a grant to JWH by the Netherlands Organisation for the Advancement of Pure Research (Z.W.O).

REFERENCES

de Boer, E. and Kuyper, P. "Triggered correlation", IEEE Trans.Biomed. Eng. 15, 169-179, 1968.

Goldstein, J. L. "Auditory nonlinearity." J.Acoust.Soc.Am. 41, 676-689, 1967.

Horst, J. W., Javel, E., and Farley, G. R. "Coding of spectral fine structure in the auditory nerve. I. Fourier analysis of period and inter-spike interval histograms." J.Acoust.Soc.Am., submitted, 1985a.

Horst, J. W., Javel, E., and Farley, G. R. "Extraction and enhancement of spectral structure by the cochlea." J.Acoust.Soc.Am., submitted, 1985b.

Houtgast, T. "Lateral Suppression in Hearing." Doctoral dissertation Free University Amsterdam, The Netherlands, 1974.

Kim, D. O., Molnar, C. E., and Matthews, J. W. "Cochlear mechanics: Nonlinear behavior in two-tone responses as reflected in ear-canal sound pressure.". J.Acoust.Soc.Am. 67, 1704-1721, 1980.

Møller, A. R. "Frequency selectivity of single auditory-nerve fibers in response to broad-band noise stimuli." J.Acoust.Soc.Am. 62, 135-142, 1977.

Pfeiffer, R. R., and Kim, D. O. "Response patterns of single cochlear nerve fibers to click stimuli: Descriptions for cat." J.Acoust.Soc.Am. 52, 1669-1677, 1972.

Rhode, W. S. "Some observations on two-tone interaction measured with the Mössbauer effect." In E. F. Evans and J. P. Wilson (eds.), Psychophysics and Physiology of Hearing (New York, Academic Press), pp. 27-38, 1977.

Sellick, P. M., and Russell, I. J. "Two-tone suppression in cochlear hair cells." Hearing Res. 1, 227-236, 1979.

WIDEBAND ANALYSIS OF OTOACOUSTIC INTERMODULATION

David T. Kemp and Ann M. Brown
Department of Audiology, Institute of Laryngology and Otology
330-332 Gray's Inn Road, London WC1X 8EE

ABSTRACT

Nonlinear interaction between close tones in the cochlea has been examined in gerbil via the multiple otoacoustic emissions this interaction generates. Tonal stimuli of near equal level yield a wide spread of up to 20 otoacoustic sidebands. The amplitude and relative phase of these components have been measured and these have provided new insights into cochlear mechanical nonlinearity and data on the reverse transmission of distortion products. Otoacoustic intermodulation and suppression characteristics differ in detail from those of the simple saturating nonlinear model presented. Demodulation of the wideband emissions reveals temporal attributes of the cochlear interaction to be the major difference from the model.

1. INTRODUCTION

Otoacoustic re-emissions occur via a nonlinear transmission path. This manifests itself in the non-proportional growth of stimulus frequency acoustic re-emission with stimulus level (Kemp and Brown 1983); in the suppression of stimulus frequency emissions by a second stronger tone (Kemp and Chum 1980); in the production of combination tones during two tone stimulation (Kim et al 1980) and in the suppression of combination tones by the presence of a third and stronger tone (Brown and Kemp 1984). These same phenomena are found in simple nonlinear systems such as in figure 1, and can be understood physically as being due to the modulation each input component by the summed amplitude of all components. For multiple tone inputs to nonlinear systems the output is highly complex but unique to the particular form of nonlinearity. Cochlear nonlinearity can thus be explored through distortion product generation but otoacoustic studies of the generation and/or suppression of just one output component, be it a stimulus frequency emission, or a single distortion product such as 2F1-F2 provide a very narrow "window" with which to test hypotheses. In this paper we develop further a multicomponent, wideband analysis technique for the interpretation of otoacoustic intermodulation products (Kemp and Brown 1984).

2. INTERMODULATION SIDEBANDS

A close tone-pair signal passing through a nonlinear system acquires a distinctive pattern of sidebands described by (N+1)f1-Nf2. This set of intermodulation products is examined here in gerbil otoacoustic responses and, for comparison, in a simple nonlinear model, described in figure 1.

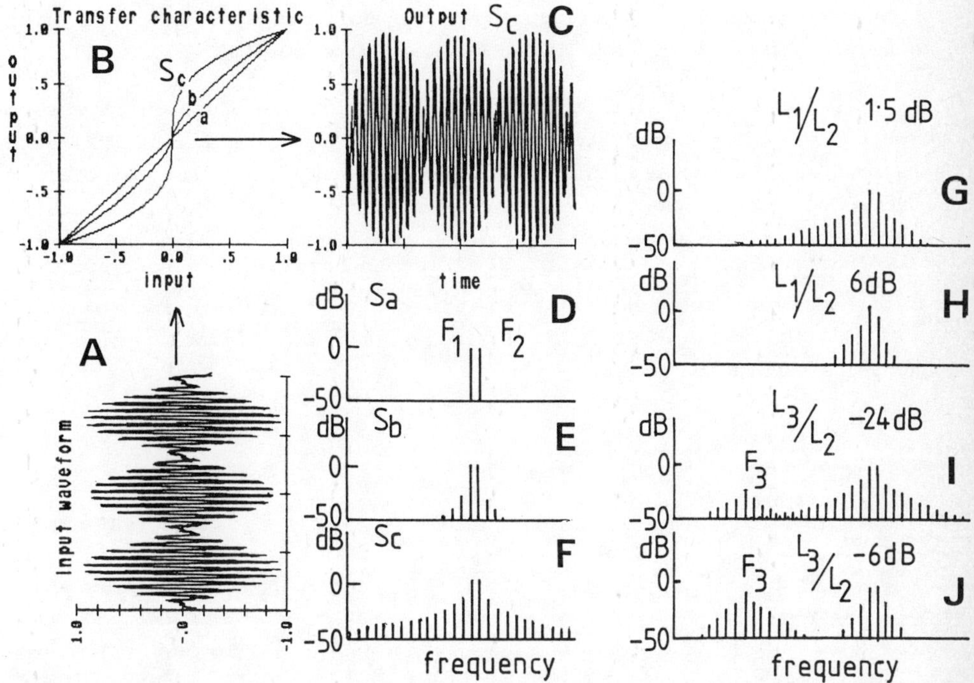

Figure 1. Some features of intermodulation found otoacoustically are here derived from a simple saturating nonlinear model. (B) shows the model input-output transfer characteristic. The instantaneous gain of the system is determined by the Sth root of the input signal modulus. Data is given for S=1, 1.4 and 3, labeled a,b,c respectively. A signal consisting of two equal pure tones (F2/F1=1.1) added together (A) and passed through the model for S=3, (B,c) results in an output (C). Some compression is noticable but spectral analysis of this output (F) shows symmetrical wide sidebands of frequencies satisfying (N+1)F1-NF2. Harmonics and other components are present but not shown. For the more linear transfer (S=1.4) the sideband range (E) is much reduced by the greater fall off with N. For linear transmission (S=1) only F1 and F2 emerge (D). Symmetrical sidebands occur only if L1=L2. (G) shows the suppression of upper sidebands due to only a 1.5 dB increase in L1 (S=3 for G through J). Greatly reduced sideband range is caused by a 6dB excess of L1, (H) although the level of the lower order sidebands still exceeds those in (E). Also in (H), L1 has suppressed the output of L2 by 4dB. The addition of a third tone F3, to the input signal produces interesting effects. For low levels (L3 24dB below L1=L2), F3 has wide sidebands (I). As L3 is increased F2,F1 sidebands are suppressed. With L3 still 6dB below L1,L2, this effect is very strong on high order components (J). With L3=L2=L1 (not shown), 2F1-F2 and 2F2-F1 are suppressed by 8dB, and higher orders by much more, whilst F1 and F2 output is reduced by only 2dB. This is relevant to the interpretation of otoacoustic suppression tuning curves for 2F1-F2 and stimulus frequency components.

Experimental data was collected from 5 Mongolian gerbils (Meriones unguicualatus) using a meatal acoustic probe previously described. Real time computer based spectral analysis was performed on the meatal sound field during (mainly) 2 tone stimulation. Frequency domain averaging (phase and amplitude) was employed to achieve a noise per 10Hz bandwidth of down to -20dbSPL. Stimulus tones were asynchronously generated but phase alignment was accomplished in software prior to sample summation. Individual data sample spectra were phase aligned to a time point when F1 and F2 where out of phase. ie the absolute time reference was the null/minimum in the stimulus beat pattern. Stimuli used from 2000-6000Hz and from 50-70dB spl.

Figure 2 shows spectral data from gerbils. Extensive sidebands are seen below the two stimuli, especially when level L1 is just greater than L2. Such a wide spread of almost equal intensity lower sidebands could not be obtained from the model even with infinite clipping saturation. Model sidebands always showed a decrease in intensity with order number. Also unlike the model the upper sidebands were never as extensive as the lower sidebands. Upper sidebands were most extensive at the higher L1 (70dB) and with the lowest F2 (2530Hz). With L1 and L2 dissimilar, the gerbil data became more similar to the saturation model.

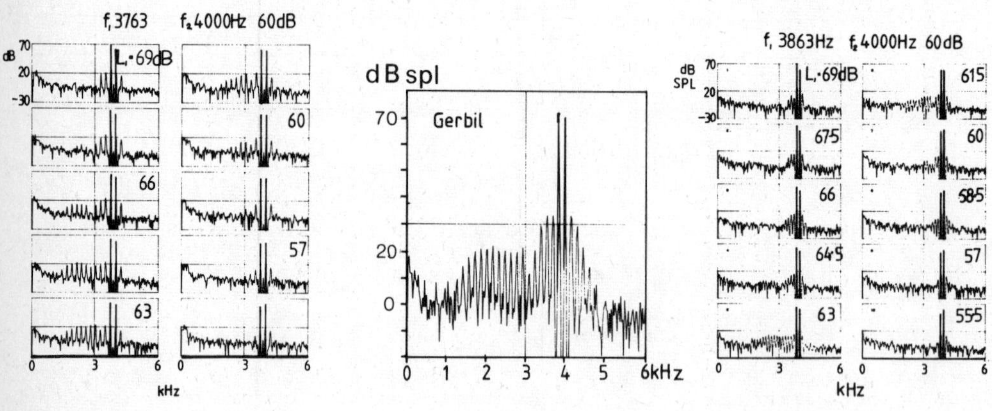

Figure 2. Spectral analysis of ear canal sounds in gerbil during two tone stimulation. The enlarged example (b) shows 17 lower and 5 upper sidebands clearly visible, covering a 3.5kHz range. Unlike the model there is no strong fall-off with order number. Stimuli were both 70dB spl with F1=3850Hz and F2=4000Hz, (F2/F1=1.039). For 60dB spl stimuli, the upper sideband spread is less. The relative level of L1 and L2 affects the lower sideband spread. In (a) and (c), L1 is changed from 69-56.5dB in 1.5 dB steps, with L2 fixed at 60dB, for F2/F1 ratios of 1.063 and 1.035 respectively. Note the limited L1 range for wide lower sideband spread.

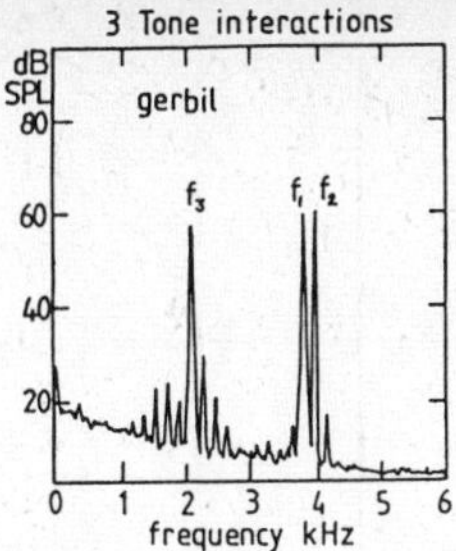

Figure 3. An example in gerbil of the effect of a third tone, F3. F2 is 4000Hz and F2/F1 is 1.047 with L1=L2=60dB spl. L3 was 3dB below L1 yet strong suppression of the F2,F1 sidebands is present. Double sidebands also surround F3. The model showed a very similar response (Fig 1j). F1 and F2 are here closer than Brown and Kemp 1984 used to obtain suppression tuning curves for 2F1-F2. The new data illustrates that suppression of 2F1-F2 does not equate to suppression of distortion generation.

The effects of a third tone F3 on gerbil intermodulation products are shown in figure 3. These are- suppression of sidebands around F1 and F2 by F3 and the appearance of a new double sideband distortion family around F3, as in the model.

Observations were made over a range of F2/F1 from 1.029 to 1.7. In the model, increasing this ratio simply spread the sidebands proportionately wider. In the gerbil the lower sidebands remained confined in the same broad band (figure 4). The upper sidebands tended to expand proportionaly as in the model.

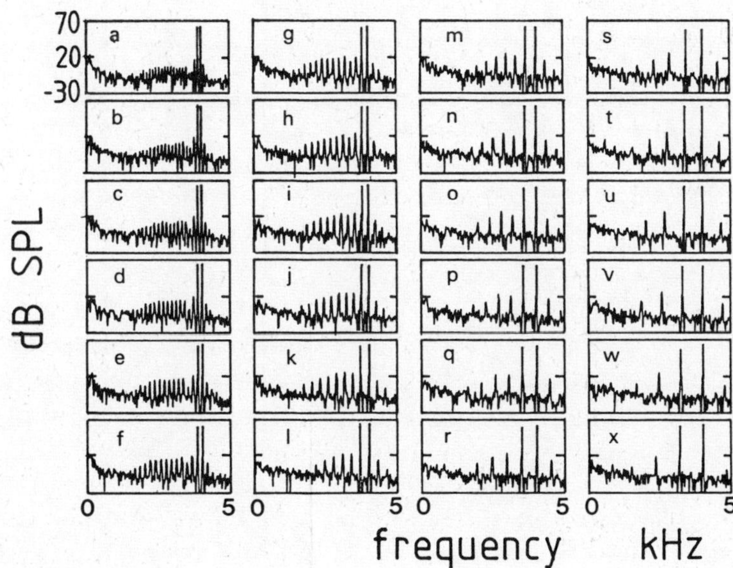

Figure 4. Intermodulation sidebands in a gerbil as a function of F2/F1. The stimulus frequency ratio R, is here changed from 1.0292 to 1.262 in (a) through (x) by 10% increments in the frequency difference (F2-F1). F2 remained fixed at 4000Hz and L2=L1-3dB=60dB. Note the near constant lower sideband range and the variable upper sideband range.

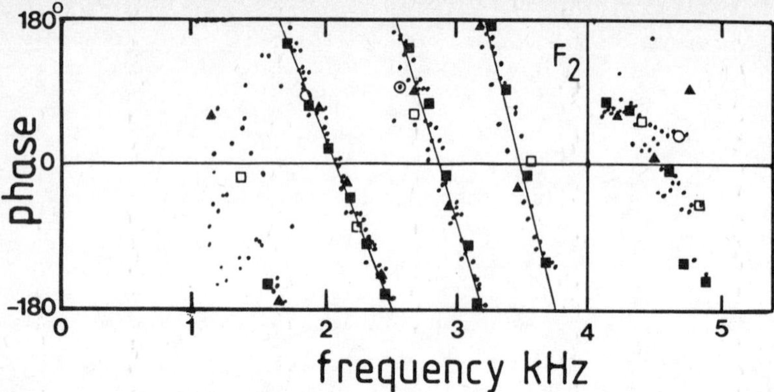

Figure 5 Phase information for all the spectral peaks shown in figure 4 except the stimuli. Time was referenced to the node of the stimulus beat in the meatal signal. Highlighted data points are from particular spectra in fig 4. (d)- filled squares, (j)- filled triangle, (p)- square, (v)-circle. Note the adherence to a near linear progression between 2 and 4 kHz, by all lower sideband orders from all ratios. The phase trend slope in fact decreases towards 2kHz. The same pattern was observed in all of 18 data sets taken at levels from 50 to 70 dB and F2's from 2-7kHz, in 4 gerbils.

Each sideband component is characterised by a phase relationship to F1,F2. Phase data corresponding to figure 4, is shown in figure 5. Data for all orders and all ratios are superimposed. A common phase trend exists for all lower sidebands for all ratios which is remarkable for such a wide range of stimulus conditions. Clearly the phase of a distortion component depends mostly on its frequency not on its order number N. Otoacoustic sideband generation is thus largely phase independent of frequency ratio, even though the cochlear excitation pattern must change substantially.

From the common phase trend a single latency frequency function can be obtained for the nonlinear channel involved; applicable to the whole F2/F1 range. Latency functions for several conditions are shown in figure 6. The lowest frequency lower sidebands generally have about one half the latency of those near to the stimulus. The latency of 2F1-F2 has previously been obtained from the phase gradient by a dynamic method ie sweeping the frequency of F1 with F2 fixed (Kemp and Brown 1983). Since the excitation pattern in the cochlear changes during this determination, the absolute validity of this could be questioned. The new data provides a "static" determination of latency from each spectrum. The coherence of phase data for allow orders and ratios shows that static and dynamic methods give the same latency. The latency of upper sidebands is not straightforward and will be discussed elsewhere.

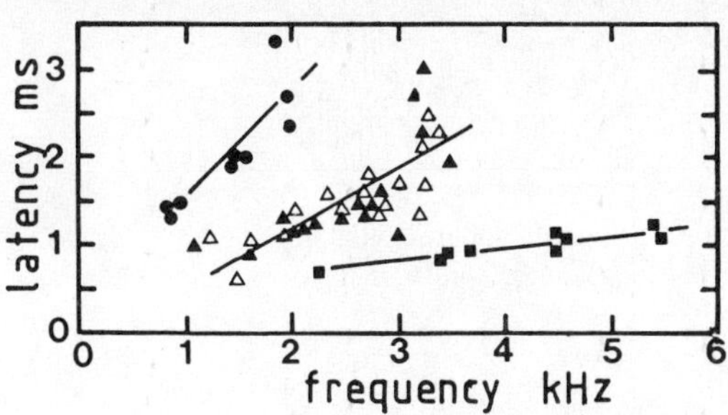

Figure 6 Sideband group latency was computed from the gradients of the near linear phase trends with frequency in figure 5 and from 18 comparable data sets. Group sideband latency was obtained as a function of frequency, from each 360 degree section. Circles are for F2 2520Hz (50-70dB), triangles F2 4000Hz (filled 60dB, unfilled 50 and 70 dB), squares 6350Hz (50-80dB). Note the decreased group latency towards lower sideband frequencies for a given F2, in contrast to increased latency with decreasing F2.

It is difficult to appreciate the physical significance of the sideband pattern looking at the frequency domain data. Inverse fourier transformation allows us to translated data back intc the time domain. This is shown in figure 7 and it is clear from the model and gerbil data that distortion is produced mainly in short bursts at the nodes in the beating input signal, ie when the input envelope is changing most rapidly.

2. DISCUSSION AND INTERPRETATION

We have found wide intermodulation sidebands to be a general feature of gerbil otoacoustic emissions in response to two tone stimuli. For near equal L1 and L2, the lower sideband is wider and has many more components than the upper. We propose that internally a double sideband distortion set is generated mostly at a place in the F1,F2 cochlear region where L1=L2 locally, and takes the form of a series of short pulses at each excitation beat node. We attribute the single sideband tendancy of otoacoustic emissions primarily to the reverse transmission path from the F2,F1 place region since the F2 place cannot be the source of reverse travelling waves above F2. (Asymmetrical sidebands can be produced by unequal stimulus intensities but this effect alone could not explain the data.) Components near to F1 and F2 in frequency would suffer considerable reverse propagation delay (equal to the forward delay). The negative frequency dispersion within the lower sideband (fig. 6) is strong evidence for this. Basic transform theory dictates that the shortness/sharpness of the emission bursts determines the sideband width. A simple saturating model could not be made to give a gerbil-like lower sideband pattern, with a flat, N independent region, although in figure 7d,e the distortion envelopes of both model and gerbil look similar.

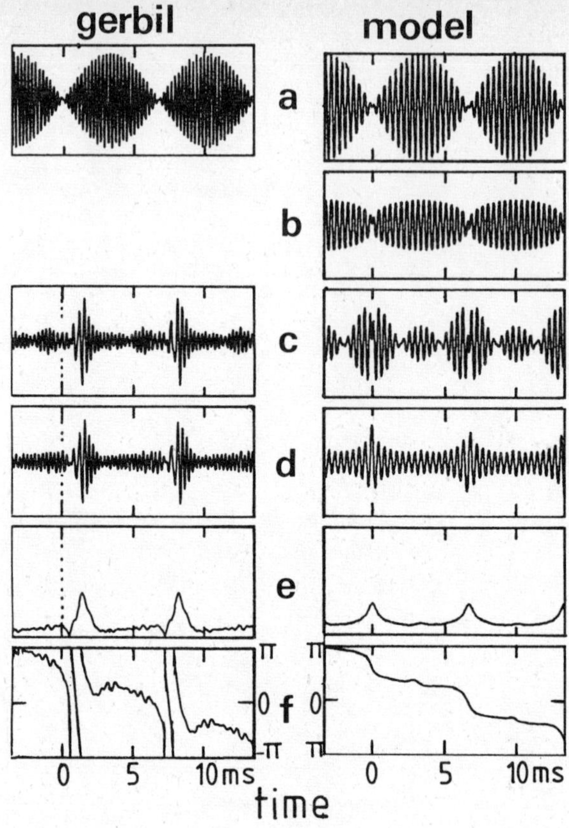

Figure 7. A time domain study of intermodulation in both the gerbil and the model. For gerbil the data corresponds to the spectrum in figure 4d. (a) shows the input, consisting of two equal sinusoids of similar frequency added (as in fig. 1a). (b) The signal emerging from the nonlinear model system, with S=3). Corresponding gerbil data is not available from this experiment since the stimuli obscured the output of stimulus frequency component from the ear. (c) shows amplified meatal and model "output" signals after both the stimulus frequency components have been filtered out. This was achieved by frequency domain editing followed by inverse fourier transformation. In the model the non-stimulus waveform burst is centered on input nodes. In the gerbil there is a 1-2ms delay, in agreement with fig 6. Further editing produces (d), the waveform of the lower sideband signal only. For the gerbil (c) and (d) are similar since the upper sideband is small. Note the wave period for gerbil changes from long to short during the emission burst- ie negative frequency dispersion. In the model, the period is more steady. Demodulation of the lower sideband signals with respect to F1 yields a complex modulating function. The modulus of this function (e) follows the envelope of the single sideband emission bursts and is similar in model and gerbil. The arguement (f) relates to phase modulation. A steep gradient implies a frequency shift. The gerbil data has steep gradients corresponding to substantial period changes through the burst, accounting for the wide lower sidebands. The model shows only small phase modulation. The negative frequency dispersion in gerbil could be due to reverse cochlear transmission from a point. If so the internal burst would be very short indeed.

The true difference between gerbil and model is revealed by the vector angle of the demodulated sideband (7f). It shows rapid change during the burst. This is another manifestation of the negative frequency dispersion and causes a temporal broadening of the otoacoustic burst. Traced back to the origin in the cochlea , we predict a very sharp burst (<1ms) of mechanical energy is transmitted at deep excitation nulls during two tone stimulation- hence the wide otoacoustic sidebands. It should be possible to use the technique here to test proposed nonlinear models against experimental data.

ACKNOWLEDGEMENTS

This work was supported by the MRC and the Iron and Steel Trades Insurers Group.

REFERENCES

Brown, A. M. and Kemp D.T., "Suppressability of the 2f1-f2 stimulated acoustic emissions in gerbil and man." Hearing Research. 13,29-37, 1984

Kemp D.T. and Chum R., "Observations on the generator mechanism of stimulus frequency acoustic emissions- two tone suppression", in Psychophysical, physiological and behavioural studies in hearing, Eds G. van den Brink and F.A.Bilsen., Delft University Press, pp 34-41 1980

Kemp, D.T. and Brown A.M., "A comparison of mechanical nonlinearities in the cochleae of man and gerbil form ear canal measurements". in Hearing-Physiological Bases of and Psychophysics, Ed's R.Klinke and R. Hartman, Springer-Verlag, New York, NY, pp. 82-88 1983

Kemp, D.T. and Brown A.M., "The interpretation of acoustic distortion responses", Abstract, British Journal of Audiology 18, p256 1984. (Also submitted in full).

Kim,D.O, Molnar,C.E. and Mathews J.W., "Cochlear mechanics: Nonlinear behavior in two tone responses as reflected in cochlear nerve fibre responses and in ear canal sound pressure", J. Acoust. Soc. Am. 67, 1704-1721, 1980

CHARACTERIZATION OF CUBIC INTERMODULATION DISTORTION PRODUCTS IN THE CAT EXTERNAL AUDITORY MEATUS

P.F. Fahey* and J.B. Allen

AT&T Bell Laboratories, Murray Hill, NJ 07974
*University of Scranton, Scranton, PA 18510

1. INTRODUCTION

We have made extensive measurements of the lower frequency cubic intermodulation distortion product sound pressure in the external auditory meatus of cats. This distortion product has a frequency, $f_D=2f_1-f_2$, where f_1 is the lower frequency of two pure tone input sounds. A_1 and A_2 are the amplitudes of the pure tone inputs at frequencies f_1 and f_2 respectively. The evidence is substantial that this distortion product is generated in the cochlea and that it propagates back out into the ear canal (e.g., Kim, 1980). Our data is presented here as Thevenin equivalent source pressures rather than as raw ear canal pressures. The Thevenin equivalent source pressure is a measurement of the properties of the source (assuming that a system is approximately linear in its overall characteristic) while the raw ear canal pressure depends upon not only the source but also the impedances of the sound delivery system and the animal's input impedance at the place of pressure measurement. (In animal studies, this is frequently the input impedance at the tympanic membrane.)

The measurements presented here have been organized in such a way as to present both new details in the distortion product database and also to show older observations in a new light. Generally, distortion product Thevenin equivalents (or the Thevenin equivalents normalized by A_1) measured as a function of f_D with f_2 held constant were found to be the most fruitful organization of most of our data.

In particular, our new observations are that:

(a) The envelopes of the distortion product versus f_D generally show saturation in the peak region at the higher input levels and as the input level decreases the envelopes scale approximately linearly when changing input sound pressure level by varying A_1 and A_2 together. The approximation to linearity is more exact if A_2 only is varied.

(b) There are very sharp nulls (Kim, 1980) that are within the overall envelope of the distortion product generation process and that these nulls shift to higher frequency of f_D as A_1 decreases.

The depths of the nulls are sensitive to ratio of A_2 to A_1. These nulls are responsible for the appearance of the low frequency tails and for the sharpening of the low frequency side of the bandpass characteristic. As A_2/A_1 becomes smaller so do the depths of the nulls. If A_2 only is varied, the nulls are stable in frequency.

(c) If the phase of the distortion product is plotted versus f_D after subtracting out a constant time delay, the phase slope will be approximately zero until f_D is near f_2. Near f_2 the phase goes through a pi phase shift. The time delay that must be subtracted to obtain the flat phase curve decreases as f_2 increases. This simple phase behavior was not observed in data acquired while f_1 was held constant.

(d) It is generally observed that if A_1 is varied such that the sum of A_1 and A_2 is held constant at fixed frequencies, f_D, f_1, f_2, the distortion product is a maximum near $A_2/A_1 = 0.5$. This behavior would be expected of a cubic law generator.

Older observations that we report with new detail are as follows:

(e) We show in a much more complete and detailed way than has previously been shown that distortion product generation is very sensitive to localized cochlear damage (Zurek, et al, 1982).

(f) We also show that the overall impression of the distortion product generation process is that it is bandpass in nature for f_2 greater than 3 kHz and that the maximum occurs for f_2/f_1 greater than 1.0. The maximum in the distortion product generation is $1.1<f_2/f_1<1.5$ (e.g., Wilson, 1980). Generally, as f_2 becomes greater the f_2/f_1 ratio at the maximum of the distortion product becomes smaller.

(g) The unusual level dependence of distortion product generation at a fixed frequency of f_D that others have observed (Kim, et al, 1980) is explained by the sharp moveable nulls that we mentioned above in (b).

2. METHODS

The data in this report were selected from a database containing measurements on more than 20 healthy adult cats. These animals were free of middle and external ear infections. Surgery was either as described in Allen (1983) for auditory nerve experiments or was limited to removal of the pinna and very wide opening of the bulla and septum. The experimental apparatus and the laboratory facilities were as described in Allen (1983). The ear canal pressure was measured with a Bruel and Kjaer one-half inch microphone of sensitivity 45 microvolts/Pascal that was at the end of a 2 cm probe tube. We take the ear canal pressure amplitude at a given frequency to be the amplitude of the Fourier transform at that frequency.

The Thevenin equivalent source pressure (at frequency, f_D), P_{th}, can be determined from a measurement of the ear canal pressure, P_{ec}, at frequency f_D, using the following expression:

$$P_{th} = (Z_{in} + Z_s)P_{ec}/Z_s$$

Where Z_{in} is the input impedance at the place of the pressure measurement and Z_s is the impedance of the sound delivery system. The measurement of these impedances is detailed in the paper by Allen in this volume. In that paper are figures showing the impedance of the sound delivery system during one set of experiments and also the input impedance of one of our cats. Since we usually measure the input impedance within 4 mm of the tympanic membrane, the input impedance is approximately the input impedance at the tympanic membrane. While the cat auditory periphery is obviously a nonlinear sound system, the input impedance is linear to about one part in one hundred, so that we approximate the system as a linear system for the purpose of calculating the Thevenin pressure. Matthews (1983) has shown the importance of taking the sound delivery system impedance into account when interpreting measurements of ear canal pressure. Features in the ear canal pressure can be due to either the impedance of the animal or the impedance of the sound system or both. By expressing the distortion product data as Thevenin equivalent pressures, we compensate for the impedance of the sound system and of the animal preparation. This kind of compensation can be important to resolve details in the distortion product itself and is important for comparison of data across animals and across laboratories.

3. RESULTS

Fig. 1 shows a family of Thevenin equivalent source pressure curves as a function of f_D for several different values of f_2. The most obvious features are the bandpass nature of the generation mechanics and the low frequency tails (for the higher values of f_2). Fig. 2a shows a family of curves for a given f_2 as a function of f_D, where the changing parameter is the input sound pressure level (A_2=A_1). Notice that there are sharp nulls within the overall bandpass envelopes and that generally these nulls move to higher frequencies as the input level decreases. This movement of the nulls has been observed in all animals studied. If one imagines a line drawn through this set of

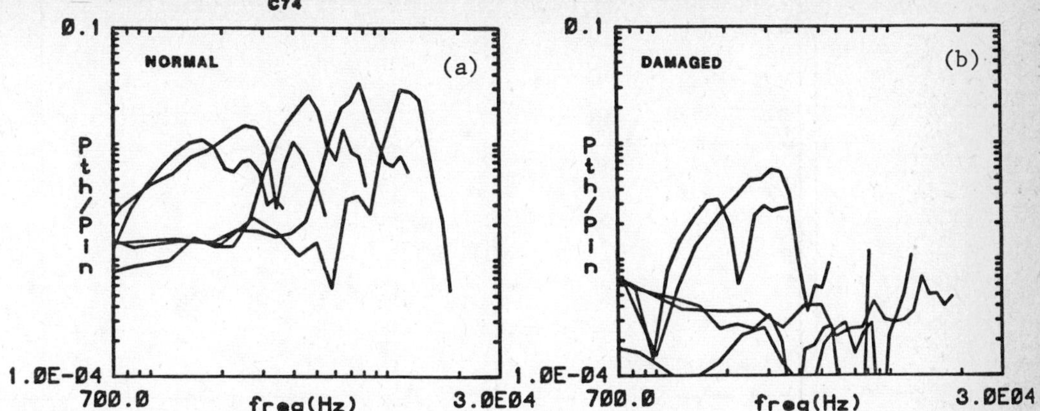

Figure 1. The Thevenin equivalent pressure, Pth, normalized by the input sound pressure level, Pin, is plotted as a function of fD for several different values of f2. (a) shows the response of an intact cochlea and (b) shows a cochlea damage with Na-pentobarbitol injection. A1=A2=94 dB SPL (re 20 micro Pascal).

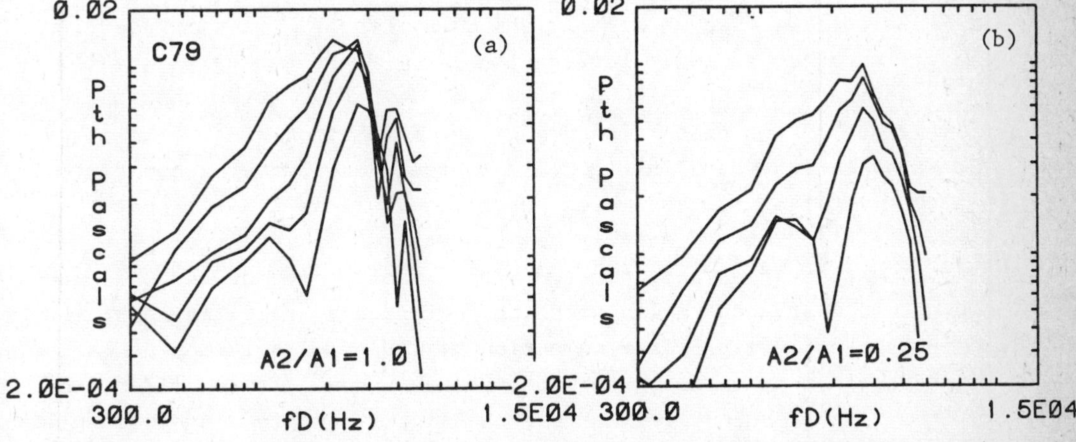

Figure 2. (a) Pth in Pascals is plotted vs fD for several different levels of input pressure. The highest curve corresponds to A1=A2=94 dB SPL. In each succeeding curve, the level of A1 and of A2 is 3 dB less than in the preceding curve. A2/A1=1. (b) Same as (a) except A2/A1= 0.25.

curves at a constant f_D, one can see that the Thevenin equivalent pressure could have unusual level dependence if the line intersected one or more nulls. In Fig. 2b, we change the ratio of A_2 to A_1 and plot the same kind of data as Fig. 2a. Decreasing the A_2/A_1 ratio removes the highest frequency null. Generally, decreasing A_2/A_1 also decreases the depths of the nulls. Fig. 3 details the change in P_{th} vs f_D when only the amplitude of A_2 is decreased (A_1 held constant).

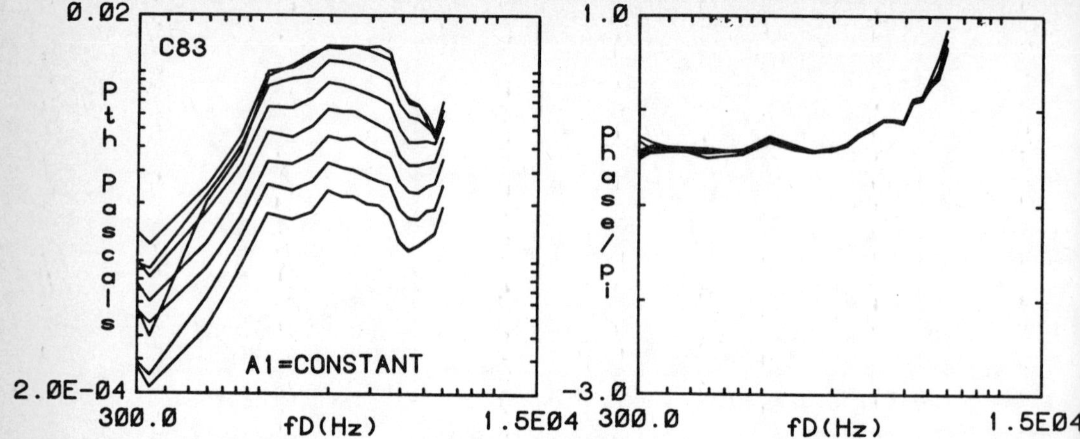

Figure 3. Same as Fig. 2 except A2 only is decreased by 3 dB SPL from one curve to the next. The highest input was A1=A2=94 dB SPL.

Figure 4. The phase data of Fig. 3 plotted as a function of fD after the subtraction of a time delay of 0.7 ms.

Here we observe that at high levels of A_2, the P_{th} is saturated in level; whereas, at lower levels of A_2, P_{th} decreased very linearly in A_2. Notice that the nulls do not shift. In Fig. 4, we show the phase dependence of the distortion product as a function of f_D at f_2 equal constant for the data in Fig. 3. In the phase curve, a constant delay time of .7 msec has been subtracted. It is evident that the phase is flat until f_D approaches the maximum in the level plot. If the same type of data is plotted with f_1 held constant the phase curve is not so flat, especially at the higher levels. Fig. 1b shows the effects of a 40 microliter injection of 1.0 gm/ml solution of Na-pentobarbitol in water through the round window. It is evident that in the high frequency region that the distortion product has completely disappeared. The distortion product in the high frequency region is at the level of the noise. Both the bandpass part and the low frequency tail part of the distortion product generator have been compromised. The cochlear microphonic (CM) measured with a round window electrode is also markedly decreased after Na-pentobarbitol injection. The most striking feature of the damage study is the sensitivity of the distortion product generator to localized damage. Indeed, we have been able to do damage that diminishes the highest frequency P_{th} vs f_D curve only.

4. DISCUSSION

Since a distortion product of frequency equal to $2f_1-f_2$ can be generated by a cubic or higher order odd nonlinearity, one would expect that the distortion product signal should exhibit at least a cubic level dependence, i.e., A_d would be expected to be proportional to $(A_1)^2A_2$. It has been observed in cat ear canal pressure (Kim, 1980) that the distortion product at fixed f_D shows a level dependence that is more linear than cubic and that has an unexpected nonmonotonic level dependence when both primaries are varied in level. The nonmonotonic feature may be interpreted in terms of our Fig. 2. That is, since the nulls shift frequency with input level, data taken at one frequency will show unusual level dependence as nulls move into or out of that frequency. Note that the envelopes of the distortion product curves scale almost linearly with sound pressure level. Hence, the cubic distortion product does not display cubic power law growth but rather an almost linear growth. This approximation to linearity is well known both in human psychophysical data (Goldstein, 1967) and animal auditory nerve data (e.g., Fahey and Allen, 1985). Indeed, if the data of Fig. 3 is normalized by the input A_2, the linearity is almost exact.

The reason for the apparent linearity of the cubic distortion product is certainly not known at present, but it has been suggested that the source of the distortion product is an essential nonlinearity (Goldstein, 1967) that is in place even at very low levels and necessarily that this nonlinearity is not expandable as a power series since the nonlinearity both exists at low level and that its relative strength is constant with level. An alternative to this view would be that the nonlinearity that generates the distortion product is indeed cubic but that it follows the nonlinearity that generates two tone suppression. Given what is known about two tone suppression in cats, this alternative is also plausible. It should be possible to distinguish these two possibilities by studying the level dependence on A_2 of the distortion product at $3f_1-2f_2$. Preliminary results shown in Fig. 5 of the dependence of this distortion product vs f_D with A_1= constant and A_2 varied in 3 dB steps indicate that the distortion product scales in level as $(A_2)^2$.

We have found some adherence to a cubic scaling rule in one aspect of our data. If we change the ratio of A_2/A_1 while leaving A_1+A_2=constant, then the maximum in distortion product production should occur at A_2/A_1=0.5 (i.e., the distortion product envelopes should be greatest at this ratio) if the mechanics are cubic. Indeed, our data on these ratio changes is consistent with cubic mechanics.

It is also evident from our level dependent data that the nulls in envelopes have frequencies that are dependent upon both the

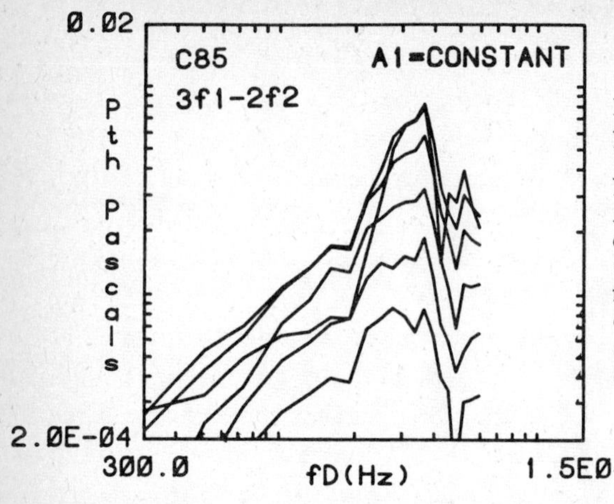

Figure 5. The Thevenin equivalent pressure, Pth, of the 3f1-2f2 distortion product vs fD. The highest curve corresponds to A1=A2=94 dB SPL. In this family of curves, A2 decreases in 3 dB step while A1 is constant. The maximum occurs at the same value of fD as the maximum in the 2f1-f2 distortion product and not at the same value as the f2/f1 ratio.

overall level and on the ratio of A_2/A_1. Zwicker (1981) attributes the nonmonotonic behavior of the distortion product to an interference pattern that results from the superposition of wavelets (produced at the many points of generation on the basilar membrane). The extent of this wavelet production is level dependent in his model of distortion product production. Hall (1974), on the other hand, suggests that the nonmonotonic behavior of the distortion product is due to reflections within the auditory periphery -- these would not be expected to be level dependent.

5. CONCLUSION

We have measured distortion products detected in the ear canal that are rich in features. The phase data and the lability to cochlear insult affirm again what has already been shown by others -- that the distortion products originate within the cochlea. The cochlear damage studies show that the distortion products are correlated with local cochlear damage. The ear canal distortion products are substantial in experimental animals, and since they are sensitive to cochlear health, the potential exists for their use as a control and as a diagnostic measure in studies that involve the opening of the cochlea and its subsequent manipulation.

Finally, we believe that the use of Thevenin equivalent pressures rather than ear canal pressures will constitute a useful protocol in ear canal pressure measurements. Thevenin equivalent pressures are much closer to fundamental properties of the cochlea than the raw ear canal pressures and they provide a ready means of comparing data across laboratories and across different preparations.

REFERENCES

Allen, J.B., "Magnitude and Phase-frequency Response to Single Tones in the Auditory Nerve." J. Acoust. Soc. Am., 73, 2071-2092, 1983.

Fahey, P.F. and Allen, J.B., "Nonlinear Phenomena as Observed in the Ear Canal and at the Auditory Nerve." J. Acoust. Soc. Am., 77, 599-612, 1985.

Goldstein, J.L., "Auditory Nonlinearity." J. Acoust. Soc. Am., 41, 676-689, 1967.

Hall, J.L., "Two-tone Distortion Products in a Nonlinear Model of the Basilar Membrane." J. Acoust. Soc. Am., 56, 1818-1828, 1974.

Kim, D.O., "Cochlear Mechanics: Implications of Electrophysiological and Acoustical Observations." Hearing Res., 2, 297-317, 1980.

Kim, D.O., Molnar, C.E., and Matthews, J.W., "Cochlear Mechanics: Nonlinear Behavior in Two-Tone Responses as Reflected in Cochlear-Nerve-Fiber Responses and in Ear-Canal Sound Pressure." J. Acoust. Soc. Am., 67, 1704-1721, 1980.

Matthews, J.W., "Modeling Reverse Middle Ear Transmission of Acoustic Distortion Signals." In Mechanics of Hearing, pp. 11-18. Editors: E. deBoer and M.A. Viergever. Martinus Nijhoff Publishers, Delft University Press, 1983.

Wilson, J.P., "The Combination Tone 2f1-f2 in Psychophysics and Ear Canal Recordings." In Psychophysical, Physiological and Behavioral Studies in Hearing, pp. 43-58. Editors: G. Van den Brink and F.A. Bilsen. Delft University Press, Delft, 1980.

Zurek, P.M., Clark, W.W., and D.O. Kim, "The Behavior of Acoustic Distortion Products in the Ear Canal of Chinchillas with Normal and Damaged Ears." J. Acoust. Soc. Am., 72, 774-780, 1982.

Zwicker, E., "Cubic Difference Tone Level and Phase Dependence on Frequency and Level of Primaries." In Psychophysical, Physiological and Behavioral Studies in Hearing, pp. 268-273. Editors: G. Van den Brink and F.A. Bilsen. Delft University Press, Delft, 1981.

ACOUSTIC OVERSTIMULATION REDUCES $2f_1-f_2$ COCHLEAR EMISSIONS AT ALL LEVELS IN THE CAT

Michael L. Wiederhold[1,2,3], Judy W. Mahoney[3] and Dean L. Kellogg[2]
Division of Otorhinolaryngology[1] and Department of Physiology[2],
University of Texas Health Science Center, and
Audie L. Murphy Memorial Veterans' Hospital[3], San Antonio, TX 78284

ABSTRACT

The cubic distortion product, $2f_1 - f_2$ (DP), was recorded in the cat ear canal as the levels of two primary tones (f_1=4.0, f_2=5.2 kHz here) were varied from approximately 40 to 90dB SPL. The DP level grows as the cube of the primary level at low and high levels, and usually exhibits a dip near L_1 = 65dB SPL. DP growth functions and N_1 amplitude-level functions were compared before and after exposure to either 4 kHz pure tones or to 500 Hz octave bands of noise. Pure-tone exposure produced a dose-dependent shift in the N_1 growth whereas the low-frequency noise exposure produced widely varying shifts in N_1 functions. In either case, however, changes in DP were correlated with changes in the neural responses. The level of the DP was compared before and after exposure at 2 levels below and one above the dip at L_1 = 65dB SPL. For all of these measures, the decrease in DP level was significantly correlated with the shift in the N_1 growth function. At both low and high primary levels, decrease in DP grows approximately as the square of the N_1 shift for shifts greater than 15dB.

INTRODUCTION

Several recent studies have shown that nonlinearities in cochlear mechanics, evidenced by acoustic intermodulation-distortion products in the external ear canal, can be altered by exposure to high-level acoustic stimuli(2,5,6). If the changes in distortion products (DP) caused by acoustic trauma are highly correlated with the degree of sensorineural hearing loss produced by the exposure, measurement of DP in the external ear canal might be considered as an objective and relatively noninvasive diagnostic test for noise-induced hearing loss. The data presented here were obtained to test the feasibility of such an application. In anesthetized cats, level and frequency content of acoustic exposures were varied, and changes in neural responses (N_1) and ear canal DP after these exposures were compared.

METHODS

Healthy adult cats, free of external- or middle-ear pathology, were anesthetized with Dial in urethane (85 mg/kg). Both pinnae were removed and approximately 5 mm of cartilagenous ear canal retained to form an acoustic seal over the tip of the hollow earbar portion of a closed acoustic system. For recording sessions, the animal was inside a double-walled, electrically-shielded sound-attenuating chamber. Electrical responses to filtered clicks (FC) were recorded with a wire electrode adjacent to the round-window in the opened bulla. Filtered clicks were generated by passing a square pulse through a 1/3-octave filter (48dB/octave roll-off). The duration of the pulse was one-half of the period of the filter's center frequency. Cochlear microphonic and N_1 components were measured from the average of 500 to 1,000 responses. The sound source was a Bruel & Kjaer 4144 1" condenser microphone. The sound field near the tympanic membrane (TM) was measured with a B & K 4134 1/2" microphone through a probe tube projecting to within 3 mm of the TM. Acoustic transmission properties of the probe tube were determined by comparing the spectrum of the sound pressure measured through the probe tube with that at the end of the earbar in an acoustic coupler when the source was driven with white noise. All measurements of SPL were made with a Nicolet 660B fast-Fourier-transform spectrum analyzer. For determination of primary-tone levels (L_1 and L_2) and DP above approximately 15dB SPL, the analyzer displayed 400 lines (with 50 Hz spacing) from 50 Hz to 20 kHz. For measurements at lower levels, the expansion mode was employed, in which an 80 Hz band, with 0.2-Hz line spacing, including the spectral component of interest, was analyzed. This decreased bandwidth reduced the noise floor of our measurements to approximately -25dB SPL.

RESULTS

When two tones are presented simultaneously, we are able to detect the cubic distortion product, $2f_1 - f_2$, in the external ear canal for primary-tone levels above 35-50dB SPL. For primary tones 4.0 and 5.2 kHz, as for most of the data presented here, and of equal SPL in the ear canal ($L_1 = L_2$), the growth of the DP at 2.8 kHz with increases in L_1 and L_2 varies between animals, as illustrated in the left panel of Figure 1. An intermodulation-distortion product can only be generated if both primary tones produce a significant response at the same point along the basilar membrane (BM). Since the amplitude of BM displacement in response to a single tone decreases from its maximum more rapidly in the apical direction, the location of significant interaction between two tones will be closer to the characteristic location of the higher-frequency tone (f_2) and the amount of displacement produced by f_1 at that point will be less than that produced by f_2 (3). In an attempt to more nearly equalize the mechanical responses produced by f_1 and f_2 at the point of maximum interaction, L_1 was increased 5dB relative to L_2. Under these conditions, the shape of the DP growth functions are more repeatable, as illustrated in the right panel of Figure 1. Note that at low levels the curves for these five normal animals all grow with a slope of 3, reach a local maximum between 55 and 60dB SPL, exhibit a sharp dip between 60 and 65dB SPL and then resume growth with a slope of 3 before approaching saturation at and above 80dB SPL.

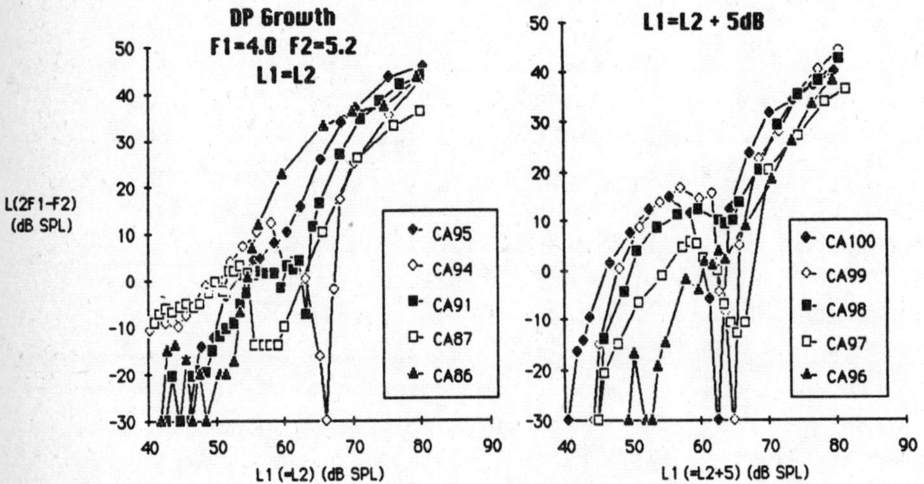

Figure 1. Growth of $2f_1-f_2$ distortion product level with increase in level of primary tones for ten normal cats. $L_1=L_2$ in left-hand panel and $L_1=L_2+5$dB in right-hand panel. NOTE: For this and other figures, DP levels plotted as -30dB SPL represent measurements for which no DP spectral peak could be distinguished from noise.

Figure 2 illustrates an example of N_1 and DP level functions before and after exposure to a 500 Hz octave band of noise at 110db SPL for 60 min. For 500 Hz filtered noise exposure, N_1's were measured with 4 kHz FC's, since previous work has indicated discrete lesions near the 4 kHz location with similar exposures (1). At low FC levels, where the pre- and post-exposure N_1 amplitude-level functions are approximately parallel, the post-exposure function is shifted to higher levels by an average of 36dB in the case of Fig. 2. This average shift is the measure we have used to characterize effects of exposure on auditory-nerve responses. The right panel of Figure 2 illustrates the effect of the exposure on the DP level function. Note that for all L_1 below 68dB SPL, no DP could be measured after exposure. For L_1 from 70 to 80dB SPL, the level of the DP is substantially reduced after exposure. We did not measure DP for primaries above 80dB SPL before exposure, to avoid overstimulation effects from the measurement itself, since each DP level determination required approximately 1 min of exposure.

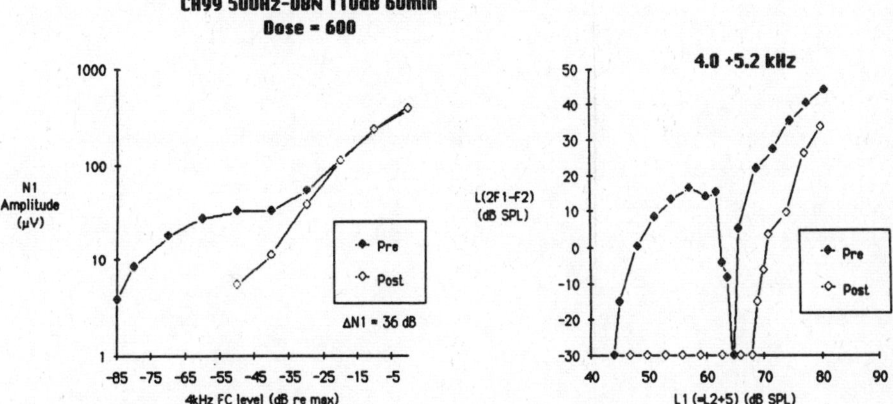

Figure 2. Round-window recorded N_1 amplitude in response to 4 kHz filtered clicks (FC) vs. FC level (in dB attenuation from maximum level) before (filled diamonds) and after (open diamonds) exposure to low frequency noise (left panel). Average shift at low levels caused by exposure is 36dB. Right panel: level of DP vs. level of primary tones (L_1=L_2+5dB) before and after exposure. Cat CA99.

In order to test for correlations between the effects of overstimulation on neural responses and on DP, we exposed cats to different levels and durations of 4 and 8 kHz pure tones as well as to 500 Hz octave bands of noise. Since exposure to pure tones produces maximum damage at frequencies up to a half-octave above the exposure, effects of 4 kHz exposure were tested with 5 kHz FC's and 8 kHZ exposure with 10 kHz FC's. Figure 3 illustrates the shifts in the N_1 amplitude-level functions for 9 cats exposed to 4 kHz, 5 exposed to 8 kHz tones, and 9 exposed to 500 Hz octave bands of noise. The exposure "dose" is expressed in "energy" units, with unit exposure being 90dB SPL for 10 min. For every 3dB increase in total SPL (for all spectral configurations) or doubling in exposure duration, the exposure dose is doubled. Note that for the two pure tone conditions, the N_1 shift grows approximately linearly with exposure dose and that equal doses at 4 and 8 kHz produce approximately equal shifts. For the 500 Hz octave bands of noise, however, within a limited range of exposures (dose from 128 to 600), N_1 shifts vary from 0 to 36dB.

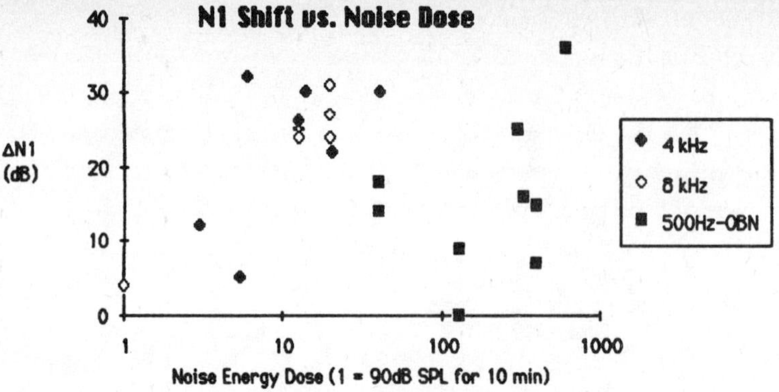

Figure 3. Shift in N_1 level functions produced by exposure to 4 kHz (closed diamonds), 8 kHz (open diamonds) and 500 HZ octave band of noise (closed squares). N_1 measured with 5 kHz FC's for 4 kHz exposure, 10 kHz FC's for 8 kHz exposure and with 4 kHz FC's for 500 Hz filtered-noise exposure. Measurement of N_1 shift and exposure dose discussed in text.

The data shown in Figure 2 are among the largest effects observed in this series on both N_1 and DP responses. The example in Figure 4 shows a much smaller (7dB) N_1 shift and correspondingly smaller effect on DP. Note that even with this reduced effect of exposure, DP level is reduced by approximately 5dB both below and above the dip. In this case the dip itself is greatly exaggerated after exposure. In other cases the dip has been shifted by up to 5dB and consequently, at some primary levels, DP after exposure can be increased in level over the pre-exposure value.

As can be seen in Figure 3, the N_1 shifts produced by our different exposure configurations ranged from 0 to 36dB. To test for correlation between the effects of exposure on neural responses and on DP, several measures of changes in DP were employed. For those DP growth functions for which a clear local maximum or "hump" was apparent below the dip, the decrease in DP level was measured at the level of the hump before exposure. The average primary level (L_1) at which the hump occurred, was 57dB SPL, so the decrease at this fixed level was measured and, to evaluate the effect well above the dip, DP decrease was measured at 75dB SPL. For those cases in which the DP was decreased to the point that it could no longer be detected, the postexposure value was arbitrarily set to -25dB SPL, the noise floor of our measurements. All three of these measurements correlate well with the N_1 shifts. Figure 5 illustrates scatter plots for the DP decrease at the level of the hump (left panel) and at L_1=75dB SPL in the right panel. Of the three measures tested, DP decrease at the hump correlates best with N_1 shifts.

For the data in Figure 5, the correlation coefficient for DP reduction at the hump and N_1 shift is 0.81 while for L_1=75dB SPL, it is 0.75. In both cases consistent DP decreases occur for N_1 shifts greater than 15dB. The amount of DP decrease is not greatly lower for measurements at 75dB SPL, compared to those measured at the hump. A similar relationship, with a correlation coefficient of 0.76 was obtained at L_1=57dB SPL. Thus it appears that overexposure produces a somewhat more consistent decrease in DP below the level of the dip, but that the decrease achieved at 75dB SPL is similar to that seen at low levels. Note also that regardless of whether the exposure was a pure tone or low-frequency noise, the amount of DP decrease predicts the amount of neural shift reasonably well.

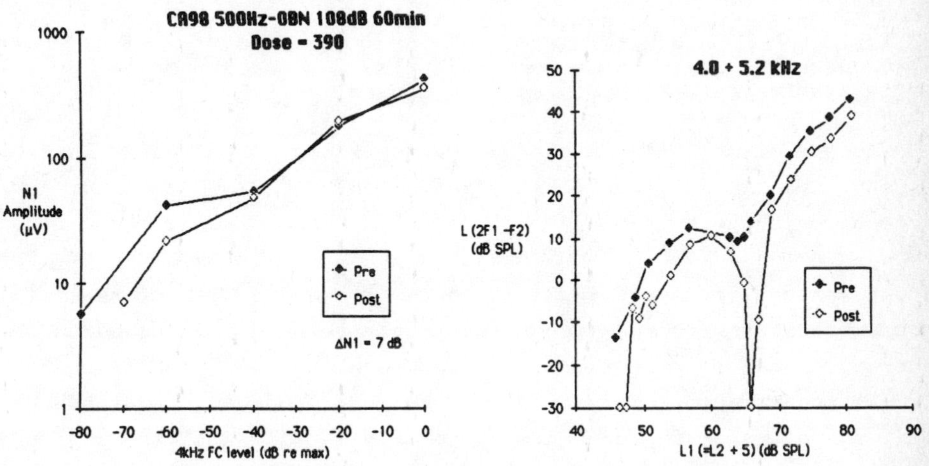

Figure 4. N_1 and DP level functions before and after exposure to 500 Hz octave band of noise. Same as Figure 2, except noise at 108dB SPL. Cat CA98.

DISCUSSION

The growth in the cubic DP with increasing level of primaries bears some similarities to that seen by Rosowski et al., in that DP level grows as the cube of primary levels at low and high levels with a transition region between the two ranges. They modelled their results with several components producing cubic distortion (6). If such a model applies to our cat data, the sharp dips we observe suggest that two components must be of nearly equal amplitude and opposite phase to produce cancellation for primaries near 65dB SPL. Whereas Rosowski's data indicate that the dominant nonlinearity above the plateau is not vulnerable to moderate

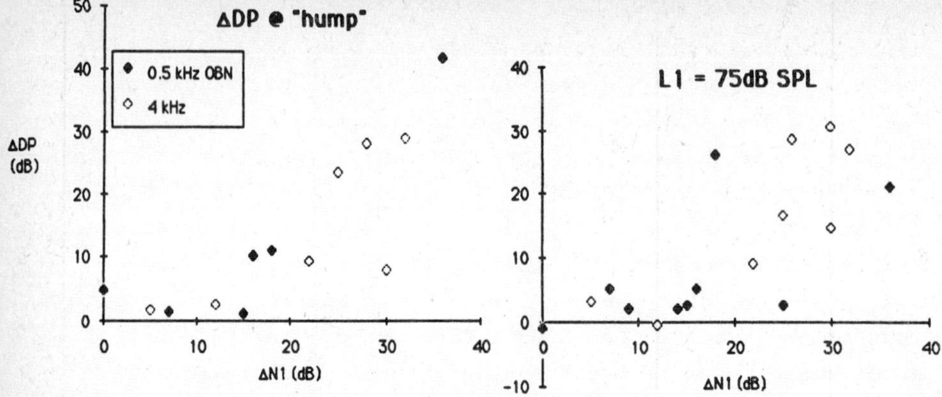

Figure 5. Scatter plot of decrease in DP level caused by exposure to 500 Hz octave band of noise (closed diamonds) or 4 kHz tones (open diamonds) plotted against shift in N_1 amplitude-level function. Left panel: decrease measured at level of primaries which produced hump in DP growth curve before exposure. Right panel: DP at L_1=75dB SPL.

acoustic overstimulation in the alligator lizard, our data indicate that similar exposures do significantly reduce both the low- and high-level $2f_1-f_2$ DP in the cat.

The effects of overstimulation on N_1 responses illustrated in Figure 3 are compatible with the suggestion that threshold shift produced by continuous tones is proportional to the total energy of the exposure (7). However, the effects (measured with 4 kHz FC) produced by the 500 Hz octave bands of noise do not correlate nearly as well with the exposure energy. This is compatible with the finding of discrete lesions near the 4 kHz region of the chincilla BM after similar exposures (1). Presumably in those cases in which we find large N_1 shifts, a lesion was produced at the 4 kHz location. However, in those cases where high-level and long durations of noise only produced a small N_1 shift (e.g., Fig. 4), if a lesion was produced, it was somewhat remote from the 4 kHz place. It is noteworthy that those high-dose exposures which failed to produce large N_1 shifts also failed to produce large decrements in DP, indicating that local damage at the site of the primaries is needed to strongly affect the cochlear nonlinearities. Although the correlations depicted in Figure 5 exhibit more scatter than would be desired for a diagnostic tool, they do suggest that with further improvement in technique such an application might be feasible. These data do not contradict cautionary conclusions noting a lack of correlation between the physiological state of the cochlea and DP levels (4,8). We too, find that the DP decays very slowly after death, especially when high-level primaries are used. Similarly, with high-level primaries, distortion in the sound-generating system can obscure DP of cochlear origin so that DP level can remain high after noise trauma. Thus, it seems clear that if ear-canal DP recording is to be considered for diagnostic uses, the most useful measurements are likely to be at low stimulus levels. Signal processing techniques such as the digital expansion used here to achieve a narrow bandwidth and new, low-noise microphones may make this realizable for human use.

[Supported by Veterans Administration Medical Research funds.]

REFERENCES

1. Bohne, B. A., and Clark, W. W., "Growth of Hearing Loss and Cochlear Lesion with Increasing Duration of Noise Exposure." In New Perspectives in Noise-Induced Hearing Loss, Ed. by R. P. Hamernik, D. Henderson and R. Salvi, Raven Press, New York, pp. 283-302, 1982.

2. Dolan, T. G., and Abbas, P. J., "Changes in the $2f_1 - f_2$ Acoustic Emission and Whole-nerve Response following Sound Exposure: Long-term Effects." J. Acoust. Soc. Amer. 77, pp. 1475-1483, 1985.

3. Fahey, P. F., and Allen, J. B., "Nonlinear Phenomena as Observed in the Ear Canal and at the Auditory Nerve." J. Acoust. Soc. Amer. 77, pp. 599-612, 1985.

4. Schmiedt, R. A. and Adams, J. C., "Stimulated Acoustic Emissions in the Ear Canal of the Gerbil." Hear. Res. 5, pp. 295-305, 1981.

5. Siegel, J. H., and Kim, D. O., "Cochlear Biomechanics: Vulnerability to Acoustic Trauma and Other Alterations as Seen in Neural Responses and Ear-Canal Sound Pressure." In New Perspectives in Noise-Induced Hearing Loss, Ed. by R. P. Hamernik, D. Henderson and R. Salvi, Raven Press, New York, pp. 137-151, 1982.

6. Rosowski, J. J., Peake, W. T., and White, J. R., "Cochlear Nonlinearities Inferred from Two-Tone Distortion Products in the Ear Canal of the Alligator Lizard. Hear Res 13, pp. 141-158, 1984.

7. Ward, W. D., and Turner, C. W., "The Total Energy Concept as a Unifying Approach to the Prediction of Noise Trauma and its Application to Exposure Criteria." In New Perspectives in Noise-Induced Hearing Loss, Ed. by R. P. Hamernik, D. Henderson and R. Salvi, Raven Press, New York, pp. 423-435, 1982.

8. Zurek, P. M., Clark, W. W., and Kim, D. O., "The Behavior of Acoustic Distortion Products in the Ear Canals of Chinchillas with Normal or Damaged Ears. J. Acoust. Soc. Amer. 72, pp 774-780, 1982.

HARMONIC ACOUSTIC EMISSIONS IN THE EARCANAL GENERATED BY SINGLE TONES: EXPERIMENTS AND A MODEL

Richard A. Schmiedt
Department of Otolaryngology
and Communicative Sciences
Medical University of South Carolina
Charleston, SC 29425

ABSTRACT

Harmonics of moderate-level single tones can be measured in the earcanals of gerbils with healthy ears. The same harmonic components measured in a cavity are as much as 20 to 40 dB below those measured in the earcanal. Earcanal harmonics, specifically the second and third, are generated largely within the cochlea and are physiologically vulnerable. Harmonic magnitudes rapidly decrease for primary frequencies above about 3 kHz. High-frequency masking tones can have dramatic suppression effects on the harmonic structure of low frequency primaries. For example, a 2 or 4 kHz masker at moderate levels can totally suppress the third harmonic of a 200 Hz tone. Many of these results can be simulated by a vector-sum model. In this model harmonic emissions can be thought of as very gross potentials that convey little information concerning specific regions of the cochlea.

1. INTRODUCTION

Acoustic correlates of some aspects of cochlear processing are present in the earcanal in the form of delayed or spontaneous emissions, or distortion products seen with continuous tones (Kemp, 1978; Kim, 1980; Zurek, 1981). It has been shown that these emissions require a metabolic substrate and decline in magnitude with cochlear injury (Anderson and Kemp, 1979; Kim et al., 1980; Schmiedt and Adams, 1981; Siegel and Kim, 1982).

The harmonic structure of single tones is an appealing choice to investigate, given the simplicity of the procedure relative to those in which an intermodulation distortion product, like $2f_1-f_2$, is measured. Only one primary tone is involved, and the frequency of the harmonic is always an integral multiple of the primary frequency. Moreover, if the distortion generators are localized to the region of the cochlea associated with the primaries, then it would seem that much of the information obtained with two tones could be had with one. As will be shown, this is not the case. A single primary seems to excite many harmonic generators within the cochlea.

2. METHODS

Mature gerbils (Meriones unguiculatus) with healthy ears were used in these experiments. All the animals were born and raised in a sound-controlled environment where the ambient sound levels were less than 35 dB SPL (A weighted, SPL referred to 20 μPa). Most of the surgical procedures and acoustic methods followed those of Schmiedt and Adams (1981). Briefly, gerbils were anesthetized with either sodium pentobarbital or a ketamine-xylazine mixture, their pinnas removed on the right or left side, and the bulla opened widely. The acoustic source was a Beyer DT-48 earphone with a passive electrical equalizer and an acoustic impedance matching section. When sealed to the earcanal, the variation in SPL was ± 5 dB over a frequency range of 100 Hz to 22 kHz. A Knowles microphone (EA-1842) with a 1 cm probe tube formed an integral part of an assembly holding the earphone. The assembly in turn attached directly to the head holder. All experiments were done in a heated, sound-isolated booth, and a round-window electrode was used to monitor to state of the cochlea and nerve.

Tonal stimuli were generated with low-distortion oscillators, attenuators, and gates, all under computer control. An acoustic calibration was done for each experiment so that earcanal SPL variations across frequency could be kept to within 1 dB. Harmonic components were measured on a wave analyzer with a 90 dB dynamic range. The analyzer was externally controlled by means of a phase-locked loop configuration that allowed accurate tracking at frequencies that were integer multiples of the primary frequency. A voltage from the analyzer proportional to the logarithm of the stimulus (or harmonic) intensity was digitized and corrected for the probe-tube characteristics to yield the actual earcanal SPL. Thus, a system was devised to sweep a primary tone across frequency keeping its intensity constant, while measuring a harmonic component over a 90 dB dynamic range.

3. RESULTS

A. EXPERIMENT

Of main interest are the second and third harmonics of the primary tone. It is these harmonics that are most intimately related to the difference (f_2-f_1) and the cubic difference ($2f_1$-f_2) tones, respectively, assuming a simple polynomial nonlinearity. The second and third harmonic structures of 70 dB primary tones as measured in the earcanal of a gerbil are shown in Fig. 1. Note that the levels of the harmonics are plotted at the frequencies of the respective primary tones. The magnitude of the third harmonic is generally greater than that of the second, and the levels of both harmonics show peaks and dips as a function of primary frequency. The absolute harmonic levels are about 40-50 dB below the primary. The harmonics decline to the system noise floor (thin line) above about 3 kHz. Our noise

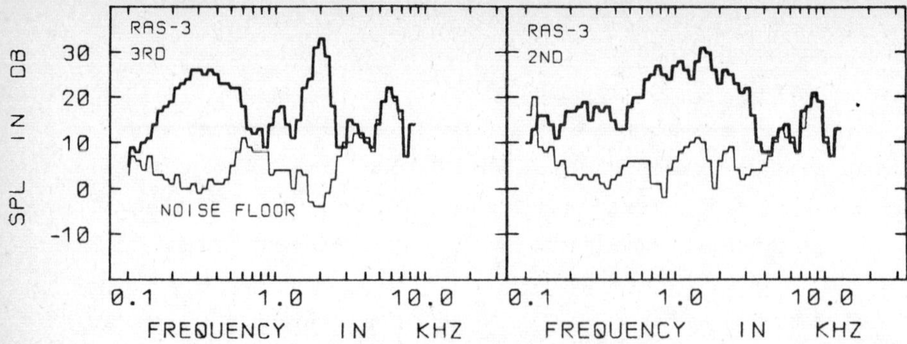

Fig. 1. Magnitudes of second and third harmonics of 70 dB SPL primaries in the earcanal of a gerbil. Harmonic magnitudes are plotted at the respective primary frequencies with bold lines. Thin lines indicate harmonic levels obtained from the same animal with a dead cochlea.

floor is always defined here as those harmonics levels existing in the earcanal of a live gerbil with a dead cochlea. The cochlea was poisoned by means of a KCl solution perfused into scala tympani.

Harmonic "signatures" like those in Fig. 1 vary somewhat between animals with regard to absolute levels and overall shape; however, the data shown are typical of the 25 gerbils we have studied. The magnitude of the third harmonic is often bimodal with a minimum at around 1 kHz, and peaks around 0.3 and 1.8 kHz. The second harmonic is more unimodal, peaking at around 1.8 kHz. It is of interest to note that the magnitude transfer function of the middle ear of the gerbil has its maximum between 1 and 2 kHz (Schmiedt and Zwislocki, 1977).

In hearing, one of the properties of the intermodulation distortion seen with two primary tones is that it seems to arise largely from the cochlear region specifically associated with the primaries. This is indicated, for example, in the pyschophysical studies of Smoorenburg (1972) and in the physiological studies of Siegel et al. (1982). Does this property also apply to the harmonic structure of single tones as measured acoustically in the earcanal? In other words, do the harmonics produced by a primary tone arise primarily from the cochlear place associated with that tone?

Tone-on-tone masking was used to ascertain some features of the frequency specificity of the harmonic structure of a single tone. Results from one such experiment are shown in Fig. 2. In this paradigm, the harmonic structure across frequency is measured with and without a second (masking) tone added to the primary tone in the earcanal. The masking tone was continuous, and its intensity was 70 dB SPL. The arrow in Fig. 2 indicates the

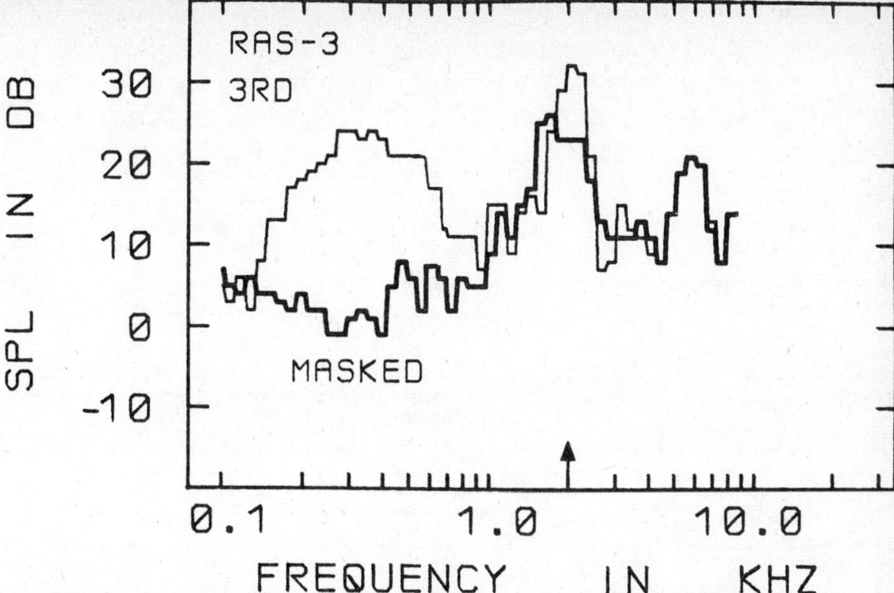

Fig. 2. Magnitudes of the third harmonic in the same animal as in Fig. 1 obtained with (bold line) and without (thin line) a continuous 2 kHz masker. The masker and primary levels were both 70 dB SPL. Note the downward spread of masking; i.e., the suppression of harmonics of primary frequencies much lower than that of the masker.

frequency of the masker. The thin line is the control (no masker) condition, the bold line is the masked condition. Note the suppression of the harmonics of low-frequency primaries by the high-frequency masker. Instances of such a downward spread of masking are unusual in hearing, especially at the moderate levels used in this study.

B. MODEL

A family of models that can explain many of the phenomena associated with acoustic emissions is one that assumes a distribution of nonlinearities spread over the length of the basilar membrane. A given nonlinearity generates distortion components as a function of the strength of the primary tone present at its input. The distortion components generated by a number of nonlinearities are allowed to add vectorially, and the resulting sum is proportional to what is measured acoustically in the ear canal. For convenience, we shall refer to this family as "vector-sum" models.

Vector-sum models attempting to describe nonlinear phenomena in hearing have been proposed and investigated by Schroeder (1969) and, most recently, by Zwicker (1980). Using a such model, Zwicker was able to successfully reproduce the shape of the intensity function of the cubic difference tone measured pyschophysically.

A block diagram of the model used in this study is shown in Fig. 3. The middle-ear of the gerbil is assumed to be a simple bandpass filter with regard to the transform between earcanal SPL and stapes velocity (Schmiedt and Zwislocki, 1977). The output of the middle ear drives an array of distortion generators, each with an assumed amplitude and phase

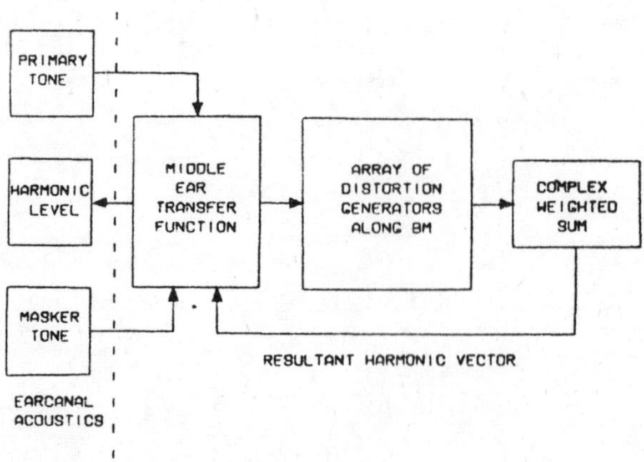

Fig. 3. Block diagram of vector-sum model. Middle-ear transfer function is assumed to be the same for signals to and from the cochlea.

output depending on its cochlear location and the primary frequency. The generators are independent of one another. A weighted complex sum is formed at the appropriate harmonic frequency and is retransformed by the middle ear to arrive at the harmonic SPL measured in the earcanal. The weighting factor is due to losses along the basilar membrane as the output of each generator travels to the point of summation. It is assumed that the loss is the same for forward and backward traveling waves and is proportional to the distance between the source and the summation point (see Zwicker, 1980). The summation point is

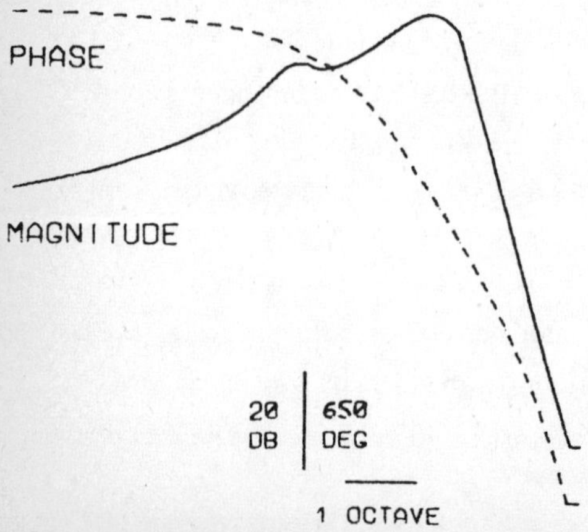

Fig. 4. Assumed magnitude and phase of an array of 256 generators of third harmonic distortion along the length of the cochlear duct. The place of lowest frequency is 125 Hz at the right, the highest is 32 kHz at the left. Maximum amplitude of the third harmonic is assumed to occur at the place of primary frequency (500 Hz in this case). The effects of a 2 kHz masker have been added to the magnitude and phase. The presence of the masker is indicated by the secondary maximum and an inflection in the magnitude and phase curves, respectively.

taken to be the point on the basilar membrane corresponding to the harmonic frequency. Figure 4 illustrates the magnitude and phase distribution of an array of 256 harmonic generators. The magnitude of each point is calculated from a complex pole-pair system as driven by the middle ear. It is assumed that the largest contribution of harmonics corresponds to the cochlear place associated with the driving primary tone. In the model the steepness of the apical slope can be varied from that of a normal resonance curve to one being a straight line slope of N dB/octave where N ranges from 25 to 1000 dB/octave. In Fig. 4 the slope is 50 dB/octave, and the curves represent a system with a single pair of complex poles having a normalized damping factor of 0.1. An apical plateau is added to reduce computation time, and the number of pole-pairs as well as the damping of the system can be changed. The weighting of the backward and forward traveling waves as well as the effects of adding a 2 kHz masking tone are included in the magnitude and phase curves shown in Fig. 4.

The phase profile of the generators assumes a minimum phase system of an appropriate number of pole-pairs. Also included in the phase calculations is a multiplier for the given harmonic number, the middle-ear phase, and the travel times of the primary tone as well as those of the resulting harmonic "wavelets" from each source.

The main effect of the masker is to disrupt the phase continuity of the harmonic generators. A specified fraction of the masker magnitude can be added to that of the primary-tone, but the magnitude effect is minimal on the efficiency of the masker with regard to suppressing the harmonics of low-frequency primaries. It is the addition of the masker phase to the primary phase profile that largely determines the amount of masking.

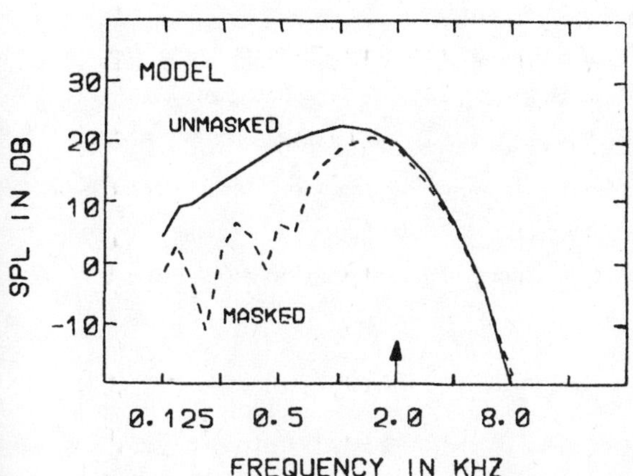

Fig. 5. The third-harmonic output of the vector-sum model with and without a 2 kHz masker. Masker frequency is indicated by the arrow. Model output peaks at the best frequency of the middle-ear and rapidly falls off above 3 to 4 kHz. The masker is most effective for low-frequency primaries, an effect primarily due to increased phase cancellation occuring among the distortion generators.

The output of the model follows the experimental results in terms of the general shape of the harmonic structure of a single tone swept in frequency. The model also qualitatively predicts the overall effects of adding a tonal masker. The output of the model with and without a masker is shown in Fig. 5. The model produces harmonic levels that have a maximum at the best frequency of the middle ear and fall off rapidly above about 3-4 kHz. Addition of the tonal masker typically suppresses the harmonics associated with primary tones lower in frequency to the masker. The above phenomena are robust in that changing the damping, the number of pole-pairs, or the weighting factor changes the shape of the curves, but not the general trend as shown in Fig. 5. Interestingly, shallow apical slopes of the distortion array yield the best suppression of harmonics at low frequencies.

4. DISCUSSION

The vector-sum model can successfully explain some of the more puzzling aspects of acoustic emissions. For example, the rapid decline in emissions magnitude above 3 kHz can be attributed not only to the frequency response of the middle ear, but also because higher frequencies are more apt to phasically cancel in such a model. A similar phase cancellation model was proposed by Whitfield and Ross (1965) to explain some of the phasic interactions apparent in the cochlear microphonic (CM).

The fact that all classes of acoustic emissions, evoked and spontaneous, are strongest for frequencies below about 3 kHz is consistent with a phasic model. The 3 kHz cut off seems stable irrespective of animal species and the attendent middle-ear transfer function, suggesting that phase cancellation predominates at frequencies above 3 kHz.

Another appealing aspect of this model is its explanation of the downward spread of masking seen with acoustic emissions. As shown here, a high-frequency masker can strongly suppress emissions associated with tones two or three octaves lower in frequency. This is hard to explain in terms of the usual traveling wave mechanics of the cochlea. A downward spread phenomenon is also seen with the cubic distortion product and its response to noise fatigue (Schmiedt, 1984, Fig. 10). After a high-frequency noise exposure the $2f_1-f_2$ component of the acoustic emissions is suppressed for low-frequency primaries even though the whole-nerve response to tone pips is normal in the same frequency range. Again, such an effect can be simply explained by a model incorporating a summation of activity from many sources.

5. Acknowledgements

I thank J.C. Adams, J.H. Mills, C.E. Smith, and J.J. Zwislocki for their helpful comments on this manuscript and their insights with regard to cochlear processing. N. Topham is responsible for the excellent job of typing the manuscript. This work was supported in part by NSF grant BNS 82-10233.

6. REFERENCES

Anderson, S. D. and Kemp, D. T. "The evoked cochlear mechanical response in laboratory primates." Arch. Otorhinolaryngol. 224, pp. 47-54, 1979.

Kim, D. O. "Cochlear mechanics: implications of electrophysiological and acoustical observations." Hear Res. 2, pp. 297-317, 1980.

Kim, D. O., Molnar, C. E., Matthews, J. W. "Cochlear mechanics: nonlinear behavior in two-tone responses as reflected in cochlear-nerve-fiber responses and in ear-canal sound pressure." J. Acoust. Soc. Am. 67, pp. 1704-1721, 1980.

Kemp, D. T. "Stimulated acoustic emissions from within the human auditory system." J. Acoust. Soc. Am. 64, pp. 1386-1391, 1978.

Schmiedt, R. A. "Acoustic injury and the physiology of hearing." J. Acoust. Soc. Am. 76, pp. 1293-1317, 1984.

Schmiedt, R. A. and Zwislocki, J. J. "Comparison of sound-transmission and cochlear-microphonic characteristics in mongolian gerbil and guinea pig." J. Acoust. Soc. Am. 61, pp. 133-149, 1977.

Schmiedt, R. A. and Adams, J. C. "Stimulated acoustic emissions in the ear canal of the gerbil." Hearing Res. 5, pp. 295-305, 1981.

Schroeder, M. R. "Relation between critical bands in hearing and the phase characteristics of cubic difference tones." J. Acoust. Soc. Am. 46, pp. 1488-1492, 1969.

Siegel, J. H. and Kim, D. O. "Cochlear biomechanics: vulnerability to acoustic trauma and other alternations as seen in neural responses and ear-canal sound pressure." In New Perspectives on Noise-Induced Hearing Loss, Ed. by R. Hamernik, D. Henderson, and R. Salvi, Raven press, New York, pp. 137-152, 1982.

Siegel, J. H., Kim, D. O., and Molnar, C. E. "Effects of altering organ of Corti on cochlear distortion products f_2-f_1 and $2f_1$-f_2." J. Neurophysiol. 47, pp. 303-328, 1982.

Smoorenburg, G.F. "Combination tones and their origin." J. Acoust. Soc. Am. 52, pp. 615-632, 1972.

Whitfield, I. C. and Ross, H. F. "Cochlear-microphonic and summating potentials and the outputs of individual hair-cell generators." J. Acoust. Soc. Am. 38, pp. 126-131, 1965.

Zurek, P. M. "Spontaneous narrowband acoustic signals emitted by human ears." J. Acoust. Soc. Am. 69, pp. 514-523, 1981.

Zwicker, E. "Nonmonotonic behaviour of ($2f_1$-f_2) explained by a saturation-feedback model." Hearing Res. 2, pp. 513-518, 1980.

STEADY-STATE RESPONSE DETERMINATION FOR MODELS OF THE BASILAR MEMBRANE

Irwin W. Sandberg and Jont B. Allen
AT&T Bell Laboratories
Murray Hill, NJ 07974

ABSTRACT

Electrical networks consisting of linear passive elements and many nonlinear resistors are often used to model the basilar membrane. The inputs to these networks are typically a sum of sinusoids switched on at $t=0$, and the resulting quantities of interest because of their interpretation as analogs of experimental observables are the steady-state response components of a certain current and of certain voltages. In this paper, recently obtained mathematical results concerning the input-output representation of nonlinear systems are used to give, for the first time, a locally convergent expansion for all of the steady-state quantities of interest. Also given is a good deal of information concerning general properties of the expansion, and this establishes important properties of the nonlinear network's response. Of particular practical interest is a term in the expansion that contains a component whose frequency is $(2f_1-f_2)$ when the network's input consists of a sum of two sinusoids, with frequencies f_1 and f_2. One of our main results is an explicit expression for this $(2f_1-f_2)$ component.

1. INTRODUCTION

Electrical networks of the type shown in Fig. 1, together with sophisticated frequency-domain measurement techniques, play a central role in the modeling and analysis of the peripheral auditory system [1-9]. In the figure, which shows a one dimensional lumped-element transmission line model of the basilar membrane, the inductors and capacitors are linear, the box at the upper left contains lumped elements and, as indicated, the resistors are nonlinear. The voltage e_0 applied to the network is typically a finite sum of sinusoids (often a sum of just two sinusoids) switched on at some finite time that we take to be $t=0$. The resulting quantities of interest, because of their interpretation as analogs of experimental observables, are the steady-state response components of the current i_0 and of one or more of the voltages $s_1, \ldots, s_q$.

In models of interest today the number q of nonlinear resistors is typically taken to be between two hundred and five hundred. The resistors are assumed to be current controlled, with each current-voltage relationship often represented by the sum of linear and cubic terms [1,2,3].

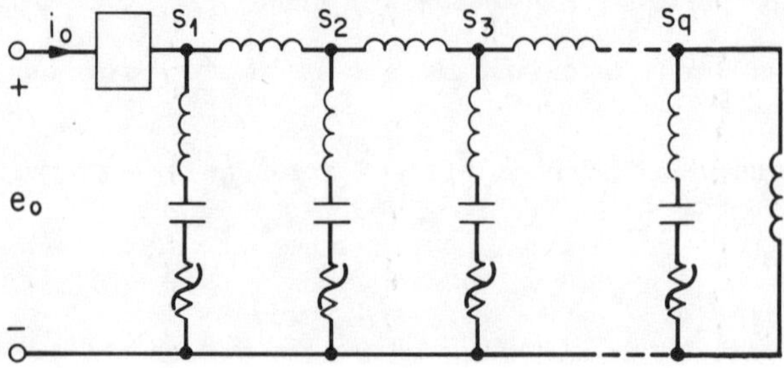

Fig. 1 — Network model.

The purpose of this paper is to use recently obtained results [10] concerning the input-output representation of nonlinear systems to give, for the first time, an expansion for all of the steady-state quantities of interest in Fig. 1. The expansion is in terms of e_0 and is locally convergent.

By this we mean that whenever the sum of the Fourier coefficients of e_0 is sufficiently small, and some reasonable additional conditions are met, the steady-state quantities exist and are given by the sum of the terms in the expansion, with each term dependent on the frequencies and Fourier coefficients of e_0. We emphasize that the expansion provides an exact representation of the response; it is not merely an approximation or a formal expansion whose convergence has not been proved. However, in this paper we do not give lower bounds on the *size* of the region of convergence. Questions of this type are the subject of ongoing studies [11].

In Section 2 it will become clear that the terms in the expansion are defined by a certain recursive process. Of particular practical interest at the present time is the term we call the third order term which contains a component whose (radian) frequency is $(2\omega_1-\omega_2)$ when e_0 consists of a sum of two sinusoids, one of frequency ω_1 and another of frequency ω_2. One of our main results is an explicit expression for this $(2\omega_1-\omega_2)$ component, under some very reasonable assumptions.

2. EXISTENCE, PROPERTIES, AND EVALUATION OF THE STEADY-STATE QUANTITIES

2.1 Formulation

In order to enable attention to be more sharply focused on the concepts of importance to us, it is helpful to generalize our problem. Thus, we consider instead of Fig. 1 the network of Fig. 2 in which $\mathcal{N}$ is a linear time-invariant network and $s_1, \ldots, s_p$ are voltages in $\mathcal{N}$, measured with respect to the ground terminal, where p is any positive integer.

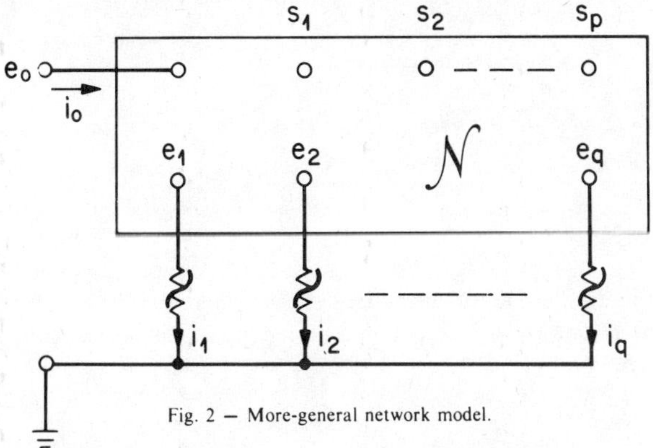

Fig. 2 — More-general network model.

Let i and e, respectively, denote the transpose of the current and voltage row vectors $(i_1, \ldots, i_q)$ and $(e_1, \ldots, e_q)$. Assume that $\mathcal{N}$ has the representation

$$i(t) = \int_0^t h_a(t-\tau)e_0(\tau)d\tau + \int_0^t h_c(t-\tau)e(\tau)d\tau + u_1(t), \quad t \geqslant 0 \tag{1}$$

in which h_a and h_c are $q \times 1$ and $q \times q$ matrix-valued impulse response functions and u_1 (which takes into account initial conditions) is a bounded continuous function that approaches zero as $t \to \infty$.* Similarly, let r stand for the transpose of the response $(s_1, \ldots, s_p, i_0)$ and suppose that there are $(p+1) \times 1$ and $(p+1) \times q$ matrix-valued impulse response functions h_d and h_b, respectively, for which

$$r(t) = \int_0^t h_d(t-\tau)e_0(\tau)d\tau + \int_0^t h_b(t-\tau)e(\tau)d\tau + u_2(t), \quad t \geqslant 0 \tag{2}$$

where u_2 is also a bounded continuous function that approaches zero as $t \to \infty$.

---

* We could have assumed that u_1 and the transient functions u_2 and u_3 to be introduced, are all zero functions. However, we wish to establish that the steady-state responses are *robust* with respect to these functions in the strong sense that, under the conditions to be described, they are independent of them.

Each element of h_a, h_b, h_c, and h_d is assumed to be an absolutely integrable real-valued function on $[0,\infty)$ with possibly an impulse at $t=0$. We use H_a to denote the Fourier transform of h_a, i.e.,

$$H_a(\omega) = \int_0^\infty h_a(t) e^{-j\omega t} dt, \quad -\infty < \omega < \infty .$$

Similarly, H_b, H_c, and H_d stand for the Fourier transforms of h_b, h_c, and h_d, respectively. Of course $H_a(\omega)$, $H_b(\omega)$, $H_c(\omega)$, and $H_d(\omega)$ are also matrices. Notice that, from (1) and (2), each of these matrices has a natural transfer-function interpretation. For example, from (1) we see that the elements of $H_a(\omega)$ are the voltage to current transfer functions from the system input e_0 to the "inputs" i of the nonlinear resistors, when these resistors are replaced with short circuits.

The nonlinear resistors in Fig. 2 are assumed to be represented by

$$e_k(t) = R_k[i_k(t)], \quad (k=1,\ldots,q) \tag{3}$$

with each R_k an analytic function in some neighborhood of the origin of the complex plane, such that $R_k(z)$ is real when z is real, $R_k(0) = 0$, and $dR_k(z)/dz = 0$ at $z=0$. (In particular, the R_k can be polynomials with real coefficients.) In Fig. 1 the nonlinear resistors typically have a relatively large linear part. These linear parts can be taken into account in Fig. 2 in $\mathcal{N}$. Using known properties of networks with positive elements, it is not difficult to show that the assumptions made above concerning h_a, h_b, h_c, h_d, u_1, and u_2 are satisfied for the network of Fig. 1 when put in the form of Fig. 2, as long as the linear part of each resistor has positive resistance, all linear elements are passive, the impedance of the two-terminal box is not zero at zero frequency, and each s_k in Fig. 2 is an s_k in Fig. 1.

2.2 Steady-State Responses: Properties and Evaluation

We now assume that e_0 is given by

$$e_0(t) = \sum_{k=-\infty}^{\infty} a_k e^{j\omega_k t} + u_3(t), \quad t \geqslant 0$$

in which the sum of the $|a_k|$ is finite, $j=(-1)^{\frac{1}{2}}$, the ω_k are real, and u_3, like u_1 and u_2, is bounded, continuous, and approaches zero as $t \to \infty$. We do not require that the ω_k are multiples of some fixed constant. Thus, the input is assumed to be the sum, for $t \geqslant 0$, of a so-called "almost-periodic" signal

$$\sum_{k=-\infty}^{\infty} a_k e^{j\omega_k t}, \quad -\infty < t < \infty \tag{4}$$

and a transient part u_3. Although all almost-periodic signals have a generalized Fourier series of the form (4), the sum of the magnitudes of the Fourier coefficients need not be finite. We shall use (AP) to denote the subset of almost-periodic functions for which this sum *is* finite.

At this point we are able to state our main result, which is: Under the assumptions already discussed, and for $\sum_{k=-\infty}^{\infty} |a_k|$ as well u_1, u_2, and u_3 sufficiently small,*

(i) there are unique bounded functions i, e, and r that satisfy (1), (2), (3), and (regarding uniqueness) a certain very reasonable neighborhood condition† concerning i,

(ii) there is a $(p+1)$-vector-valued function r_{ss} defined on $(-\infty,\infty)$, with each of its $(p+1)$ components belonging to (AP), such that

---

* By "small" for u_1, u_2, and u_3 is meant small in the reasonable sense of the $\mathscr{B}_0$ norm of [10, p. 692].

† The condition is simply that the function i must lie in a certain neighborhood of the origin. See the first of the two footnotes in the appendix.

$$r(t) - r_{ss}(t) \to 0 \text{ as } t \to \infty$$

(i.e., the response r approaches the steady state r_{ss} as $t \to \infty$), and

(iii) r_{ss} is independent of u_1, u_2, and u_3. It is given by

$$r_{ss}(t) = \sum_{m=1}^{\infty} [r_{ss}(t)]_m \quad , \quad -\infty < t < \infty \tag{5}$$

in which the $[r_{ss}(\cdot)]_m$ are $(p+1)$-vector-valued functions, with components belonging to (AP), defined by

$$[r_{ss}(t)]_1 = \sum_{k=-\infty}^{\infty} H_d(\omega_k) a_k e^{j\omega_k t}$$

and

$$[r_{ss}(t)]_m = \sum_{k_1=-\infty}^{\infty} \cdots \sum_{k_m=-\infty}^{\infty} \mathscr{R}_m(\omega_{k_1}, \ldots, \omega_{k_m}) a_{k_1} \cdots a_{k_m} e^{j(\omega_{k_1}+\ldots+\omega_{k_m})t} \tag{6}$$

for $m \geqslant 2$, where the $\mathscr{R}_m(\omega_{k_1}, \ldots, \omega_{k_m})$, which depend on H_a, H_b, H_c, and the derivatives of the R_k at the origin, but not on the coefficients a_k, are defined by the recursive relations (10), (11), and (12) in the appendix. The infinite sum in (5) converges uniformly in t.

Notice that a fundamental property of the class of network models considered is that, with excitations as indicated, each component of any steady-state response r_{ss} belongs to (AP). In particular, the r_{ss} are well behaved; any r_{ss} is continuous in t, has a Fourier series, and the Fourier series converges to $r_{ss}(t)$ for each t.

It is shown in the appendix that the result described above follows from the main theorem in [10]. In addition, bounds in [10, Section 2.4.3] show that the following can be added to (i)-(iii).

(iv) There are positive constants α and β such that, with $([r_{ss}(t)]_m)_k$ the kth component of $[r_{ss}(t)]_m$,

$$\sum_{m=(M+1)}^{\infty} \max_k |([r_{ss}(t)]_m)_k| \leqslant \alpha \left(\beta \sum_{k=-\infty}^{\infty} |a_k|\right)^{(M+1)} \quad , \quad -\infty < t < \infty$$

for any positive integer M (which provides useful information concerning the error in discarding all terms in (5) beyond the Mth).

2.3 The $(2\omega_1 - \omega_2)$ Component of $[r_{ss}(t)]_3$

Each $[r_{ss}(t)]_m$ in (5) is of order m in the sense that the effect of multiplying all of the Fourier coefficients of e_0 by a constant λ is to cause $[r_{ss}(t)]_m$ to be replaced by $\lambda^m [r_{ss}(t)]_m$. Of particular interest in applications is an explicit expression for $T(\omega_1, \omega_2, a_1, a_2)$, the component at the frequency $(2\omega_1 - \omega_2)$ of the *third* order term in (5), when

$$e_0 = a_1 e^{j\omega_1 t} + a_{-1} e^{-j\omega_1 t} + a_2 e^{j\omega_2 t} + a_{-2} e^{-j\omega_2 t}$$

(where a_{-1} and a_{-2} are the complex conjugates of a_1 and a_2, respectively), $0 < \omega_1 < \omega_2 < 2\omega_1$,* and $\alpha_k(\ell) = 0$ $(k = 1, \ldots, q)$ for $\ell = 2$, where here and in the appendix $\alpha_k(\ell)$ denotes $d^\ell R_k(z)/dz^\ell |_{z=0}$.

Under the condition on the $\alpha_k(2)$ indicated, the expression (12) for the $\mathscr{R}_m$ yields

---

* For ω_1 and ω_2 that meet these conditions, $(2\omega_1 - \omega_2)$ is not equal to ω_1, ω_2, $3\omega_1$, $3\omega_2$, or $(2\omega_2 - \omega_1)$ which are the only other positive frequencies at which $[r_{ss}(t)]_3$ can have components. However, higher order terms of odd index *can* possess components at $(2\omega_1 - \omega_2)$. For example, $(\omega_{k_1} + \ldots + \omega_{k_5}) = (2\omega_1 - \omega_2)$ if $\omega_{k_1} = \omega_{k_2} = \omega_1$, $\omega_{k_3} = -\omega_2$, and $\omega_{k_4} + \omega_{k_5} = 0$.

$$\mathscr{R}_3(\omega_{k_1}, \omega_{k_2}, \omega_{k_3}) = \frac{1}{6} H_b(\omega_{k_1}+\omega_{k_2}+\omega_{k_3}) \operatorname{diag}[\alpha_1(3),$$

$$\ldots, \alpha_q(3)] \hat{\chi}[H_a(\omega_{k_1}), H_a(\omega_{k_2}), H_a(\omega_{k_3})],$$

where $\hat{\chi}[H_a(\omega_{k_1}), H_a(\omega_{k_2}), H_a(\omega_{k_3})]$ denotes the q-vector whose kth element is the product $[H_a(\omega_{k_1})]_k\ [H_a(\omega_{k_2})]_k\ [H_a(\omega_{k_3})]_k$ of kth elements for each k. Thus, using (6) with $m = 3$, $a_0 = 0$, and $a_k = 0$ for $|k| > 2$, as well as the observation that $(\omega_{k_1}+\omega_{k_2}+\omega_{k_3}) = (2\omega_1-\omega_2)$ only if one of the ω_{k_ℓ} is $-\omega_2$ and the other two are equal to ω_1, it easily follows that the sum of the coefficients of $\exp[j(2\omega_1-\omega_2)t]$ in $[R_{ss}(t)]_3$ is

$$\frac{1}{2} H_b(2\omega_1-\omega_2) \operatorname{diag}[\alpha_1(3), \ldots, \alpha_q(3)] \hat{\chi}[H_a(\omega_1), H_a(\omega_1), H_a(-\omega_2)] a_1^2 a_{-2}.$$

This shows that

$$T(\omega_1,\omega_2,a_1,a_2) = \operatorname{Re}\{H_b(2\omega_1-\omega_2) \operatorname{diag}[\alpha_1(3), \ldots,$$

$$\alpha_q(3)] \hat{\chi}[H_a(\omega_1), H_a(\omega_1), H_a(-\omega_2)] a_1^2 a_{-2} \exp[j(2\omega_1-\omega_2)t]\},$$

where Re { } stands for the real part of { }. Since H_a and H_b have a direct interpretation in terms of the structure of the network of Fig. 2, so does $T(\omega_1,\omega_2,a_1,a_2)$.

As a matter of convenience we have chosen to let $\mathscr{N}$ take into account the linear parts of the nonlinear resistors. We could have assumed instead that $\mathscr{N}$, without these linear parts, has sufficient damping that our conditions on h_a, h_b, h_c, h_d, u_1 and u_2 are satisfied. Under some very reasonable assumptions (see [10, comments on p. 694 concerning H.3]), our expression for $T(\omega_1,\omega_2,a_1,a_2)$ would then explicitly exhibit its dependence on these linear parts, and this may be of interest in some cases. It can be shown, using a result in [10, Section 2.4.3], that the alternative expression for $T(\omega_1,\omega_2,a_1,a_2)$ that we would have obtained is

$$\operatorname{Re}\{H_b(2\omega_1-\omega_2) F(2\omega_1-\omega_2) \operatorname{diag}[\alpha_1(3), \ldots, \alpha_q(3)] \hat{\chi}[E(\omega_1) H_a(\omega_1), E(\omega_1) H_a(\omega_1),$$

$$E(-\omega_2) H_a(-\omega_2)] a_1^2 a_{-2} \exp[j(2\omega_1-\omega_2)t]\},$$

in which, with 1_q the identity matrix of order q, $F(\omega) = \{1_q - \operatorname{diag}[\alpha_1(1), \ldots, \alpha_q(1)] H_c(\omega)\}^{-1}$, and $E(\omega) = \{1_q - H_c(\omega) \operatorname{diag}[\alpha_1(1), \ldots, \alpha_q(1)]\}^{-1}$ (with both inverses existing for $-\infty < \omega < \infty$).

2.4 Discussion

In this paper we have derived and discussed a general expansion for the response of a cochlear model having a nonlinear membrane. The nonlinearities of the model take into account the membrane's nonlinear damping. Of particular interest is the third order term in the expansion for the case described in Section 2.3, in which the input is a sum of sinusoids at frequencies ω_1 and ω_2. This term is the first term in the expansion that gives rise to a component at the frequency $(2\omega_1-\omega_2)$.

The expression for the third order term is seen to depend on two transfer-function matrices H_a and H_b, where H_b relates the output response vector r to the voltages across the resistors in Fig. 2 under the condition that e_0 is zero, and H_a relates the currents through the resistors to the input voltage e_0 under the condition that the resistors are replaced by short circuits.

In the expression, the transfer function H_b is evaluated only at $(2\omega_1-\omega_2)$. The function H_b has the interpretation that it corresponds to a filter that alters the distortion products after their generation on the basilar membrane.

The terms $\alpha_1(3), \ldots, \alpha_q(3)$ are measures of the generator strength of the nonlinear distortion as a function of position, in the sense that each $\alpha_k(3)$ is proportional to the coefficient of the cubic term in the power series expansion of the resistor function R_k. Cubic nonlinearities have been used previously in basilar membrane models to model the generation of distortion products.

The transfer function H_a enters the expression for T in a particularly interesting way. Notice that any element of T, say the ℓth, is the real part of

$$\sum_{k=1}^{q} [H_b(2\omega_1-\omega_2)]_{\ell k}\ \alpha_k(3)[H_a(\omega_1)]_k^2[H_a(-\omega_2)]_k\ a_1^2 a_{-2} e^{j(2\omega_1-\omega_2)t} ,$$

which is a linear combination of q terms with, so to speak, H_a appearing three times in each term, twice for ω_1 and once for ω_2.

3. APPENDIX

Proof of the Main Steady-State Response Result; Recursive Relations for the $\mathcal{R}_m$

Theorem 3 of [10] would be directly applicable to the network governed by (1), (2), and (3) if $h_a(t-\tau)$, $h_b(t-\tau)$, $h_c(t-\tau)$, and $h_d(t-\tau)$ were *square* matrices of the *same* size. Since this condition is not met, we proceed to construct a suitable related set of system equations.

Let $n = (q+p+2)$, and define K_a, K_b, K_c, and K_d to be the convolution operators associated with h_a, h_b, h_c, and h_d, respectively. Let v, x, y, and w be given by $v = (e_0, u_1, u_2)^{\text{tr}}$, $x = (i, x^{[p+2]})^{\text{tr}}$, $y = (e, y^{[p+2]})^{\text{tr}}$, and $w = (r, w^{[q+1]})^{\text{tr}}$, where "tr" denotes transpose, and $x^{[p+2]}$, $y^{[p+2]}$, and $w^{[q+1]}$ are unspecified vector-valued functions (on $t \geqslant 0$) of the indicated dimensions. Notice that v, x, y, and w are all n-vector valued. Finally, let η_k $(k=1, \ldots, n)$ be the functions defined by $\eta_k = R_k$ $(1 \leqslant k \leqslant q)$, with η_k equal to the zero function for $(q+1) \leqslant k \leqslant n$.

Consider the equations

$$x = Av + Cy \tag{7}$$

$$w = Dv + By \tag{8}$$

$$y = Nx \tag{9}$$

in which by (9) we mean $y_k(t) = \eta_k[x_k(t)]$ for each k and t, and in which A, C, D, and B are given in partitioned form by

$$A = \begin{bmatrix} K_a & I(q) & Z(q,p+1) \\ Z(p+2,1) & Z(p+2,q) & Z(p+2,p+1) \end{bmatrix}, \quad C = \begin{bmatrix} K_c & Z(q,p+2) \\ Z(p+2,q) & Z(p+2,p+2) \end{bmatrix},$$

$$D = \begin{bmatrix} K_d & Z(p+1,q) & I(p+1) \\ Z(q+1,1) & Z(q+1,q) & Z(q+1,p+1) \end{bmatrix}, \text{ and } B = \begin{bmatrix} K_b & Z(p+1,p+2) \\ Z(q+1,q) & Z(q+1,p+2) \end{bmatrix},$$

where, for any positive integers q and s, $I(q)$ denotes the identity operator on the space of q-vector valued functions on $t \geqslant 0$, and $Z(q,s)$ is the zero operator from the space of s-vector-valued functions on $t \geqslant 0$ into the corresponding space of functions whose values are of dimension q.

We see that if (7), (8), and (9) are satisfied, then (1), (2), and (3) are met, and that if the latter set of equations are satisfied and $x^{[p+2]}$, $y^{[p+2]}$, and $w^{[q+1]}$ are zero functions, then (7), (8), and (9) are satisfied. Using the fact that A, B, C, D, and N meet the conditions of theorem 3 of [10], it follows from that theorem that statements (i) and (ii) of Section 2.2 hold.* It also follows from the theorem that r_{ss} is independent of u_1, u_2, and u_3, and using the relation $w = (r, w^{[q+1]})^{tr}$, that $r_{ss}(t)$ can be written in the form (5) with the components of each $[r_{ss}(\cdot)]_m$ elements of (AP), with

$$[r_{ss}(t)]_1 = \sum_{k=-\infty}^{\infty} H_d(\omega_k) a_k e^{j\omega_k t}$$

and each $[r_{ss}(t)]_m$ for $m \geqslant 2$ specified as follows (after some straightforward analysis involving partitioned matrices).

With $c_1, c_2, \ldots$ arbitrary n-vectors, and $\beta_1, \beta_2, \ldots$ arbitrary real numbers, let q-vector-valued functions $Q_1, Q_2, \ldots$ and $(p+1)$-vector-valued functions $P_2, P_3, \ldots$ be defined by $Q_1(c_1, \beta_1) = H_a(\beta_1)(c_1)_1$, $Q_m(c_1, \ldots, c_m, \beta_1, \ldots, \beta_m) = H_c(\beta_1 + \ldots + \beta_m) S_m$ for $m \geqslant 2$, and $P_m(c_1, \ldots, c_m, \beta_1, \ldots, \beta_m) = H_b(\beta_1 + \ldots + \beta_m) S_m$ for $m \geqslant 2$, in which†

$$S_m = \sum_{\ell=2}^{m} (\ell!)^{-1} \sum_{\substack{k_1+\ldots+k_\ell = m \\ k_j > 0}} \operatorname{diag}[\alpha_1(\ell), \ldots, \alpha_q(\ell)] \hat{\chi}[Q_{k_1}(c_1, \ldots, c_{k_1}, \beta_1, \ldots, \beta_k), \ldots,$$

$$Q_{k_\ell}(c_{(m-k_\ell+1)}, \ldots, c_m, \beta_{(m-k_\ell+1)}, \ldots, \beta_m)],$$

$(c_1)_1$ is the first component of c_1,

$$\alpha_k(\ell) = \frac{d^\ell R_k(z)}{dz^\ell}\Big|_{z=0} \quad (k=1, \ldots, q)$$

for each ℓ, "diag" indicates a diagonal matrix, and $\hat{\chi}$ is defined by the condition that $\chi[c_1, \ldots, c_\ell]$ is the q-vector with kth element $(c_1)_k \cdots (c_\ell)_k$ $(1 \leqslant k \leqslant q)$. In terms of these P_m, we have

$$[r_{ss}(t)]_m = \sum_{k_1=-\infty}^{\infty} \cdots \sum_{k_m=-\infty}^{\infty} P_m(d_{k_1}, \ldots, d_{k_m}, \omega_{k_1}, \ldots, \omega_{k_m}) e^{j(\omega_{k_1} + \ldots + \omega_{k_m})t},$$

where $d_k = (a_k, 0, \ldots, 0)^{tr}$ for each k.

Observe that for any m and k with $1 \leqslant k \leqslant m$, each Q_m and each P_m is linear in c_k and independent of $(c_k)_\ell$ for $\ell \geqslant 2$. Thus, each $Q_m(c_1, \ldots, c_m, \beta_1, \ldots, \beta_m)$ is equal to $Q_m(u, \ldots, u, \beta_1, \ldots, \beta_m)(c_1)_1 \cdots (c_m)_1$, where $u = (1, 0, \ldots, 0)^{tr}$, and similarly for the P_m. Therefore, if $\mathscr{S}_m$ and $\mathscr{R}_m$ are defined by

$$\mathscr{S}_1(\beta_1) = H_a(\beta_1), \tag{10}$$

$$\mathscr{S}_m(\beta_1, \ldots, \beta_m) = H_c(\beta_1 + \ldots + \beta_m) \sum_{\ell=2}^{m} (\ell!)^{-1} \sum_{\substack{k_1+\ldots+k_\ell=m \\ k_j > 0}} \operatorname{diag}[\alpha_1(\ell), \ldots,$$

$$\alpha_q(\ell)] \hat{\chi}[\mathscr{S}_{k_1}(\beta_1, \ldots, \beta_{k_1}), \ldots, \mathscr{S}_{k_\ell}(\beta_{(m-k_\ell+1)}, \ldots, \beta_m)] \tag{11}$$

for $m \geqslant 2$, and

$$\mathscr{R}_m(\beta_1, \ldots, \beta_m) = H_b(\beta_1 + \ldots + \beta_m) \sum_{\ell=2}^{m} (\ell!)^{-1} \sum_{\substack{k_1+\ldots+k_\ell=m \\ k_j > 0}} \operatorname{diag}[\alpha_1(\ell), \ldots,$$

---

* The "neighborhood condition" of statement (i) is inherited from [10, part (iib) of Theorem 3] via the relationship between (1), (2), and (3) and (7), (8), and (9).

† In the expression for S_m, $\sum_{\substack{k_1+\ldots+k_\ell=m \\ k_j > 0}}$ denotes a sum over all positive integers $k_1, \ldots, k_\ell$ that add to m.

$$\alpha_q(\ell)]\hat{\chi}[\mathscr{S}_{k_1}(\beta_1, \ldots, \beta_{k_1}), \ldots, \mathscr{S}_{k_\ell}(\beta_{(m-k_\ell+1)}, \ldots, \beta_m)] \tag{12}$$

for $m \geqslant 2$, we have

$$[r_{ss}(t)]_m = \sum_{k_1=-\infty}^{\infty} \cdots \sum_{k_m=-\infty}^{\infty} R_m(\omega_{k_1}, \ldots, \omega_{k_m}) a_{k_1} \cdots a_{k_m} e^{j(\omega_{k_1}+\ldots+\omega_{k_m})t}, \quad -\infty < t < \infty$$

for $m = 2, 3, \ldots$. This completes the appendix.

REFERENCES

1. J. L. Hall, "Observations on a Nonlinear Model for Motion of the Basilar Membrane." In *Hearing Research and Theory*. Volume 1, pp. 1-61, Academic Press, New York, 1981.
2. J. L. Hall, "Two-tone Distortion Products in a Nonlinear Model of the Basilar Membrane." J. Acoust. Soc. Am. *56*, pp. 1818-1828, 1974.
3. J. W. Matthews, "Modeling Reverse Middle Ear Transmission of Acoustic Distortion Signals." In *Mechanics of Hearing,* Ed. by E. de Boer and M. A. Viergever, Martinus Nijhoff Publishers, Delft University Press, 11-18, 1983.
4. E. Zwicker, "Cubic Difference Tone Level and Phase Dependence on Frequency and Level of Primaries." In *Psychophysical, Physiological and Behavioral Studies in Hearing.* (van den Brink and Bilsen, eds.). Delft Univ. Press, Netherlands, 1981.
5. J. L. Goldstein, "Auditory Nonlinearity." J. Acoust. Soc. Am. *41*, pp. 676-689, 1967.
6. J. L. Goldstein, G. Buchsbaum and M. Furst, "Compatibility Between Psychophysical and Physiological Measurements of Aural Combination Tones." J. Acoust. Soc. Am. *63*, pp. 474-485, 1978.
7. J. L. Goldstein, and N. Y. S. Kiang, "Neural Correlates of the Aural Combination Tones." Proc. of the IEEE *56*, pp. 981-992, 1968.
8. D. O. Kim, C. C. Molnar, and J. W. Matthews, "Cochlear Mechanics: Nonlinear Behavior in Two-Tone Response as Reflected in Cochlear-Nerve-Fiber Responses and in Ear-Canal Sound, Pressure," J. Acoust. Soc. Am. *67*, pp. 1704-1721, 1980.
9. J. B. Allen and P. F. Fahey, "Nonlinear Behavior at Threshold Determined in the Auditory Canal and on the Auditory Nerve," in *Hearing-Physiological Bases and Psychophysics*, Ed's. R. Klinke and R. Hartman, Springer-Verlag, New York, NY, pp. 128-134, 1983.
10. I. W. Sandberg, "Existence and Evaluation of Almost Periodic Steady-State Responses of Mildly Nonlinear Systems," IEEE Trans. Circuits and Systems, Vol. 31, No. 8, pp. 689-701, August 1984.
11. I. W. Sandberg, "Criteria for the Global Existence of Functional Expansions for Input-Output Maps," to appear in the AT&T Technical Journal, 1985.

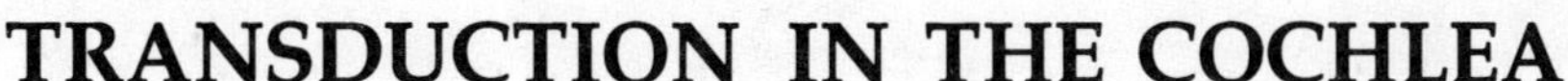

TRANSDUCTION IN THE COCHLEA

TRANSDUCTION IN COCHLEAR HAIR CELLS

I.J. Russell and A.R. Cody
MRC Neurophysiology Group
School of Biological Sciences
University of Sussex
Brighton, BN1 9QG, UK

The hair cells in the cochlea are divided into two morphologically distinct populations: a single row of inner hair cells (IHC) and three or four rows of outer hair cells (OHC). On the basis of differences in morphology, IHCs and OHCs have been attributed with different roles in mechano-electric transduction in the cochlea. The aim of this paper is to review recent studies on the electrophysiological properties of hair cells in the mammalian cochlea to see if there is any basis for this suggestion.

The responses of cochlear hair cells to tones

In response to low frequency tones, IHCs produce receptor potentials that are always asymmetrical in the depolarizing phase regardless of stimulus frequency and intensity. In contrast, the OHC responses to the same stimuli, are almost symmetrical at low intensities and become asymmetrical in the hyperpolarising direction with increasing intensity. Finally, at stimulus intensities above about 90 db SPL, the OHC receptor potential polarity is reversed becoming asymmetrical in the depolarising direction. This is seen in figure 1 where the responses of an OHC to 600Hz tones (figure 1 B,D) can be compared with the depolarising responses of an adjacent IHC (figure 1 A,C).

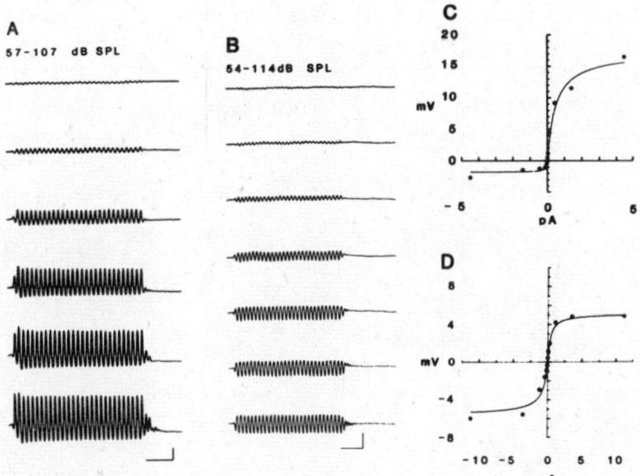

Figure 1. Voltage responses of A, an IHC, and B, an OHC in the basal turn of the same cochlea to 600Hz tones. C and D represent the relationships between peak voltage response and peak sound pressure for the IHC and OHC respectively. Sound pressure rarefaction is positive, compression is negative. The smooth curves through the points are rectangular hyperbolae whose constants are empirically derived. Scales: 5mV, 10 msec.

IHCs in the basal turn of the cochlea are characterised by their large dc voltage responses to tones in the high frequency tip region of their response areas (Russell and Sellick, 1978). In contrast, dc receptor potentials have not been recorded from OHCs in response to high frequency tones, unless the intensities exceed 90 db SPL. Examples of the dc voltage responses of an OHC to intense, high frequency tones are shown in figure 2B, where they can be compared with the dc receptor potentials from an adjacent IHC to tones of the same frequency (figure 2A). The dc responses of OHCs are much smaller than those of IHCs (less than 5mV in amplitude), invariably depolarising at these intensities, and the rise times are level dependent and slower than those in IHCs.

A comparison between intracellular and extracellular receptor potentials in the cochlea

The cochlear microphonic (CM), and the summating potential (SP) are the extracellular sound evoked responses of the cochlea. Since these must be a reflection of the intracellular responses of hair cells, one may ask how the extracellular and intracellular responses compare. It has long been held that the CM is the product of the OHCs (Dallos, Billone, Durrant, Wang and Raynor, 1972), and this concept has received substantial support from intracellular recordings of the receptor potentials from OHCs in the third (Dallos, Santos-Sacchi, and Flock, 1982) and basal turns (Russell and Cody, 1984) of the guinea-pig cochlea. These recordings show that the intracellular waveforms are almost identical to the extracellular CM. Furthermore, Dallos (1984) has found a close correspondence between the SP, recorded in the organ of Corti in the third turn, and the dc components of the OHC voltage responses which reverse their polarity from hyperpolarising at low frequencies, to depolarising in response to tones close to their best frequencies. In the basal turn, stimulation by high frequency tones produces large intracellular responses from IHCs and SPs can be recorded either close to the IHCs (Russell and Sellick, 1978), or differentially across the cochlea (Cheatham and Dallos, 1984). No intracellular dc responses are produced by OHCs to

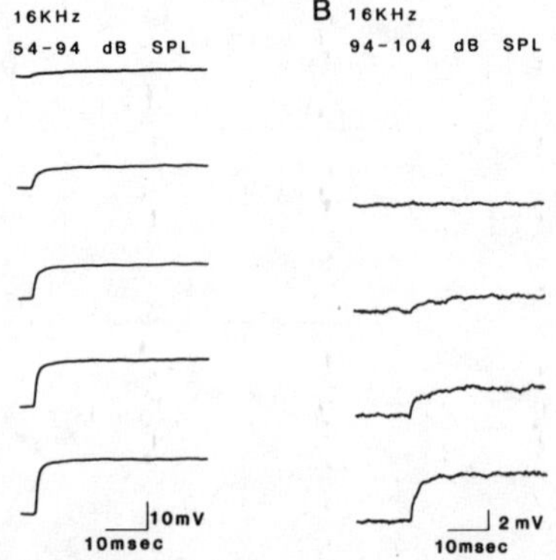

Figure 2. The responses of A, an IHC and B an OHC from the same cochlea to tones close to their best frequency (19kHz).

such low intensity stimuli. However, the extracellular SPs to high intensity tones (Honrubia and Ward, 1969; Cody and Russell, 1985), closely resemble the dc responses of OHCs to these stimuli. It appears, therefore, that the CM recorded from the organ of Corti in the basal turn of the guinea-pig cochlea is produced by the responses of the OHCs, and because OHCs do not produce responses to high frequency tones, other than at high intensities, the SP is presumably produced by IHCs at low and moderate intensities, and dominated by the voltage responses of the OHCs at high intensities.

Transfer functions of hair cells

The relationship between the peak-to-peak voltage responses and sound pressure for IHCs and OHCs from the same cochleas can be fitted empirically with pairs of rectangular hyperbolae (figure 1 C,D). This relationship has been used to describe the voltage responses of hair cells in other acoustico-lateralis receptors (Boston, 1981; Crawford and Fettiplace, 1981; Russell and Sellick, 1983), but it is not generally applicable to our data, only to selected examples. However, it is interesting that these relationships fit both IHC and OHC responses recorded from the same cochlea. We have never encountered a cochlea where the rectangular hyperbolae fit the responses of one class of hair cell and not the other. This would suggest that the voltage responses of adjacent IHCs and OHCs in the basal turn reflect a common process.

The idea of a process common to both sets of hair cells is supported by the Lissajous figures (figure 3 A,B) of IHC and OHC responses plotted against the sound pressure measured in the auditory meatus; the origin represents the resting potential of the hair cell and zero sound pressure (Russell and Cowley, 1983). These figures show that the non-linearities in the responses of adjacent IHCs and OHCs to tones of the same frequency and amplitude are very similar. If Lissajous figures of OHC versus IHC responses are plotted (figure 3C), then these non-linearities are greatly reduced, indicating again that they have a common origin.

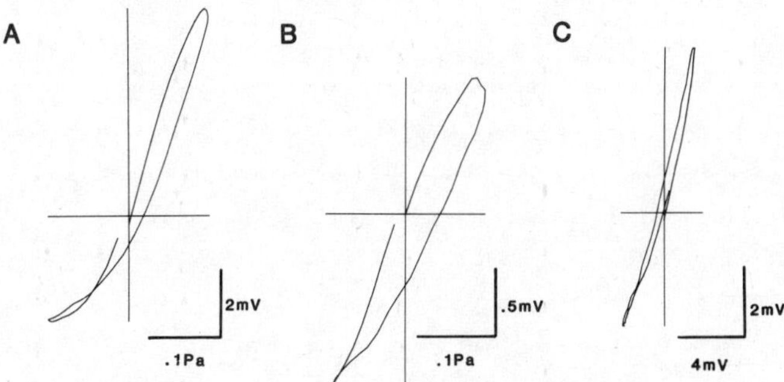

Figure 3. The relationship between sound pressure (abscissa) and voltage response to an 80 db SPL tone (ordinate) of A, an IHC and B, an OHC from the same cochlea. C, the relationship between the IHC (ordinate) and OHC (abscissa) voltage responses. The Lissajous figures are derived from averaged waveforms sampled 36 times per cycle.

The relationship between the voltage responses of adjacent OHCs and IHCs shown in figure 3C is almost linear in the depolarising phase with a sharp discontinuity in slope at the origin (resting potentials of the hair cells). The slope of the hyperpolarising phase is less steep and reflects the tendancy of the IHC response to saturate in this phase at a rate greater than that of the OHC. It has been proposed that the property of IHCs to generate dc voltage responses, even at low intensities, is due to a sharp discontinuity of slope at the origin of their transfer functions (Russell and Sellick, 1978; Russell, 1979). This idea is supported by the Lissajous figures in figure 3 where it can be seen that this discontinuity of slope is a property of the IHC (3A) and not the OHC (3B).

An analysis of the harmonic components in the responses of adjacent IHCs and OHCs to low frequency tones, also supports the hypothesis that their distortion has a common source. The harmonic composition of the IHC and OHC responses are similar. The dominant components are the fundamental, the 1st-order even (2f), and 2nd-order odd (3f) harmonics. For low intensity tones, the fundamental and 3f component are proportional to sound pressure level, while the 2f component is proportional to stimulus intensity. That is, the fundamental and the 3f component grow in proportion to the peak to peak amplitude of the voltage responses of IHCs and OHCs, while the 2f component grows in proportion to the dc component (Goodman, Smith and Chamberlain, 1982; Patuzzi and Sellick, 1983; Cody and Russell, 1985). Furthermore, at low frequencies, the 2f component is in phase with the fundamental of the IHC response. Thus, it leads the fundamental of the OHC response by about 90° (Russell and Cody in preparation; see below). Consequently, the 2f component contributes towards the depolarising phase of the voltage responses of IHCs and the hyperpolarising phase of the OHCs.

The phase relationships between IHC and OHC voltage responses

For low frequency tones, the responses of OHCs and the CM phase lead the sound pressure in the external auditory meatus by about 90° (Cody and Russell, 1985). Thus, at low frequencies, OHCs respond to basilar membrane displacement (Wilson and Johnstone, 1975), while IHCs, which phase lead sound pressure by about 180°, respond to the velocity of the basilar membrane (Russell and Sellick, 1983). At stimulus frequencies above a few hundred Hz the interpretation of phase measurements from intracellular recordings is difficult. For tones between about 500Hz and 1.5kHz, the phase lead of the IHCs, relative to the OHCs, is reduced from about 90° to 45° (Russell and Cody in preparation). This may indicate that IHCs change their responses from basilar membrane velocity to displacement at these frequencies, or that the time constant of the recording conditions had changed by about 0.1msec between the two recordings. These conclusions are consistant with the morphological relationships between the stereocilia of the hair cells and the tectorial membrane as revealed by electron microscopy (Lim, 1980); the stereocilia of the OHCs are firmly attached to the tectorial membrane and are likely to respond to basilar membrane displacement, while the stereocilia of the IHCs stand free in the fluid of the sub-tectorial space and are likely to respond to viscous drag (Dallos et.al., 1972).

The generation of receptor potentials during high frequency tones

The asymmetry of the sound-pressure voltage-response curve is important in signal transmission because it enables hair cells to produce voltage responses to tones above the cut-off frequency set by their electrical time constants (Russell and Sellick, 1983). For tones with periods considerably longer than the time constant, modulation of the receptor current by the displacements of the stereocilia bundle cause the hair cells to be alternately depolarised and hyperpolarised. This modulates the release of transmitter by the hair cells and produces phase locked patterns of activity in the afferent nerve fibres which synapse on them. If the period of the tone is much shorter than the time constant, then the receptor current which is modulated by the tone will be integrated by the capacitance of the hair cell. Thus, if the charge carried by the inward current equals that carried by the outward current, as it would if the transducer conductance was symmetrical, then the net change in transducer current would be zero. No receptor potential would be developed, and consequently the release of afferent transmitter would be unchanged. However, in IHCs the depolarising asymmetry of the transducer conductance ensures that there is a net inward current during each cycle of the tone and this is integrated by the membrane capacitance to produce a steady depolarisation. This is the dc potential (intracellular +SP), which lasts, without adaptation for the duration of the stimulus. The dc potential then leads to a steady increase in the release of chemical transmitter and an increase in the firing rate of the afferent fibres which is not phase locked to the stimulus. The fact that OHCs do not generate voltage responses to high frequency tones, except at high intensities, is an indication that the net transfer of current across their membranes is zero at these frequencies, and they cannot, therefore, signal responses to high frequency tones.

A mechanism for producing the asymmetrical voltage responses of hair cells

A mechanism for producing the metabolically labile asymmetry of the IHC responses has been proposed (Russell and Ashmore, 1983). This model is similar to the shear-excitation model of Davis (1958), but in addition, takes into account the firm attachment of the OHC stereocilia to the tectorial membrane (Lim, 1980). According to this model shown in figure 4, the OHC stereocilia restrict the shear displacement of the tectorial membrane when the basilar membrane moves towards the scala media, causing the tectorial membrane to buckle near its attachment to the spiral lamina, which reduces the subtectorial space (figure 4A). This amplifies the angular displacement of the IHC stereocilia in the depolarising direction, thus producing the depolarising asymmetry of the IHC voltage responses. The angular displacement of the IHC stereocilia could be further amplified if the OHCs actively reacted against the shear force imposed on them, thereby increasing the buckling. This would be the case if mechano-electric transduction was a bi-directional process (Weiss, 1982). The hyperpolarizing asymmetry of the OHCs might arise if the tectorial membrane incorporated an elastic element, as proposed by Allen (1980). This element, under tension, would exert a radial force on the OHC stereocilia bundles which would increase when the basilar membrane moved towards the scala tympani (figure 4).

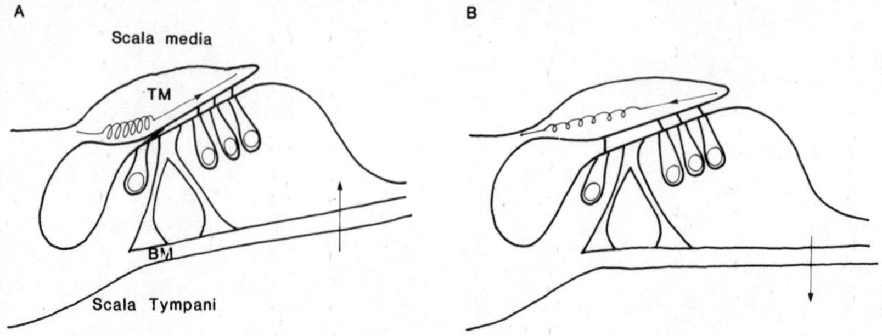

Figure 4. Proposed mechanism to explain the asymmetry of the IHC and OHC voltage responses. A, schematic of organ of Corti showing bending of TM and compression of spring and subtectorial space when BM is displaced towards scala media. B, shows expansion of spring and subtectorial space when BM is displaced towards scala tympani. Horizontal arrows, direction of force acting on OHCs. Vertical arrows, direction of BM displacement. (Modified from Russell and Ashmore, 1983).

Any change in the mechanical properties of the stereocilia, tectorial membrane, and/or their coupling, will alter the buckling of the tectorial membrane and, consequently, the mechanical input to the hair cells. Miller et.al. (1985) found that the stiffness of OHC stereocilia, in *in vitro* preparations of the guinea-pig cochlea, was reduced following exposure to intense sounds. It is suggested that if such a change occurred *in vivo* it would reduce the buckling of the tectorial membrane during the depolarising phase of the basilar membrane motion. This would give rise to the observed changes in symmetry of the IHC and OHC responses following intense tones (Cody and Russell, 1985); the IHC voltage responses become more symmetrical and those of the OHCs become asymmetrical in the depolarising direction.

The responses of hair cells to intracellularly injected current

Preliminary reports of the voltage responses of IHCs to injected current showed them to be relatively linear (Russell, 1983; Brown and Nuttall, 1984). However, re-examination of the original data and recent evidence (Nuttall, 1985; Russell and Cody, 1985), shows that the conductances of IHCs and OHCs are voltage sensitive. The conductances of the IHCs, in these experiments, ranged between 13 and 40nS at their resting membrane potentials (-25 to -45mV) and decreased to nearly half their resting value when the IHCs became more hyperpolarised than -60mV (figure 5A). In the case of OHCs, which have resting membrane potentials between -75 and -100mV, this rectification is more pronounced (figure 5B) and their conductances increase from about 20nS at -75mV to more than 100nS at -70mV. OHCs also show an inward rectification when their membrane potentials are hyperpolarised below about -90mV. Thus, the current-voltage responses of OHCs are similar to those in lower vertebrate hair cells (Corey and

Hudspeth, 1979).

The functional significance of the non-linear current-voltage relationships of the OHCs has yet to be explored, but it may underpin the observations of Mountain, Geisler, and Hubbard (1980), who showed that voltage dependent elements are involved in the generation of the CM and the sound induced resistance changes measured in the scala media of the guinea-pig cochlea. The OHC non-linearities may also provide a basis for the observation by Nuttall (1985) that current injected into the scala media, but not intracellularly into the IHCs, cause non-linear changes in the DC components of the receptor potentials and the frequency tuning of IHCs. These changes may be mediated by non-linear, current sensitive, mechanical elements in the cochlea (Hubbard and Mountain, 1983), and it would be interesting to discover if the non-linearity of these elements was primarily due to the voltage sensitive conductances of the OHCs.

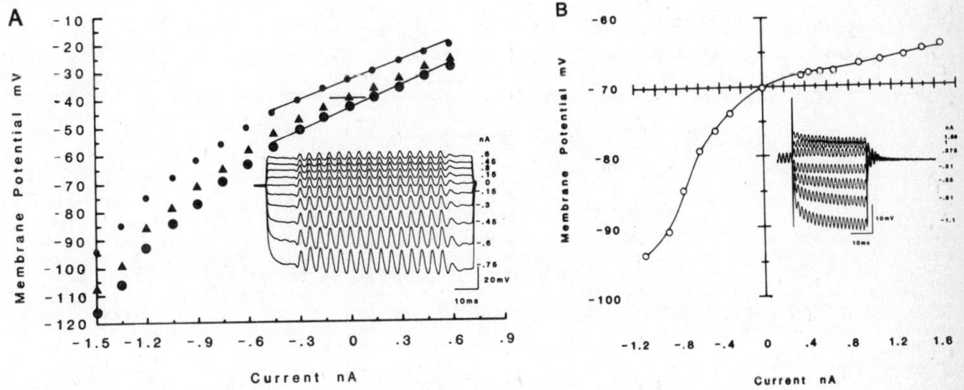

Figure 5. The relationship between voltage and injected current for an IHC (A) and an OHC (B) when stimulated by 600Hz tone at 84 dB SPL. A, • positive and ● negative responses to tones, ▲ membrane potential in the absence of a tone. B, o membrane potential in the absence of a tone. Insets: Voltage responses to injected current (amplitude shown by each trace.)

It remains to be seen which of the conductances, recently characterised in isolated hair cells in the frog and chick vestibular systems, contribute to the non-linear current voltage responses of cochlear hair cells (Lewis and Hudspeth, 1983; Ohmori, 1984). IHCs might be expected to have voltage dependent calcium conductances associated with their afferent synapses, because calcium plays an important role in transmitter release (Baker, 1972). At their resting potentials (-25 to -50mV), IHCs are more depolarised than the activation voltage of -60mV reported for the non-inactivating calcium channels of hair cells in the chick and frog vestibular systems. If calcium conductances contribute towards the non-linear current-voltage relationships of IHCs, then the steady influx of calcium ions would result in a continuous release of transmitter from their presynaptic membranes and the spontaneous activity in the afferent fibres. The functional significance of this would be to set the operating point of each afferent synapse in the steep, central region of its transfer function, thereby increasing its sensitivity to changes in the membrane potential of the IHC (Russell, 1979).

On the basis of their ultrastructures, OHCs have few, if any, functional afferent synapses (Spoendlin, 1978). Furthermore, given their large resting potentials (-75 to -100mV) and small receptor potentials, (Dallos et.al., 1982; Russell and Cody, 1984; Cody and Russell, 1985), OHCs are not likely to approach the activation voltages (-50 to -60mV) of the calcium channels which have so far been discovered in hair cells, even when excited by intense tones. In view of the doubts about the efficacy of their synapses, it remains to be seen if OHCs do relay sensory signals to their afferent innervation. Indeed, in the one report of intracellular recording from an identified OHC afferent fibre (Robertson, 1984), it was found to be silent.

The transducer conductance

The most widely accepted theory of sensory transduction in the cochlea is Davis' resistance microphone theory (Davis, 1958). This was prompted by the unusual extracellular environment of cochlear hair cells whose apical sensory surfaces are exposed to the potassium rich endolymph of the scala media, the composition of which is maintained by an electrogenic pump in the cells of the stria vascularis. The pump generates the endocochlear potential (EP) of about 80mV, with respect to the perilymph of the scala tympani. According to Davis' theory, the EP and the hair cell resting potential combine to produce a driving voltage of about 120 to 160mV for the receptor current across the sensory membrane of the hair cells. The transducer conductance behaves as a passive resistor which is by changed sound. This, in turn, modulates the flow of transducer current and, consequently, the membrane potential of the hair cell. Davis proposed that the receptor current is carried by potassium ions travelling down their electrical gradient into the hair cells and down their chemical gradient into the scala tympani (Johnstone and Sellick, 1972). If the receptor current is carried by potassium ions then the receptor current should reverse when the membrane potential of the hair cells is depolarised to the EP. Estimates of the reversal potential of the receptor current in IHCs (Russell, 1983, Russell and Cody, 1985) show that this is indeed the case, with a very close correspondence between the measured EP and reversal potentials.

Although our experiments indicate that the receptor current in IHCs is probably carried by potassium, it does not necessarily follow that the transducer channels are selective for this ion given that those in other vertebrate hair cells are non-selective (Corey and Hudspeth, 1979). On the basis of extracellular current measurements, Hudspeth (1982) has shown that the transducer channels of hair cells are probably located at the tips of the stereocilia, and there is support for this idea in ultrastructural studies of the stereocilia (Pickles, Osborne and Comis, 1984). Estimates of the conductances of single transducer channels in hair cells range from

12pS (bullfrog sacculus: Holton and Hudspeth, 1985) to 50pS (chick vestibular system: Omohri, 1984). In the guinea-pig, the total transducer conductances of IHCs and OHCs are estimated to be about 2.5nS and 3.1nS respectively (Russell, 1983; Russell and Cody, 1985). Thus, there are probably between 1 and 5 channels per stereocilium.

The roles of inner and outer hair cells

The suggestion that hair cells behave as receptors and effectors in sensory transduction was stimulated by the finding, in hair cells, of proteins normally associated with contractile processes (Flock, 1983 for a review). Direct studies of the stiffness properties of hair cells in isolated preparations show that their stiffness is susceptable to agents which induce contraction in muscle fibres (Orman and Flock, 1983) and the mechanical properties of stereocilia are changed by efferent stimulation (Ashmore, 1984). Furthermore, there is now evidence, in the turtle cochlea, that the voltage responses of the hair cells are closely reflected in mechanical movements of their stereocilia bundles; an indication that transduction and frequency selectivity in these cells involves electromechanical feedback (Crawford and Fettiplace, 1985).

In the mammalian cochlea, indirect measurements have shown that active mechanical processes underlie the efferent control of cochlear sensitivity (Mountain, 1980; Siegel and Kim, 1982). The origin of these active mechanical processes has yet to be directly investigated. However, studies on isolated cochlear hair cells show that OHCs but not IHCs have motile responses to injected current and to the application of the putative efferent transmitter acetylcholine (Brownell, Bader, Bertrand, and Ribaupierre, 1985). OHCs are, therefore, likely to be sources of the observed mechanical changes associated with inhibition in the cochlea. OHCs have also been proposed as the source of active mechanical processes responsible for acoustic emissions (Wilson, 1980). However, the frequency of oto-acoustic emissions measured at the tympanic membrane may exceed several kHz (Kemp, 1979). If OHCs are responsible for such high frequency mechanical changes, then their motile processes must respond to the flow of current through the OHC membranes and not to the voltage developed across them because, as discussed above, OHCs do not generate receptor potentials to high frequency stimuli.

In conclusion, it may be that the dual sensory-motor properties which have been reported in the hair cells of lower vertebrates are separated in the cochlea. Evidence presented in this paper shows that IHCs have a predominantly sensory role in the cochlea. In OHCs, their motile properties, the apparent lack of voltage responses to high frequency tones, and doubts about the efficacy of their afferent synapses, suggests that OHCs may be effectors rather than receptors. On the basis of the limited studies to date, it seems that OHCs may have a mainly motor role, responding to changes in their membrane conductance and to commands from their efferent innervation. If they have a sensory function, it may be to signal proprioceptive information about the tonic state of the basilar membrane. This information could be usefully integrated with the efferent commands to the cochlea.

Acknowledgements

We thank Dr. A.R. Palmer for his helpful comments on the manuscript, Mrs. E.M. Cowley for excellent assistance, and Mrs. J. Hutchings for her careful typing. This work was supported by the M.R.C. A.R.C. is a N.H and M.R.C., C.J.Martin Fellow.

REFERENCES

Allen, J. B., "Cochlear Micromechanics - A Physical Model of Transduction." J. Acoust. Soc. Am. 68, pp. 1660-1670, 1980.

Ashmore, J. F., "The Stiffness of the Sensory Hair Bundle of Frog Saccular Hair Cells." J. Physiol. 350, 20P, 1984.

Baker, P.F., "Transport and Metabolism of Calcium Ions in Nerve." Biophys. molec. Biol. 24, pp. 177-223, 1972.

Boston, J.R., "A Model of Lateral Line Microphonic Response to High-level Stimuli." J. Acoust. Soc. Am. 67, pp. 875-881, 1981.

Brown, M.C. and Nuttall, A.L., "Efferent Control of Cochlear Inner Hair Cell Responses in the Guinea-Pig." J. Physiol. 354, pp. 625-646, 1984.

Brownell, W.E., Bader, C.R., Bertrand, D. and Ribaupierre, Y., "Evoked Mechanical Responses of Isolated Cochlear Outer Hair Cells." Science 227, pp. 194-196, 1985.

Cheatham, M.A. and Dallos, P., "Summating Potential (SP) Tuning Curves." Hearing Res. 16, pp. 189-200, 1984.

Cody, A.R. and Russell, I.J., "Outer Hair Cells in the Cochlea and Noise Induced Hearing Loss." Nature (in press), 1985.

Corey, D.P. and Hudspeth, A.J., "Ionic Basis of the Receptor Potential in a Vertebrate Hair Cell." Nature 281, pp. 625-627, 1979.

Crawford, A.C. and Fettiplace, R., "Non-linearities in the Responses of Turtle Hair Cells." J. Physiol. 315, pp. 317-338, 1981.

Crawford, A.C. and Fettiplace, R., "The Mechanical Properties of Ciliary Bundles of Turtle Cochlear Hair Cells." J. Physiol. (in press), 1985.

Dallos, P., "Control of Extracellular Electrical Activity in the Cochlea. Inferences from Responses at Low Frequences." Abst. Assoc. Res. Otolaryngol. 7, pp. 34, 1984.

Dallos, P., Billone, M.C., Durrant, J.D., Wang, C-Y., and Raynor, S., "Cochlear Inner and Outer Hair Cells: Functional Differences." Science 177, pp. 356-358, 1972.

Dallos, P., Santos-Sacchi, J. and Flock, A., "Intracellular Recordings from Cochlear Outer Hair Cells." Science 218, pp. 582-584, 1982.

Davis, H., "A Mechano-Electric Theory of Cochlear Action." Ann. Otol. Rhinol. Laryngol. 67, pp. 789-801, 1958.

Flock, A., "Hair Cells, Receptors with a Motor Capacity?." In Hearing- Physiological Bases and Psychophysics, Ed. by R. Klinke and R. Hartman, Springer, Berlin, pp. 2-9, 1983.

Goodman, D.A., Smith, R.L., and Chamberlain, S.C., "Intracellular and Extracellular Responses in the Organ of Corti of the Gerbil." Hearing Res. 7, pp. 161-179, 1982.

Holton, T., and Hudspeth, A.J., "A Study of Hair-Cell Transduction Channel Using the Whole-Cell Voltage-Clamp Technique." Asso. Res. Otolaryngol. 8, pp. 48-49, 1985.

Honrubia, V., and Ward, P.H., "Properties of the Summating Potential of the Guinea Pig's Cochlea." J. Acoust. Soc. Am. 45, pp. 1443-1450, 1969.

Hubbard, A.E., and Mountain, D.C., "Alternating Current Delivered into the Scala Media Alters Sound Pressure at the Eardrum." Science, 222, pp. 510-512, 1983.

Hudspeth, A.J., "Extracellular Current Flow and the Site of Transduction by Vertebrate Hair Cells." J. Neurosci. 2, pp. 1-10, 1982.

Hudspeth, A.J., and Corey, D.P., "Sensitivity, Polarity and Conductance Change in the Response of Vertebrate Hair Cells to Controlled Mechanical Stimuli." Proc. Natl. Acad. Sci. U.S.A. 74, pp. 2407-2411, 1977.

Johnstone, B.M., and Sellick, P.M., "The Peripheral Auditory Apparatus." Q. Rev. Biophys. 5, pp. 1-57, 1972.

Kemp, D.T., "Evidence of Mechanical Non linearity and Frequency Selective Wave Amplification in the Cochlea." Arch. Oto.Rhino. Laryngol. 224, pp. 37-45, 1979.

Lewis, R.S., and Hudspeth, A.J., "Voltage and Ion-dependent Conductances in Solitary Vertebrate Hair Cells. Nature 304, pp. 538-541, 1983.

Lim, D.J., "Cochlear Anatomy Related to Cochlear Micromechanics. A Review." J. Acoust. Soc. Am. 67, pp. 1686-1695, 1980.

Miller, J., Canlon, B., Flock, A., and Borg, E., "High Intensity Noise Effects on Stereocilia Mechanics." Assoc. Res. Otolaryngol. 8, pp. 50, 1985.

Mountain, D.C., "Changes in Endolymphatic Potential and Crossed Olivocochlear Bundle Stimulation Alter Cochlear Mechanics." Science 210, pp. 71-72, 1980.

Mountain, D.C, Giester, C.D., and Hubbard, A.E., "Stimulation of Efferents Alters the Cochlear Microphonic and the Sound-induced Resistance Changes Measured in Scala Media of the Guinea Pig." Hearing Res. 3, pp. 231-240, 1980.

Nuttall, A.L., "Influence of Direct Current on dc Receptor Potentials from Cochlear Inner Hair Cells in the Guinea Pig." J. Acoust. Soc. Am. 77, pp. 165-175, 1985.

Ohmori, H., "Mechano-electric Transduction Currents in Isolated Vertibular Hair Cells of the Chick." J. Physiol. 359, pp. 189-218, 1984.

Orman, S., and Flock, A., "Active Control of Sensory Hair Cell Mechanics Implied by Susceptibility to Media that Induce Contraction in Muscle." Hearing Res. 11, pp. 261-266, 1983.

Patuzzi, R.B. and Sellick, P.M., "A Comparison Between Basilar Membrane and Inner Hair Cell Receptor Potential Input-Output Functions in the Guinea Pig Cochlea." J. Acoust. Soc. Am. 74, pp. 1731-1741, 1983.

Pickles, J.O., Cormis, S.D., Osborne, M.P., "Cross-links between Stereocilia in the Guinea Pig Organ of Corti, and their Possible Relation to Sensory Transduction." Hearing Res. 15, 103-112, 1984.

Robertson, D., "Horseradish Peroxidase Injection of Physiologically Characterised Afferent and Efferent Neurones in the Guinea Pig Spiral Ganglion." J. Hearing Res. 15, pp. 113-122, 1984.

Russell, I.J., "The Responses of Vertebrate Hair Cells to Mechanical Stimulation." In Neurones Without Impulses, Ed. by A. Roberts and B.M.H. Bush, Cambridge University Press, Cambridge, pp. 117-145, 1979.

Russell, I.J., "The Origin of Receptor Potential in Inner Hair Cells of the Mammalian Cochlea - Evidence for Davis' Theory." Nature 301, pp. 334-336, 1983.

Russell, I.J., and Ashmore, J., "Inner Hair Cell Receptor Potentials Investigated During Transient Asphyxia." In Hearing- Physiological Bases and Psychophysics, Ed. by R. Klinke and R. Hartman, Springer-Verlag, Berlin, pp. 10-16, 1983.

Russell, I.J., and Cody, A.R., "The Voltage Responses of Inner and Outer Hair Cells in the Guinea-pig Cochlea to Low Frequency Tones." Brit. Journ. Audiol. 18, pp. 253-254, 1984.

Russell, I.J., and Cody, A.R., "The Voltage Responses of Cochlear Hair Cells to Current Injection." Brit. J. Audiol. (in press), 1985.

Russell, I.J., and Cowley, E.M., "The Influence of Transient Asphyxia on Receptor Potentials in Inner Hair Cells of the Guinea Pig Cochlea." Hearing Res. 11, pp. 373-384, 1983.

Russell, I.J., and Sellick, P.M., "Intracellular Studies of Hair Cells in the Mammalian Cochlea." J. Physiol. 284, pp. 261-290, 1978.

Russell, I.J., and Sellick, P.M., "Low Frequency Characteristics of Intracellularly Recorded Receptor Potentials in Mammalian Hair Cells." J. Physiol. 338, pp. 179-206, 1983.

Siegel, J.H., and Kim, D.O., "Efferent Neural Control of Cochlear Mechanics? Olivocochlear Bundle Stimulation Affects Biomechanical Nonlinearity." Hearing Res. 6, pp. 171-182, 1982.

Spoendlin, H., "The Afferent Innervation of the Cochlea." In Evoked Electrical Activity in the Auditory Nervous System. Eds R.F.Naunton and C. Fernandez, Academic Press, New York, pp. 21-41, 1978.

Weiss, T.F., "Bidirectional Transduction in Vertebrate Hair Cells: A Mechanism for Coupling Mechanical and Electrical Processes." Hearing Res. 7, pp. 353-360, 1982.

Wilson, J.P., "Model for Cochlear Echoes and Tinnitus based on an Observed Electrical Correlate." Hearing Res. 2, pp. 527-532, 1980.

Wilson, J.P., and Johnstone, J.R., "Basilar Membrane and Middle Ear Vibration in the Guinea pig Measured by Capacitive Probe." J. Acoust Soc. Am. 57, pp. 705-723, 1975.

Furosemide affects ear-canal emissions produced by the injection of ac currents into scala media.

A. E. Hubbard+, D. C. Mountain* and E. L. LePage*. Departments of Otolaryngology and Departments of Systems, Computer and Electrical Engineering+ and Biomedical Engineering*, Boston University, 110 Cummington Street, Boston, MA 02215.

ABSTRACT: Tones produced by the injection of current into scala media were measured in the ear canal. The ear-canal emissions were substantially modified with the intravenous injection of a diuretic, furosemide. Changes in the endolymphatic potential and the cochlear microphonic were also measured.

INTRODUCTION

The effects of loop diuretics such as furosemide on cochlear function have been studied at length (Brown, 1981; Asakuma and Snow, 1980; Evans and Klinke, 1982; Pratt and Comis, 1982; Sewell, 1984 a,b,c). Such substances have been shown to depress the endocochlear potential (EP), the N1 component of the compound action potential and the cochlear microphonic (CM). Single unit studies of the effect of furosemide on neural activity have been carried out by Evans and Klinke (1982) and Sewell (1984 a,b,c), both studies in cats. Other cochlear research has exploited the potent drug interaction of loop diuretics such as furosemide in conjunction with aminoglycosides such as kanamycin (Brummett, et al. 1975,1981; Santi et al. 1982). Both teams of investigators found little or no acute effect due to administration of kanamycin alone, however, when followed approximately 2 hours later with furosemide administration, there was pronounced, permanent depression of cochlear responses, particularly in the first turn. Histologic examination showed destruction of cochlear hair cells.

This paper deals with the effect of furosemide and furosemide with kanamycin on the electrically stimulated ear-canal emissions, the EP, and the CM. In recent studies (Hubbard and Mountain,1983) it was found that when alternating current was delivered into scala media while at the same time acoustic tones were presented to the ear, acoustic emissions could be measured at the eardrum at the frequencies corresponding to the intermodulation distortion frequencies and the frequency of the injected current. The process responsible for the emissions was physiologically vulnerable. It was also shown that the production of emissions likely involved cochlear transduction.Further exploration of this phenomenon has been pursued using furosemide and kanamycin with furosemide. An effect of ethacrynic acid upon a sound-evoked cochlear mechanical response has previously been demonstrated (Anderson & Kemp, 1979).

METHODS

The method used to elicit and measure electrically-stimulated,ear-canal acoustic emissions by current injection in scala media of the gerbil has been described previously (Hubbard and Mountain,1983). In the the present series of experiments, 125 mg/kg furosemide was injected intravenously (IV) into an animal which in some cases had been given 400 mg/kg kanamycin intramuscularly (IM) approximately four hours earlier. Prior to furosemide injection, frequencies and sound levels which produced electrically-stimulated acoustic emissions with satisfactory signal-to-noise ratio were selected. All but one set of data were taken from the first cochlear turn and the other data were taken from the second turn. Because of signal-to-noise considerations for the components of interest, the acoustic frequency was always 1600 Hz or 800 Hz and the frequency of the injected current was 600 Hz or 300 Hz. Before, during, and following furosemide injection, the same stimuli were presented and data were collected repetitively at fixed time intervals varying from 45 seconds to five minutes apart, depending on the apparent rate of change of measured parameters,as observed on-line. The data taken always included the acoustic emissions, but in later experiments CM and EP were recorded interleaved with the emission signal, delayed in time by a known interval which was on the order of a few seconds. In some cases, the CM and EP were obtained by using the current-delivery electrode (which can exhibit slow dc-drift)as a recording electrode. In other experiments, the voltages were recorded by means of a second pipette in scala media.

The automated method of data collection and analysis has been described elsewhere (Hubbard and Mountain,1983). Basically, data records were time-averaged, and the final averaged time record was Fourier transformed. At successive times before and after furosemide delivery, the "raw" data considered was the frequency-domain representation of the averaged time signal. Three of these spectra are shown, but in general, quantitative consideration was achieved by plotting the magnitude and phase angle of various components versus time. The procedure used to deliver the furosemide was primitive. The injections were delivered by hand-held syringe, during stimulus presentation. The duration of the injection was approximately two minutes.

RESULTS

Figure 1 shows three spectra of the calibrated acoustic signal measured near the eardrum, the first obtained prior to furosemide injection and the other two obtained at approximately five and 25 minutes following injection. In the top panel one observes acoustic components typically obtained using a 1600 Hz delivered sound at 80 dB SPL in conjunction with 20 uA current at 600 Hz delivered inside of scala media of the first cochlear turn having an initial EP of 90 mV. The middle panel shows the same components measured at a time when EP had decreased to -30 mV. Notably, the sideband (SB) components flanking the fundamental acoustic (A1) component have decreased in size. The component at 600 Hz, which we call the direct emission (DE), because it can be produced without accompanying sound input has also been reduced in size relative to its pre-furosemide-injection level.In the bottom panel, the sideband components and direct emission are much larger than obtained prior to furosemide injection, with sideband components now visable flanking the components which are the acoustic harmonics (A2,A3). The size of the acoustic harmonics themselves, can be seen to vary from panel to panel. This latter effect has often been noticed, although not yet studied in detail. Also noted is the presence of the second harmonic of the direct emission (DE2), which was noted typically in about half of our experiments which did not involve the use of furosemide.

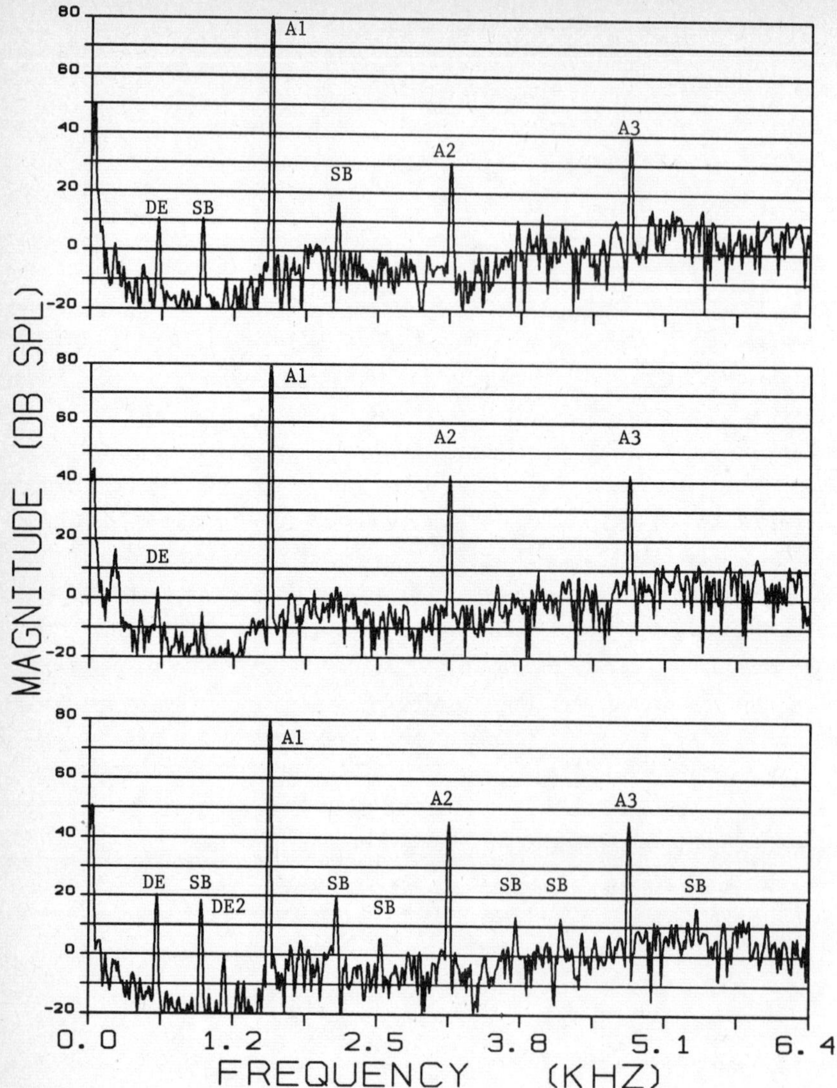

Figure 1. Spectral content of the acoustic signal measured near the eardrum during a furosemide injection experiment. The largest component (A1) in each panel is the input acoustic signal at 1600 Hz. Components at multiples of 1600 Hz are harmonics of the acoustic input signal(A2,A3). Additional components are the sideband components (SB) and the direct emission (DE). Top panel: before furosemide injection. Middle panel: EP was at a minimum. Bottom panel: about one-half hour following furosemide injection.DE2 is the second harmonic of the direct emission.

Figure 2 shows the following measured experimental data as a function of time: EP,CM, DE ,LSB (the acoustic emission which is the lower-frequency modulation product with the fundamental component of the acoustic input) and the USB (the upper-frequency modulation product). During data collection, 125 mg/kg furosemide was injected intravenously. EP and CM (bottom panel) show precipitous decline within minutes following the injection. The CM changes in a manner which is similar the change in the EP. In the top panel,LSB, USB, and the DE are seen to fall with a time-course similar to that of the CM. During the initial recovery period of the EP ;the LSB,USB, and DE overshoot their initial levels. The rate of change with time is similar for the all of the emissions.

Certainly, each quantity changes with EP in the gross sense, but many aspects defy the notion that EP-change is the single important factor relating changes with furosemide. It is obvious that the emissions recover and overshoot well before EP has even begun to recover. During this time, the CM recovers, (bottom panel) in a manner apparently, but not simply- related, to the change in EP. There is always an effect on the magnitudes of the cochlear emissions with furosemide injection. With respect to variations in effect, we feel that initial cochlear condition and the rate of furosemide injection are important factors. With injection of furosemide IV over a period of about about one minute in a fresh cochlear preparation, EP can drop from 90 mV to -30 mV within about three minutes If furosemide is infused more slowly, the rate-of-change of EP is slower.This means that more data can be collected during the transition period, however the observed effects are less dramatic. There is always an obvious enhancement of the emissions which occurs on the order of 15-30 minutes following furosemide injection, and up to three successive injections have been shown to produce successive enhancements, with some, but not much, prior reduction in the magnitude of the emissions.Normal saline IV has been injected as a control in two animals which later showed the furosemide-induced emission response,without effect.

The phase angles of the LSB,USB,DE, and CM show systematic, slow, and between one another, similar variation with time following furosemide injection. The phase angles return to pre- furosemide-injection values with time following injection.The magnitudes of the fundamental, second and third harmonic acoustic components recorded following furosemide injection are also obvserved to change . Seemingly systematic (but very small) changes in the fundamental acoustic component have been noted in only two animals. When 400 mg/kg kanamycin was injected into an animal four hours prior to the time of furosemide injection,

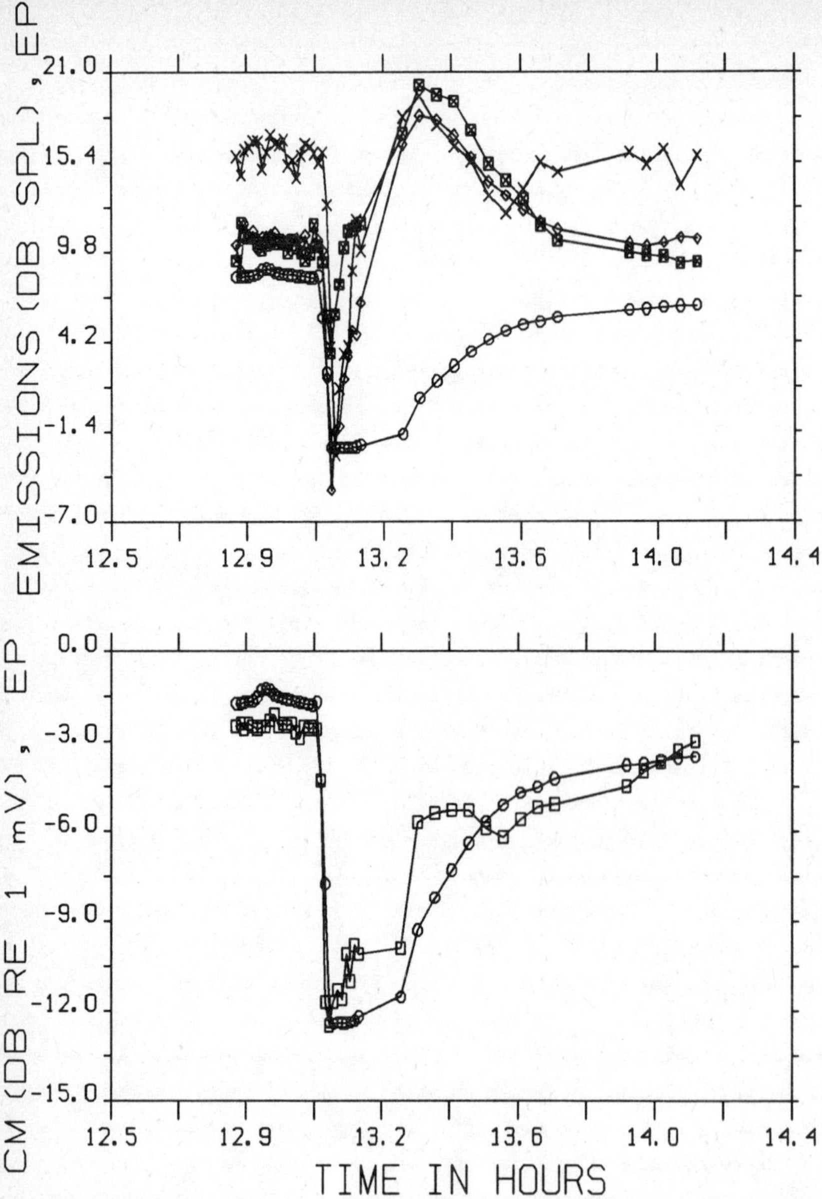

Figure 2. The time course of changes in EP,CM, and acoustic emissions following furosemide injection. Bottom panel: Scaled and shifted EP plotted along with CM. The EP (circular symbols) is scaled by a factor of 10 and shifted downwards by 10 units, i.e. the actual change in EP is from around +90 mV to -30 mV. Top panel: EP (circular symbols) scaled by a factor 10 is presented along with the direct emission (x-with-square) and the upper (x) and lower (diamond) sideband components. These figures and Fig. 1 are from the same furosemide injection run, from the same experiment.

the time-course of the EP variation was similar to that obtained without the use of kanamycin. The early time-course of the other measured quantities was also similar to that of data obtained with furosemide only. However, after about 1-hour CM, LSB, USB, and DE all dropped precipitously even though the partially-recovered EP remained approximately constant. In this preparation , the expected effect was hair-cell death (Brummett et al. 1975).

DISCUSSION

To explain the data obtained requires more than our present understanding of the phenomenon of electrically- stimulated acoustic emissions and the action of furosemide.The emissions have been shown to be related to transduction (Hubbard and Mountain,1983) . Changes in EP amount to a change in driving force (EMF) in the transducer circuit, and therefore we expected the emissions to change with EP changes. A slight growth of both EP and the emissions prior to EP- decline as well as the similar abrupt decline of the emissions with EP are,however, the only aspects of the data which tend to support that simple explanation. The emissions clearly show growth even though EP remains at a low value. When EP finally shows recovery with time, the emissions are at a level well-above pre-furosemide-injection values.During the overshoot period, changes in the emissions are similar to changes in the CM, which itself at this time changes in a way apparently unrelated to changes in the EP (cf. Fig.2).

The change in the emissions with furosemide injection might be related to a change in impedance which shunts varying amounts of current through the elements (probably the hair cells) which produce the cochlear emissions. If this is the case, changes with furosemide should resemble changes in the level of the injected current. Figure 1 could be interpreted as showing just that stimulus manipulation. The almost- parallel time courses of the emissions plotted in Fig.2 could also be produced by varying the level of injected current . The argument leads to the conclusion that first more, and then less current is shunted away from the hair cells. CM changes are apparently correlated with EP changes, but even for the CM the case is not entirely simple.

On the other hand, changes in mechanics are also a possible explanation for the observed phenomena. For example, the situation might involve shifts in basilar membrane position which are related to EP, which in turn, might affect ac-motional characteristics, i.e. the emissions. Systematic and similar phase angle changes in the emissions and CM could suggest that the phase angle of the displacement of the basilar membrane also changes. Under the assumption that the mechanical traveling-wave is governed, for the frequencies utilized,by spring and mass characteristics, it could be that the entire cochlear partition changed its stiffness characteristics due to furosemide injection.

REFERENCES

Anderson, S. D., Kemp, D. T., (1979). The evoked cochlear mechanical response in laboratory primates. A preliminary report. Arch.Otorhino- laryngol., 224(1-2). 47-54.

Asakuma. S., Snow.J.B.Jr.,(1980). Effects of furosemide and ethacrynic acid on the endocochlear direct current potential in normal and kanamycin sulfate-treated guinea pigs. Otolaryngol.Head.Neck.S., 1980, Mar-Apr. 88(2). 188-93.

Brown, R. D.,(1981). Comparisons of the acute effects of I.V. furosemide and bumetanide on the cochlear action potential (N1) and on the A.C. cochlear potential (CM) at 6 KHz in cats, dogs and guinea pigs. Scand.Audiol [Suppl]., 14 Suppl. 71-83.

Brummett, R. E., Traynor, J., Brown, R. and Himes, D., (1975). Cochlear damage resulting from kanamycin and furosemide. Acta Otolaryngol. 80, 86-92.

Brummett, R. E., Bendrick, T., Himes, D., (1981). Comparative ototoxicity of bumetanide and furosemide when used in combination with kanamycin. J.Clin.Pharmacol., Nov-Dec. 21(11-12 Pt 2). 628-36.

Evans, E. F., Klinke. R., (1982). The effects of intracochlear and systemic furosemide on the properties of single cochlear nerve fibres in the cat. J.Physiol (Lond)., Oct. 331. 409-27.

Hubbard, A.E. and Mountain, D.C., (1983). Alternating current delivered into the scala media alters sound pressure at the eardrum. Science, 222, 510-512.

Pratt, S. R., Comis, S. D., (1982). Chronic effects of loop diuretics on the guinea-pig cochlea. Br.J.Audiol., May. 16(2). 117-22.

Santi, P.A., Ruggero, M.A., Nelson, D.A. and Turner, C.W., (1982). Kanamycin and bumetanide ototoxicity: Anatomical, physiological and behavioral correlates. Hear.Res. 7, 261-279.

Sewell, W. F., (1984). The relation between the endocochlear potential and spontaneous activity in auditory nerve fibres of the cat. J.Physiol (Lond)., Feb. 347. 685-96.

Sewell, W. F., (1984). The effects of furosemide on the endocochlear potential and auditory-nerve fiber tuning curves in cats. Hear.Res. 14, 305-314.

Sewell, W.F., (1984). Furosemide selectively reduces one component in rate- level functions from auditory-nerve fibers. Hear.Res. 15, 69-72.

OUTER HAIR CELL MOTILITY: A POSSIBLE ELECTRO-KINETIC MECHANISM

William E. Brownell* and Bechara Kachar
Departments of Neuroscience and Surgery (ENT), University of Florida Gainesville, FL. and Lab of Neurobiology, NIH, Bethesda, MD.

ABSTRACT

Video enhanced microscopy is used to measure shape changes of isolated outer hair cells in response to transcellular electrical stimulation. Low frequency sinusoidal stimulation results in sinusoidal modulation of cell length about its resting value. Elongation is associated with positive potential gradients. Electrically evoked motility is unaffected by the presence of metabolic poisons that interfere with the production of ATP. Isotonic dilution of the bathing medium results in an increase in displacement magnitude. The symmetry of the displacements, their dependence on voltage gradients, independence of ATP, and evidence for an inverse relation between ionic concentration and magnitude of the movement argue against a contractile mechanism, but are compatible with electro-kinetic processes.

COCHLEAR MICROARCHITECTURE AND OUTER HAIR CELL LENGTH CHANGES

An exciting architectonic feature of the organ of Corti is its conspicuous colonnade of outer hair cells (Figure 1). The columnar effect results from the fact that outer hair cells (OHC) contact other cells only at the ends of their cylindrical cell bodies and are otherwise isolated in the spaces of Nuel. The large extracellular spaces surrounding the OHC contrast with the spaces found around the inner hair cells and the hair cells of the vestibular end organs. These other hair cells are enveloped by their supporting cells. The distances that separate cells in most organs, such as the liver or the brain, are small relative to cellular dimensions (glial cells are usually found within 30 nm of neurons in the CNS). The large extracellular spaces of Nuel surrounding outer hair cells are on the order of micrometers. *In vitro* studies (Brownell, 1984; Brownell et al., 1985a) reveal OHC to possess a surprising rigidity that must contribute to the mechanical coupling between the basilar membrane and reticular lamina and thereby maintain the architectural integrity of the organ of Corti. The same studies also

* Current address: Departments of Otolaryngology - Head & Neck Surgery and Neuroscience, The Johns Hopkins University School of Medicine, Baltimore, Maryland 21205

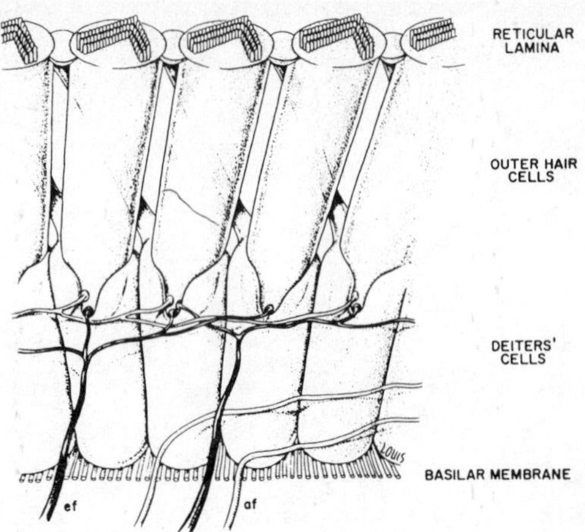

Figure 1. View of a single row of outer hair cells (OHC) and their supporting cells as it would appear if viewed from the central axis of the cochlea looking towards the bony capsule. This view emphasizes the microarchitectural features of the organ of Corti that permit free expression of OHC length changes. OHC have no cellular attachments along most of their length. They contact afferent (af) and efferent (ef) fiber terminals within an invagination of the Deiters´ cell and make a very secure attachment with the distal portion of the Deiters´ cell phalangeal process in the reticular lamina. Cytoskeletal elements are abundant in the stereociliar ends of the OHC and in the interdigitating phalangeal processes. The secure occluding junction attachments between the cells around the stereociliar end contributes to the rigidity of the reticular lamina. The proximal portion of the phalangeal processes has tensile strength but is free to rotate about its point of attachment with the Deiters´ cell body. The cochlea spirals towards the apex to the right. Note how the OHC slant towards the apex, while the phalangeal processes of the Deiters´ cells slant towards the base. The triangle formed by an OHC, the phalangeal process of the Deiters´ cell on which it rests, and that portion of the reticular lamina between the phalangeal process and the hair cell represent a structural unit within the cochlear partition. A change in OHC length will change the separation between the reticular lamina and the basilar membrane.

describe OHC length changes in response to a variety of electrical and chemical stimuli. The unique placement of OHC in an open cellular matrix would permit free expression of the length changes _in vivo_ which could effect the response of the cochlear partition to acoustic vibrations. For example, the length changes could mediate the effects of efferent fiber stimulation on cochlear mechanics (Brownell, 1984; Brownell et al., 1985a). If the OHC length changes are rapid enough they may provide the source of mechanical energy required to account for otoacoustic emmissions, low thresholds and the narrow bandpass characteristics associated with the mechanics of the cochlear partition (Davis, 1983; Zwicker and Schloth, 1984). The time constants for OHC length changes are determined by the underlying cellular mechanisms. Our current experiments suggest the basis of OHC motility is an electrokinetic mechanism that may be capable of generating force at acoustic frequencies.

RECENT EXPERIMENTS

Isolated OHC were prepared using techniques similar to those previously described (Brownell, 1984; Brownell et al., 1985a). Liebowitz L-15 culture medium with 0.5% bovine serum albumin and 1.0% HEPES added was used for dissection, dissociation and cell maintenance. A 100 ul volume of the medium containing isolated cells was placed on a polylysine coated cover slip. Cells were viewed with video enhanced microscopy using a Zeiss Axiomat configured as an inverted microscope (Kachar, 1985, Kachar et al., 1985). The experiments were carried out in the droplets at room temperature and without perfusion. OHC that maintained their characteristic cylindrical shape and cytoplasmic organization were selected for experiments. Stimulation was achieved with a pair of teflon insulated #30 silver wire electrodes separated by a distance of 150-250 um and mounted to the condenser front lens of the microscope. The electrode pair was lowered into the drop containing the cell. Sinusoidal voltage waveforms with a frequency of 5 Hz, at currents of less than 1 mA were passed across the length of the cell. The cell´s response to the electric stimulation was analysed from frame by frame play back of recorded experiments.

OHC attached to the cover slip at either their synaptic or stereociliar ends. Conspicuous stimulus evoked displacements of the unattached end were observed in 31 out of 37 OHC obtained from 11 animals. Motility was not observed in other cell types (red blood cells, Deiters´ and Hensen´s cells) that co-isolate with the OHC nor was a stimulus coupled movement of the OHC stereociliar bundle detected. Most of our observations were made 3-5 hours after the dissection procedures were initiated. The length changes of 16 cells were large enough to be analysed for the direction of movement relative to the applied electric field. Either the end of attachment changed or the microscope stage was rotated during recordings on 7 of the 16 cells so that a total of 23 observations could be made on the cells with different stimulus configurations. The most common observation (20 out of 23) was for elongation to occur when the electrode nearest the free end was positive relative to the one near the fixed end (a positive potential gradient). Shortening of the cell occurred during the negative potential gradient phase of the stimulus sinusoid.

Two of the 16 cells responded with displacements that permitted a finer analysis of their movements. The peak to peak magnitude of their displacements reached 0.5-0.6 um. Their cell length was sinusoidally modulated about its resting value at the stimulus frequency with little evidence of rectification or harmonic distortion. The phase plots of displacement against stimulus voltage for these cells resulted in narrow ellipsoidal Lissajous figures.

Once OHC were firmly attached to the cover slips the bathing medium could be changed by gently aspirating the fluid and replacing it with the desired medium. OHC responses were unaffected for up to one hour after fluid replacement with either: 1) 2 mM dinitrophenol dissolved in the L-15 culture medium; or 2) 200 ug/ml iodoacetic acid dissolved in the L-15 culture medium.

An additional set of experiments, suggested by S. McLaughlin (personal communication), involved addition of 33 ul of a dilution medium to the 100 ul droplet. The dilution medium was made of a 10 mM HEPES solution brought to the osmolarity of the cell maintenance media with sucrose (318 mOsmol). Its addition diluted the ionic concentrations of the bathing medium by 25% and resulted in an enhancement of OHC length changes. Cells were encountered that produced little or no movement in the standard bathing medium but displayed a robust movement after dilution. We were able to reverse the effect on two cells by replacing the diluted bathing medium with the original cell maintenance solution. Complete replacement of the bathing medium with the dilution medium resulted in a rapid loss of OHC turgor and their gradual deterioration over a fifteen minute time course.

LAMINATED SUBSURFACE CISTERNAE AND ELECTRO-KINETIC MECHANISMS

Most characteristics of electrically evoked length changes in isolated OHC are incompatible with conventional mechanisms of cellular motility. Contractile mechanisms generally feature a rapid contractile phase and a slow relaxation phase. The time course for both elongations and shortening of OHC is the same. The dependence of the direction of movement on the potential gradient relative to the fixed end of the cell is even more difficult to explain with conventional contractile mechanisms. Contraction in skeletal muscle results from an interaction between the proteins actin and myosin that is triggered by calcium. Calcium release in turn is governed by the muscle's membrane potential. Ion flow across most cellular membranes would not reverse if the cell were to change its orientation to an extracellular field or its point of attachment to a cover slip.

Free end elongation with a positive potential gradient is consistent with previous intracellular current injection experiments

(Brownell, 1984; Brownell et al., 1985a) in which depolarizing current injections into the synaptic end of the cell resulted in shortening while hyperpolarizing current injections caused elongation. The intracellular electrode fixed the synaptic end of the cell while the stereociliar end was free to move. The intracellular gradients causing elongation and shortening are the same sign as the extracellular gradients.

Most contractile mechanisms rely on ATP as an energy source. OHC motility appears to be independent of ATP as demonstrated by the fact that it continues after incubation with dinitrophenol (DNP) and iodoacetic acid (IAA)(Kachar et al, 1985). DNP and IAA uncouple oxidative phosphorylation and inhibit glycolytic phosphorylation respectively. The possibility that OHC motility does not require conventional energy sources follows from reports of postmortem generation of oto-acoustic distortion products (Schmiedt and Adams, 1981) for time periods corresponding to the postmortem generation of cochlear microphonics. The possibility that the energy requirements of motility are not maintained by hair cells but rather by the stria vascularis is indicated by deoxyglucose studies that it is the stria and not the OHC that increase their glucose uptake with increases in the intensity of acoustic stimulation (Ryan et al., 1984). Our dilution experiments resulting in greater responses when the bathing media contains fewer ions is extremely difficult to reconcile with conventional contractile mechanisms. A decrease in external ions would reduce skeletal muscle motility.

Finally, there is a poor representation of contractile proteins in the OHC between the cuticular plate and the nucleus (Flock, 1983; Zenner, 1985). It is in this region of the cell that the movement seems to originate and it is in this region that an anastomosis of subsurface cisternae (Saito, 1983) is found that is unique to the OHC (Figure 2). The laminated cisternal system of the mammalian OHC is not found in any other hair cell and may not be present any other cell. Its involvement in OHC motility is suggested by investigations showing that aspirin can reversibly alter outer hair cell laminated cisterae (Douek et al., 1983) and reversibly abolish spontaneous oto-acoustic emissions (McFadden and Plattsmier, 1984).

Electro-kinetic processes, including electro-osmosis and electrophoresis, are mechanisms that might provide the motive force underlying OHC motility. Electro-phoresis is a phenomena that is extensively utilized in biochemistry laboratories to separate charged molecules by driving them through a conducting medium with an applied electric field. Electro-osmosis was first described by Helmholtz and is reviewed in a

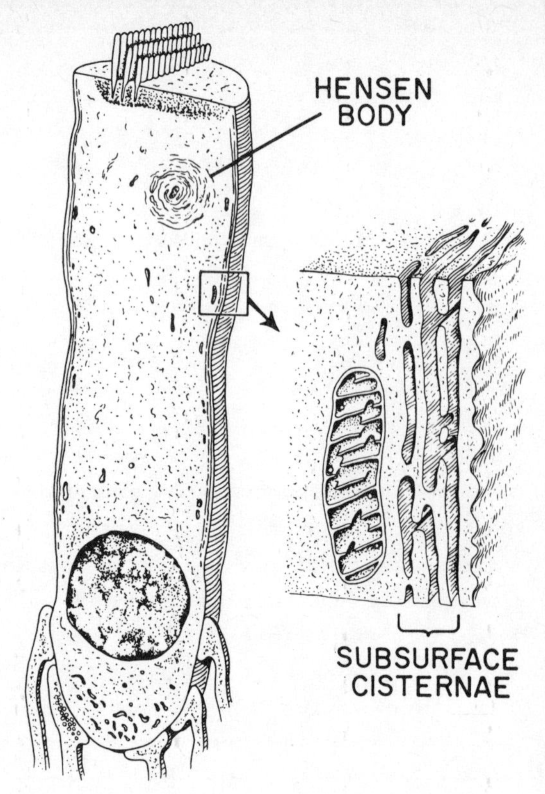

Figure 2. Morphology of outer hair cell subsurface cisternae, an anastomosis of flattened membranous sacs adjacent to the plasma membrane (blowup of the subsurface cisternae is on the right). General organization of the OHC is depicted on the left. Stereocilia insert into the cuticular plate and the nucleus of the cell is eccentrically placed towards the synaptic zone. The laminated cisternal system is found along the entire length of the OHC between the cuticular plate and the nucleus, in the same portion of the cell exposed to the spaces of Nuel. The outermost layer is parallel to and 30-35 nm from the plasma membrane. Each layer is parallel to its neighbors and separated by a gap of 15-20 nm. Filamentous material (not shown) connects the outermost layer and lateral plasma membrane. Mitochondria (see blowup on right) are associated with the cisternal system which includes the Hensen body (an annulate lammellar body often found near the cuticular plate). Cytoskeletal filaments and tubules are abundant near the cuticular plate and in the synaptic region but are not found in the cytoplasm central to the subsurface cisternae.

recent article by McLaughlin and Mathias (1985). It is a field-induced movement of fluid adjacent to a charged surface and may be viewed as the complement of electro-phoresis in that the object being moved is the fluid, while the charge is fixed to the surface. The biological equivalent of the charged surface would be the phopholipid bilayer membranes and the charged proteins that are inserted into the bilayer. Electro-osmotic fluid velocity is proportional to the applied electric field and the zeta potential associated with the charged membrane. The laminated subsurface cisternal system represents an intracellular proliferation of membrane. Our dilution experiment would have the effect of increasing the zeta potential by increasing the Debye length for the ions in the cytoplasm which, in turn, would result in greater electro-osmotically driven fluid velocity. Electro-osmotically driven cytoplasmic fluid movements need not rely on the metabolic activity of the cell but could instead be driven by the applied electric field. The absense of an effect by metabolic poisons on OHC length changes is compatible with an electro-kinetic response. The symmetry of the sinusoidal displacements of the cell about its resting length and the relation between the direction of the displacement and the sign of the potential gradient are consistent with the relation between electro-osmotic fluid velocity and the applied electric field (McLaughlin and Mathias, 1985). The elongated shape of the OHC, the presence of the laminated subsurface cisternae, and the presence of a large potential gradient across the cochlear partition that drives a substantial ionic current (Brownell et al., 1985b) provide all the necessary conditions for electro-osmosis to occur. It is the dependence on the applied electric field that increases the potential significance of an electro-osmosic basis for OHC length changes. Electrical potentials that vary at acoustic frequencies have been measured in the cochlea for over 50 years. Cochlear microphonics are believed to be generated in large part by mechano-electric transduction associated with the OHC. OHC are also electro-mechanical transducers that may use the energy of the microphonics to power the cochlear amplifier (Davis, 1983) and in certain cases produce oto-acoustic emissions.

ACKNOWLEDGEMENTS

Experiments were conducted in the Laboratories of Neurobiology and Neuro-Otolaryngology at the National Institutes of Health in Bethesda, Maryland. R.A. Altschuler, and J. Fex contributed to and participated in the experiments. We would like to thank S. McLaughlin for his helpful suggestions and N. Brownell, M. Zidanic, G.A. Spirou and P. Dulguerov for editorial comments. Work supported by research grants BNS-12174 from NSF(USA) and NS-19050 from NINCDS.

REFERENCES

Brownell, W.E., "Microscopic Observation of Cochlear Hair Cell Motility." Scan. Elec. Micro. 1984/III, pp. 1401-1406, 1984.

Brownell, W.E., Bader, C.R., Bertrand, D., and de Ribaupierre, Y., "Evoked Mechanical Responses of Isolated Cochlear Outer Hair Cells." Science 227, pp. 194-196, 1985a.

Brownell, W.E., Zidanic, M., and Spirou, G.A., "Standing Currents and Their Modulation in the Cochlea." In Neurobiology of Hearing: The Cochlea, Ed´s R.A. Altschuler, D. Hoffman, and R. Bobbin, Raven Press, N.Y., in the press, 1985b.

Davis, H., "An Active Process in Cochlear Mechanics." Hearing Res., 9, pp. 79-90, 1983.

Douek, E.E., Dodson, H.C., and Bannister, L.H., "The Effects of Sodium Salicylate on the Cochlea of Guinea Pigs." J. Laryngol. Otol. 93, pp. 793-799, 1983.

Flock, A., "Hair Cells, Receptors With a Motor Capacity?" In Hearing-Physiological Bases and Psychophysics, Ed's R. Klinke and R. Hartmann, Springer-Verlag, Berlin-Heidelberg, pp. 2-7, 1983.

Kachar, B. "Asymmetric Illumination Contrast: A Method of Image Formation for Video Light Microscopy." Science 227, pp. 766-768, 1985.

Kachar, B., Brownell, W.E., Fex, J., and Altschuler, R., "Mechanical Response of Outer Hair Cells to Transcellular Stimulation with Alternating Currents." Biophys. J. 47, pp. 442a, 1985.

McFadden, D., and Plattsmier, H.S., "Aspirin Abolishes Spontaneous Otoacoustic Emissions." J. Acoust. Soc. Am. 76, pp. 443-448, 1984.

McLaughlin, S., and Mathias, R.T., "Electro-osmosis and the Reabsorption of Fluid in Renal Proximal Tubules." J. Gen. Physiol. 85, pp. 699-728, 1985.

Ryan, A.F., Woolf, N.K., Sharp, F.R., "Deoxyglucose Uptake Patterns in the Auditory Pathway: Metabolic Response to Sound Stimulation in the Adult and Neonate." Abst. Ass. Res. Otolaryngol. 7, pp. 143-144, 1984.

Saito, K., "Fine Structure of the Sensory Epithelium of Guinea-pig Organ of Corti: Subsurface Cisternae and Lamellar Bodies in the Outer Hair Cells." Cell and Tissue Res. 229, pp. 467-481, 1983.

Schmiedt, R.A., and Adams, J.C., "Stimulated Acoustic Emissions in the Ear Canal of the Gerbil." Hearing Res. 5, pp. 295-305, 1981.

Zenner, H.P., "Structure of Hair Cells." In Neurobiology of Hearing: The Cochlea, Ed's R.A. Altschuler, D. Hoffman, and R. Bobbin, Raven Press, N.Y., in the press, 1985.

Zwicker, E., and Schloth, E., "Interrelation of Different Oto-acoustic Emissions." J. Acoust. Soc. Am. 75, pp. 1148-1154, 1984.

VISUALIZATION OF SENSORY HAIR CELLS IN AN *IN VIVO* PREPARATION

Sietse M. van Netten and Alfons B.A. Kroese
Department of Biophysics
Laboratory for General Physics, University of Groningen
Westersingel 34, 9718 CM Groningen, The Netherlands

ABSTRACT

In order to obtain visual control over micromechanical and electrophysiological measurements on the sensory hair cells of the fish lateral line *in vivo* we developed an optical method based on polarization microscopy. The thickness of the preparation prevents the use of transmitted light microscopy. The results of the investigation show that under conditions of polarized incident light illumination, visualization of the cell surface and of the hair bundles of the sensory hair cells can be realized. The image of the hair cells and bundles is formed by the light reflected back by the nerves underlying the low reflective sensory epithelium in the macula. Although at a magnification of 800 × the image of the sensory epithelium is contrast limited, under the most favourable conditions of illumination the individual stereocilia and the kinocilium of the hair bundle can be distinguished. The cupula, a jelly-like structure in which the hair bundles are embedded, does not noticeably disturb the image of the hair cells and its integrity assures physiological conditions for the apical surface of the hair cells.

1. INTRODUCTION

The progress towards understanding the transduction process in vertebrate sensory hair cells is limited at present by a lack of quantitative information on the mechanics of stereociliar motion in the physiological range. The microscope described below has been built in the course of the development of an *in vivo* vertebrate sensory hair cell preparation in which electrophysiological and micromechanical measurements on the hair cells can be combined under physiological conditions and under visual control.

The lateral line is distributed over the body surface of aquatic vertebrates and is concerned with the detection of water motion. In fish one type of organ is located in canals covered by skin and

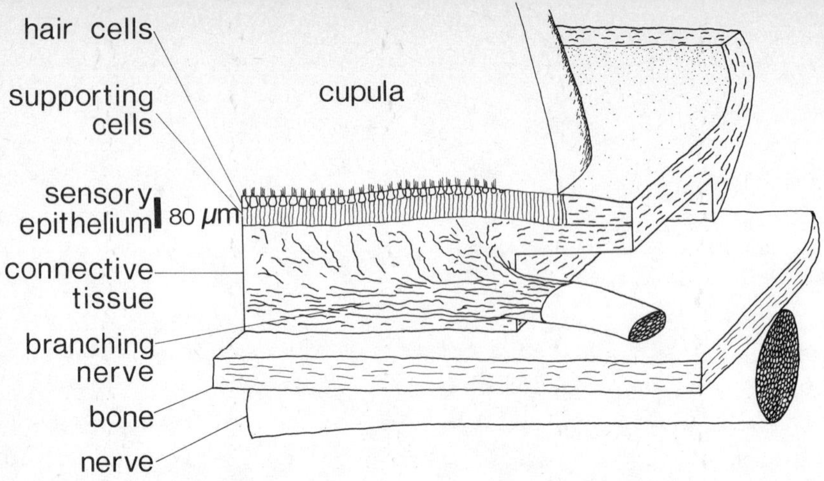

Figure 1. Schematic diagram of a cross section of a neuromast in the lateral line canal of the ruff. After Jakubowski (1963). Vascularization is not shown.

protected by a bone bridge. The sensory hair cells are grouped together in neuromasts, which in the supraorbital lateral line canal of the ruff have a diameter of about 1 mm and contain about 1000 hair cells each. The structure of these neuromasts has been studied extensively (Kuiper, 1956; Jakubowski, 1963) and will be described only briefly. The sensory epithelium rests upon an oval disc of blood capillaries and peripheral branches of the innervating nerve fibres, which penetrate the neuromast and scatter in the connective tissue under the sensory zone in the form of a brush, as illustrated in Fig.1. The sensory epithelium is covered by the cupula, a half-sphere like shaped transparent structure which couples the water movements in the canal to the sensory hairs. Each sensory hair bundle is composed of about 40 stereocilia arranged in a hexagonal pattern behind an asymmetrically located kinocilium. The hair bundle has a height of about 15 μm. Just below the thin and transparent bony bottom of the canal a thick nerve bundle in the orbit of the fish passes the neuromast.

Light microscopical techniques for imaging low reflective biological structures are commonly used in combination with transmitted light illumination. Differential interference contrast microscopy, for instance, has been used in sophisticated studies of sensory hair cell transduction on an isolated preparation of the frog sacculus (Hudspeth and Corey, 1977). However, transmitted light microscopy is precluded by thick preparations as anaesthetized fish which are intact except for the bone bridge covering the neuromast. Therefore, we investigated the feasibility of an incident light microscopical technique.

2. MICROSCOPE

The combination of incident light illumination with polarization microscopy is a common tool in the research of metals, crystals and minerals (Rabe, 1983), because of the relatively high reflection coefficients of these materials. Since living cells in general have low reflection coefficients incident light polarization microscopy is rarely used on biological preparations. We found, however, that polarization microscopy can be applied in combination with water immersion under incident light conditions for imaging low reflective biological preparations.

The microscope system consists of standard components of the Zeiss modular system. The incident light source is an UV filtered 75 Watt Xenon arc in combination with a 4 component collector system. The light is diaphragmed and directed via a polarizer to the achromat water immersion objective (40/0.75 W), which was originally designed for transmitted light and has a working distance of 1.6 mm. The use of the water immersion objective avoids steep changes in refractive index while polarization microscopy suppresses specular reflections from lens surfaces to a high degree. The Antiflex objectives (Wada, 1970) specially designed for low reflective objects have too short a working distance for use on the lateral line preparation. Contrast could further be enhanced with a low cost video camera. Photomicrographs were recorded directly from the microscope (Kodak Technical Pan 2415).

3. PREPARATION

Experiments were performed on ruff (*Acerina cernua* L.) about 11 cm long. They were anaesthetized with intraperitoneal injection (24 mg/kg) of Saffan (Glaxovet), held rigidly in place by head and body clamps and respired artificially by a flow of tap water. Canal organs in the supraorbital canal were revealed by carefully removing the overlying skin and bone bridge. The condition of the fish was monitored by visual inspection of the blood flow through the capillary vessels in the macula. Fish were usually kept in good condition for more than 6 hours.

4. VISUALIZATION OF THE HAIR CELLS

Observing the macula through a binocular operation microscope the most striking features are the branching nerve fibres and the vascularization, while the cupula and the hair cells are not visible at all. We found, however, that in the image of the sensory epithelium formed by the incident light polarizing microscope the hair cells can be distinguished. The hair cells and bundles are best visible in those regions of the epithelium which lie above a nerve bundle, where light intensity is relatively high.

In order to gain more information on the process of image formation, the amount of linearly polarized light reaching the eye pieces of the microscope was measured as a function of focal depth. While focussing down with the microscope from 800 μm above the surface of the epithelium the amount of polarized light increases gradually, until a maximum is reached. It appeared that the focal depth at which the maximum is reached coincides with the depth at which nervous tissue is present in the preparation. In all neuromasts a maximum was found at about 200 μm below the surface of the sensory epithelium where the lateral line nerve penetrates the organ and starts sending out branches in the connective tissue (see Fig.1). In a different region of the epithelium a maximum was found at the depth of the thick nerve bundle which passes underneath the transparent bony bottom of the canal.

From the results of these experiments we conclude that the image of the hair cells and bundles is formed by the light reflected from the nerves underlying the low reflective sensory epithelium. In this way a transmitted light situation is created for those regions of the sensory epithelium which are located above nerve bundles. The fact that an image of the hair cells is formed demonstrates that during the process of reflection from the nerves the linear polarization of the light is maintained to a considerable extent.

Figure 2 shows a representative image of the sensory epithelium at a location above the branching nerve bundle. The apical surface of the hair cells, the hair bundles and the supporting cells can be easily identified. Furthermore, the morphological polarization of the hair cells by the asymmetric placement of the hair bundle on the apical surface of the cell can be observed clearly. Although the images obtained by incident light polarization microscopy are light limited, the quality of the image is such that the dimensions of the hair cells and bundles can be measured. The height of the drop-shaped sensory hair cells is about 15 μm; the diameter of the cells at the basal side is

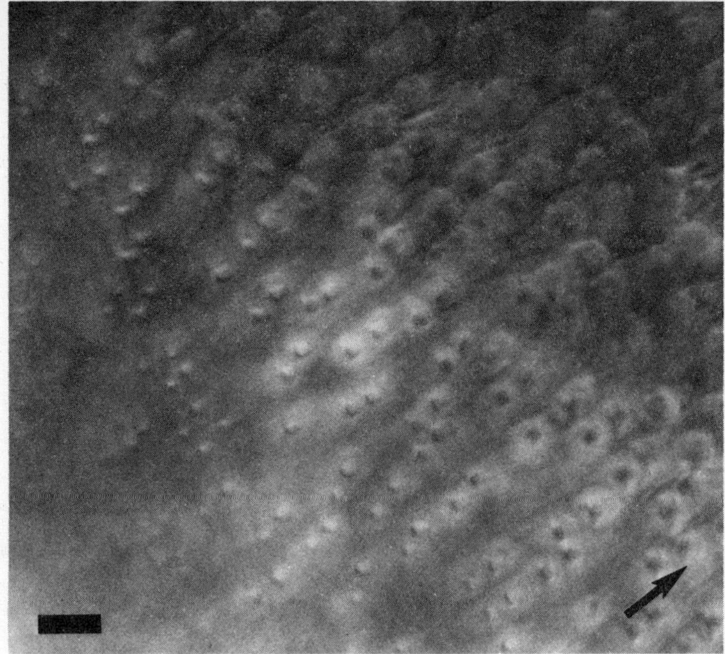

Figure 2. The sensory epithelium of a neuromast of the lateral line organ observed with incident light polarization microscopy. Focus ranges from top of hair bundles (bottom left) to just below the apical membrane of hair cells (top right). Black arrow indicates the direction of the longitudinal axis of the canal. Bar: 10 µm. Exposure time 20 s.

about 10 µm and the apical surface only 7 µm. The hair bundles have a height of about 15 µm. The supporting cells are cylindrical and about 60 µm in height and 3 µm in diameter. The smallest details in the preparation which can be distinguished in the horizontal plane of the microscope were found to be about 0.5 µm.

Sharp images of the macula up to a depth of about 100 to 150 µm can be obtained by simple focussing. Within the sensory epithelium and in the connective tissue of the macula the nerve fibres can be seen. An example of the image obtained by focussing 24 µm below the apical surface of the hair cells is given in Fig. 3. The heavy branching of the nerve fibres in between the supporting cells can be seen clearly and can be followed over long distances. In the extended vascularization of the macula erythrocytes can be seen passing through the capillary vessels. This flow was used as a simple means of monitoring the condition of the preparation.

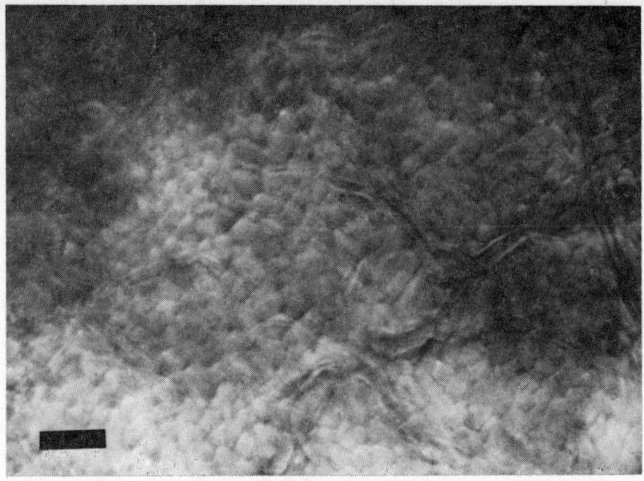

Figure 3. Image of the sensory epithelium 24 μm below the apical surface of the hair cells revealing branching nerve fibres in between supporting cells. Bar: 10 μm.

The quality of the image obtained by the microscope can be substantially improved by placing a small mirror (0.5 × 0.5 mm) parallel to the epithelial surface in the orbit of the fish directly below the bone which carries the neuromast. The mirror can be placed in position without any interference with the mechanical properties of the lateral line organ. The additional mirror brightens the image and considerably improves the contrast. Under these conditions it becomes possible to distinguish details within the sensory hair bundles such as the kinocilium and the stereocilia, which have a diameter of about 0.3 μm. An example of the image obtained in this way is given in Fig. 4. The kinocilium is slightly longer than the tallest stereocilium and has a bulbous tip. The improvement obtained by placing a mirror below the macula is thought to result from the overall increase in the amount of polarized light which passes through the sensory epithelium after reflection on the mirror. Further contrast enhancement might be obtained by electronic improvements (e.g. Inoué, 1981; Allen, 1981).

5. APPLICATION

The quality of the images of the sensory epithelium obtained with the incident light polarizing microscope permits penetration of the apical surface of the sensory hair cells with a glass microelectrode

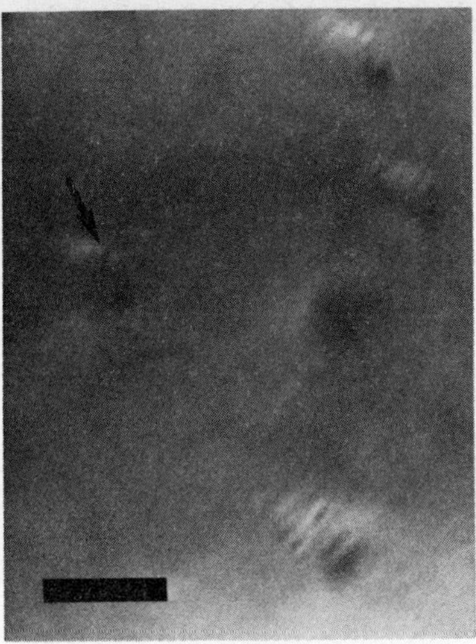

Figure 4. Image of sensory hair bundles, improved by placing a mirror in the orbit of the fish. Arrow points to kinocilium. Bar: 5 μm. Exposure time 5 s.

under visual control. The penetration can be performed on a spot chosen such as to prevent hitting with the electrode the hair bundle or the cuticular plate. Long duration intracellular recordings of the resting membrane potential of the hair cells can be made. Furthermore, in other experiments we found that the light beam (Ø 1 μm) produced by a laser-Doppler interferometer can be accurately focussed through the microscope on a single hair bundle of a selected hair cell. These results demonstrate that the incident light method is well suited for obtaining visual control over an <u>in vivo</u> sensory hair cell preparation.

The intracellular recordings from the hair cells show stable resting membrane potentials for periods of up to 30 minutes for individual cells, and for up to 6 hours for the whole preparation. So probably the amount of light from the Xenon arc does not cause immediate damage to the hair cells. This conclusion is supported by the fact that signs of decay of the hair cells, such as swelling of the cells or desintegration of the hair bundles, were never observed during normal experimental conditions, but only after slowing down of the blood flow through the macula.

We have tested the applicability of this microscopical technique on several low reflective biological objects, unsuited for the use of transmitted light microscopy because of their thickness. In the neuromasts of the epidermal lateral line organ of the clawed frog, *Xenopus laevis*, for instance, the hair cells are located superficially in the skin. It was found that the hair bundles and the supporting cells in the neuromasts of adult animals can be visualized by the incident light polarization microscopical technique.

In conclusion, the present method permits visualization of sensory hair cells and hair bundles in an in vivo preparation of the fish lateral line.

ACKNOWLEDGEMENTS

This work was supported by the Netherlands Organization for the Advancement of Pure Research (Z.W.O.). We thank Prof. H. Duifhuis for creating a productive research environment. The authors are particular grateful to Halbe H. Elsenga (Fryslân) who braved all sorts of weathers to provide us with ruff. Also we are indebted to Zeiss Nederland for the patient lending of a number of optical components used in this study.

REFERENCES

Allen, R.D., Travis, J.L., Allen, N.S., Yilmaz, H., "Video-enhanced contrast polarization (AVEC-POL) microscopy". Cell Motility 1, 275-289, 1981.

Hudspeth, A.J. and Corey, D.P., "Sensitivity, polarity, and conductance change in the response of vertebrate hair cells to controlled mechanical stimuli". Proc. Natl. Acad. Sci. USA 74, 2407-2411, 1977.

Inoué, S., "Video image processing greatly enhances contrast, quality and speed in polarization-based microscopy". J. Cell Biol., 89, 346-356, 1981.

Jakubowski, M., "Cutaneous sense organ of fishes. I. The lateral-line organs in the stone-perch (*Acerina cernua* L.)". Acta Biol. Cracoviensia, Zool. 6, 59-82, 1963.

Kuiper, J.W., "The microphonic effect of the lateral line organ", Thesis, University of Groningen, The Netherlands, 1956.

Rabe, H., "Die Erkennung und quantitative Datenerfassung optischer Hauptschnitte von stark absorbierende, anisotropen Kristallen mit dem Auflicht-Polarisationsmikroskop". Fortschr. Miner. 61, 243-281, 1983.

Wada, S., "The reflected-light microscope". Zeiss Inform. 75, 19-21, 1970.

A MODEL FOR TRANSDUCTION IN HAIR CELLS INVOLVING STRAIN-ACTIVATED CONDUCTANCE

Jonathan Bell
State University of New York
Buffalo, New York 14214

and

Mark H. Holmes
Rensselaer Polytechnic Institute
Troy, New York 12180

ABSTRACT

Based on the experimental observations of the mechano-electrical transduction in hair cells, we formulate a model of the receptor potential utilizing a simple model circuit and ideas of stretch activation. The stereociliary displacement-response relation is developed based on the cilia crosslinking, thus incorporating notions of bidirectional sensitivity, asymmetry, saturation, and adaptation naturally into the model. We then give some simulation results involving periodic stimuli to the hair bundle as well as current stimuli to study latency behavior and other qualitative properties of the model.

INTRODUCTION

We base our model on the experimental measurements of the change in the receptor potential in the hair cell (HC) in response to direct mechanical stimulation of the stereociliary bundle. Of particular importance is the dependence of the receptor potential on the angular deflection of the bundle and the current-voltage relation for the hair cell which are used to determine the qualitative and quantitative behavior of the model. The dynamics of the receptor potential are based on a circuit model. The model contains a number of nonlinear membrane features. In particular, our conductance which is associated with the transduction current is dependent on the angular displacement of the ciliary bundle. This dependence is due to the axial strain induced in the cilia membrane as the bundle deflects. Hence the bidirectionality, asymmetry, and saturation characteristics of the receptor potential are incorporated naturally into the model. We show that the model predicts the appropriate response to periodic stimuli, proper response latency behavior, and qualitatively reasonable reversal behavior (such as what appears in [6]). The model also accounts for the nonlinear nature of the steady state receptor current as well as the linear change in the receptor current in response to a periodic stimulus at a fixed receptor potential [3].

TRANSDUCTION MODEL

Based on experimental studies of the mechanical stimulation of the HC, we assume that the mechanicoelectrical transduction mechanism is associated with the angular displacement of the stereociliary bundle. However, our model does not suggest a detailed mechanism for the transduction process since further experimental evidence on the molecular movement associated with transduction is necessary. Hudspeth [9] has given strong evidence that the location of the transduction mechanism resides in the distal ends of the stereocilia. This is reinforced by the observations of Evans, et al [7] that the cross-links between the stereocilia appear to be oriented in such a way that deflection of the ciliary bundle alters the axial strain in the apical membrane of the stereocilia (figure 1). In particular, deflection towards the kinocilium increases the strain and it is decreased by deflection in the opposite direction. This dependence is strongly correlated with the asymmetry in the dependence of the receptor potential with angular displacement (figure 2). For this reason the axial strain (here denoted by ε) that is induced in the ciliary membrane as the bundle deflects is a central component in our description of the channel density of the membrane.

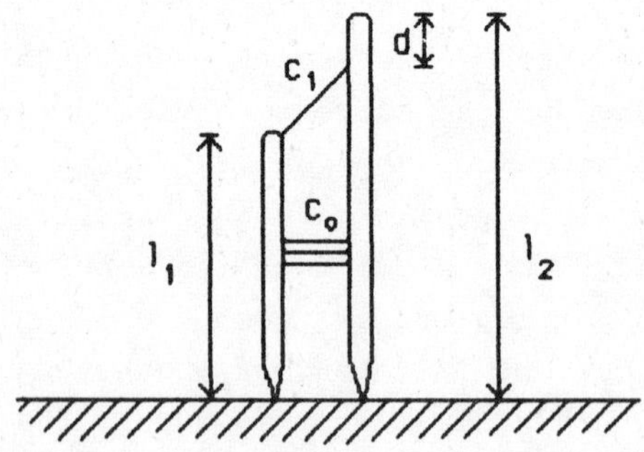

Figure 1. Schematic of the two cilia model. For the calculations discussed, $c_0 = 1/3$ μm, $l_2 = 5$ μm, $l_1 = l_2/2$, and $d_0 = l_2/3$.

We represent the effect of the mechanical stimulus on the hair cell by one variable, namely the angular displacement, θ, of the bundle from equilibrium. The angle θ is measured in the plane of symmetry with the positive direction in the direction of the kinocilium. Our model concerns the dynamics of $v(t)$, which is the receptor potential relative to the resting potential so that $v = 0$ corresponds to the resting potential of the cell.

We model the ionic current density at the cell's receptor end by an electrical circuit consisting of a capacity current density in parallel with a "transduction" current density (mostly a potassium current), and a residual current density. The model equation from the circuit has the form

$$C\,dv/dt + \overline{g}_t G(n)(v - \overline{v}_t) + \overline{g}_L H(v)(v - \overline{v}_L) = I \,. \tag{1}$$

Our model circuit is essentially that found in [6], but our representation of the conductances is quite different. Here $C, \overline{g}_t, \overline{v}_t, \overline{v}_L, \overline{g}_L$ are constants that represent, respectively, receptor membrane capacitance, maximum transduction conductance, transduction reversal potential, "residual" reversal potential, and maximum residual conductance. The I on the right hand side of equation (1) represents an applied current to the cell. The third (i.e. residual) current term in (1) represents all bundle insensitive currents at the HC's apical membrane. The conductance functions $G(n)$ and $H(v)$ are determined from constitutive laws derived from the experimental results which will be discussed below. The transduction conductance depends on a channel activation parameter n. The notational choice of n is suggestive of the potassium activation parameter in the classical Hodgkin-Huxley model of excitable squid giant axon [8]. It might be expected therefore that the transduction channel can be described by a Hodgkin-Huxley type gating mechanism This is not done here by rather it is assumed that the transduction activation process is in a quasi-steady state where the dependence of the activation parameter is only on the axial strain in the ciliary membrane. In other words, we assume that $n = n(\varepsilon)$ and exactly what this dependence is will be specified later. The adequacy of the assumption of a quasi-steady state for n depends on the time scales which are important in the observed experiments of interest. In the experiments discussed below this assumption is appropriate. Whether or not it can be used for mammalian systems at high frequencies is not clear. Also, the strain in the membrane is not homogeneous and the value used here should be considered as an average, or effective, strain that is induced in the membrane.

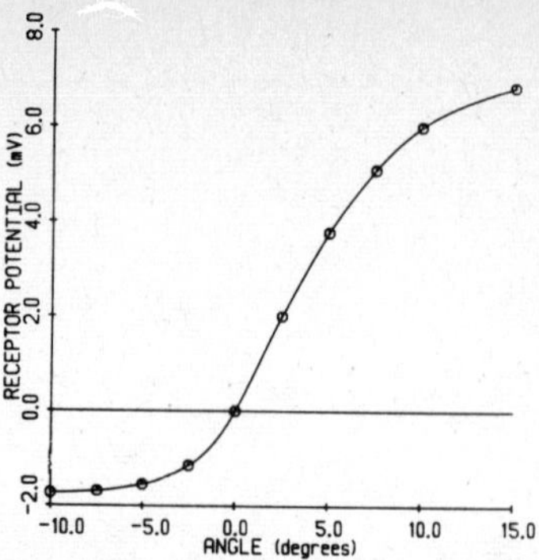

Figure 2. The steady state receptor potential as a function of ciliary deflection angle θ . Measured values from [2] are given by the "0" marks. Note that $\theta > 0$ corresponds to depolarization of the hair cell while $\theta < 0$ results in a hyperpolarization.

As represented in (1), the transduction conductance is not voltage dependent. We base this assumption on the computed receptor current voltage relation given by Corey and Hudspeth [4] for HC's in the saccule of the bullfrog and by Russell [10] for HC's in the cochlea of guinea pigs, which are linear over a large range of holding potentials. In Corey and Hudspeth's experiment all bundle displacement-insensitive currents were subtracted out. When all currents are accounted for in the full current-voltage relation, the response is nonlinear and must be taken into account through the displacement-insensitive conductance $\bar{g}_L$. This explains our formulation (1), where again we assume the pertinent residual conductance activation rate is large enough in comparison to the time scales of the relevant experiments that we can use a quasi-steady state formulation.

The nonlinear voltage dependence of the residual conductance is representative of the type of dependence seen in conduction models of other excitable membranes, while the dependence of the activation parameter n on the axial strain ε represents our approach to incorporating into the neural model the observed receptor response to direct mechanical stimulation of the ciliary bundle and subsequent behaviors to be described below. Our treatment of the strain in the membrance is a simple one, but it gives a clear interpretation for the asymmetry in the response. Just how well it describes the experimental observations will be discussed below.

$H(v)$ and $G(n)$ can be determined from a few reasonable assumptions and some experimental results from [2,4]. The residual conductance $H(v)$ is an increasing function with $0 \leq H(v) \leq 1$ and $H(v) \to 1$ as $v \to \infty$. In what follows, $H(v)$ is assumed to have the form

$$H(v) = \begin{cases} h_0 v + h_1 & \text{if } v \leq v_0 \\ (v + h_2)/(v + h_3) & \text{if } v \geq v_0 \end{cases} \tag{2}$$

where $h_0 = 0.001$, $h_1 = 0.1158$, $h_2 = 6.279$, $h_3 = 7.2335$, and $v_0 = 8.192$ mV. The coefficients in (2) are determined by fitting (1) to the measured receptor current (figure 3) using the BMDP nonlinear regression program. The particular functional form for $H(v)$ used here is simply one of convenience and other sigmoidal type functions can be used without greatly affecting the qualitative behavior of the model.

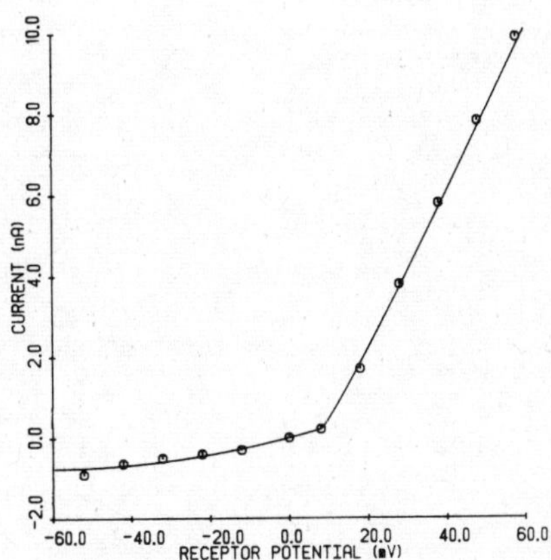

Figure 3. The steady state current obtained for a given potential in the case of no ciliary displacement ($\theta = 0$). Shown are the values from the model, solid line, and the values by Corey and Hudspeth [4], given by the "0" marks.

The activation parameter is expected to increase with strain but at the same time it must satisfy $0 \leq n \leq 1$. Therefore, it is assumed that n is a continuous increasing function of ε with $n \to 0$ as $\varepsilon \to -\infty$ and $n \to 1$ as $\varepsilon \to \infty$. In what follows we assume that

$$n(\varepsilon) = 1/(1 + n_1 \exp(-n_2 \varepsilon)) \tag{3}$$

The parameters n_1, n_2 are determined from (3) using the same nonlinear regression program mentioned above in conjunction with (2) and the measured values of the receptor potential $V(\theta)$ found from figure 2.

One finds that $n_1 = 4.978$ and $n_2 = 1.064$.

It is necessary to be able to relate n and ε with the angular deflection θ. To do this we use a two cilia model (see figure 1), where the cilia act like rigid rods and are free to rotate at their base. The cross-links, which have lengths c_0, c_1, are inextensible and are free to pivot at their attachments to the cilia. The attachments do not move relative to the cilia except for the attachment of the upper cross-link to the taller cilium. Letting $d(\theta)$ represent the distance between this attachment and the distal tip of the taller cilium, then

$$d(\theta) = l_2 - l_1 + c_0\cos(\pi/2 - \theta) - \sqrt{c_1^2 - c_0^2 \sin^2(\pi/2-\theta)}$$

where l_1, l_2 are the lengths of the cilia (with $l_1 < l_2$). The axial strain relative to the rest position is therefore given as

$$\begin{aligned} \varepsilon &= (d(\theta) - d(0))/d(0) \\ &= \frac{c_0\cos(\pi/2 - \theta) - \sqrt{c_1^2 - c_0^2 \sin^2(\pi/2 - \theta)} + \sqrt{c_1^2 - c_0^2}}{l_2 - l_1 - \sqrt{c_1^2 - c_0^2}} \end{aligned} \tag{4}$$

For small angular displacements, this reduces to $\varepsilon = c_0\theta/d_0$, where $d_0 = d(0)$.

Determination of the constant parameters in the model, other than geometric ones comes from three experiments of Corey and Hudspeth [4]. In one experiment they obtained a current-voltage relation from a voltage clamp experiment in the case when $\theta = 0$. Significantly, they found that the current is a strongly nonlinear function of the potential. The second experiment used to determine the material parameters is shown in figure 2, which gives the receptor potential $V(\theta)$ as a function of imposed angular displacement. The third experiment used is shown in figure 2 of their paper [4], where they imposed a triangular stimulus ($f = 10$ Hz) of saturating amplitude on the stereocilia and superimposed on the various voltage steps. From these experimental results we can determine the material constants $\bar{g}_L$, $\bar{V}_L$, $\bar{g}_t$, and $\bar{V}_t$. Further details are given in [1].

The qualitative description of the model is completed by specifying the constitutive law for $G(n)$. $G(n)$ should be an increasing function of the activation parameter n with $G(0) = 0$ and $G(1) = 1$. Other than this though there is at present no experiments which indicate the specific nature of the dependence of G on n. Consequently, as an ansatz we are going to let $G(n) = n$ in the calculations. Because of the observed fast response time of the receptor gating mechanism, we have no reason to suspect a power law relationship between conductance and the activation variable which is normally used to account for certain delays in activation in ionic channels of other excitable membranes.

COMPARISON BETWEEN THEORY AND EXPERIMENT

Based on the analysis of the previous section, the material parameters in (1) are $\overline{v}_t = 56.0$ mV, $\overline{v}_L = 1.8$ mV, $\overline{g}_t = 4.5$ nmho, and $\overline{g}_L = 0.204$ μmho. We also let the membrane capacitance be $3.8\mu F/cm^2$, and the surface area of the membrane is taken to be $3.0 \times 10^{-6} cm^2$. This surface area value is approximately the area of the apical membrane of the HC (see [5]).

The steady state potential-current curve that is obtained from (1) in the case when $\theta = 0$ is shown in figure 3. The agreement between this curve and the values obtained by Corey and Hudspeth [4] is very good. Of interest in this figure is the nonlinear nature of the curve in the neighborhood of the resting potential. This region contains the normal operating range of the hair cell which indicates the inappropriateness of using a linear circuit to model the transduction process over a large potential range for this preparation.

The dependence of the receptor potential on strain, and the associated measured values in [2] are shown in figure 4. It shows that for small positive strains (i.e., $0 \leq \varepsilon \leq 2\%$) the response is approximately linear, but for larger strains, or for negative ones, the dependence is distinctively nonlinear.

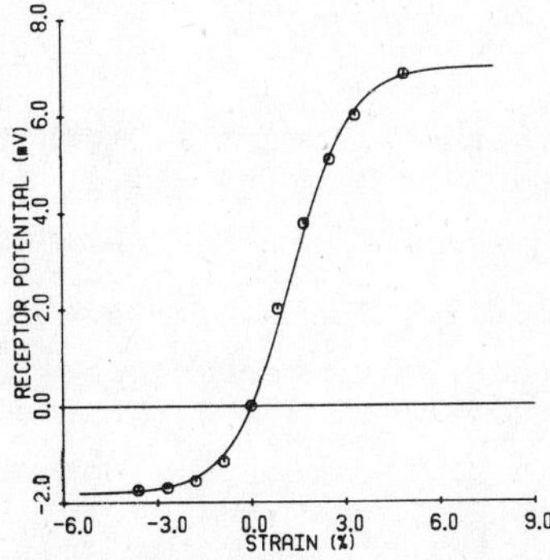

Figure 4. The steady state receptor potential as a function of strain in the ciliary membrane. Shown are the values predicted by the model, solid line, and the measured values given in [2], by the "0" marks.

As for the dynamic behavior, we computed receptor potential in response to a periodic displacement (f = 10 Hz) of the ciliary bundle. The receptor potentials for three different amplitudes of ciliary deflection were computed. The dynamic component of the receptor current in response to a triangular stimulus when the HC is held at a fixed potential was also computed. The final comparison we made between the theory and experiment concerns the response latency of the hair cell. To measure this, a short pulse stimulus with a maximum amplitude of 4°, over a duration of 150 μsec, is applied to the ciliary bundle. Agreement with analogous experiments [2,3,4] was very good, (see [1]).

ACKNOWLEDGEMENT

One author (J.B.) was supported, in part by an NSF grant MCS8301724, while the other author (M.H.H.) was supported, in part by the U.S.A.R.O., contract number DAAG29-83-K-0092.

REFERENCES

1. Bell, J., and Holmes, M.H., "A nonlinear model for transduction in hair cells", submitted to Hearing Research.

2. Corey, D.P., and Hudspeth, A.J., "Sensitivity, polarity, and conductance change in the response of vertebrate hair cells to controlled mechanical stimuli", Proc. Nat. Acad. Sci. USA 74, 2407-2411, 1977.

3. Corey, D.P., and Hudspeth, A.J., "Response latency of vertebrate hair cells", Biophys. J. 26, 499-506, 1979.

4. Corey, D.P., and Hudspeth, A.J., "Ionic basis of the receptor potential in a vertebrate hair cell", Nature 281, 675-677, 1979.

5. Corey, D.P., and Hudspeth, A.J., "Analysis of the microphonic potential of the bullfrog's sacculus", J. Neurosci. 3, 942-961,1983.

6. Crawford, A.C., and Fettiplace, R., "Non-linearities in the responses of turtle hair cells", J. Physiol. 315, 317-338, 1981.

7. Evans, P.H.R., Comis, S.D., Osborne, M.P., Pickles, J.O., Jeffries, D.J.R., "Cross-links between stereocilia in the human organ of Corti", J. Larymgol. and Otology 99, 11-19, 1985.

8. Hodgkin, A.L., and Huxley, A.F., "A quantitative description of membrane current and its application to conduction and excitation in nerve", J. Physiol. 117, 500-544, 1952.

9. Hudspeth, A.J., "Extracellular current flow and the site of transduction by vertebrate hair cells", J. Neurosci. 2, 1-10,1982.

10. Russell, I.J., "Origin of the receptor potential in the inner hair cells of the mammalian cochlea-evidence for Davis' theory", Nature 301, 334-336, 1983.

AUTHOR INDEX

PERMUTED TITLE INDEX